MATHEMATICS AND THE REAL WORLD

MATHEMATICS AND THE REAL WORLD

THE REMARKABLE ROLE OF EVOLUTION IN THE MAKING OF MATHEMATICS

ZVI ARTSTEIN

TRANSLATED FROM HEBREW BY ALAN HERCBERG

59 John Glenn Drive
Amherst, New York 14228

Published 2014 by Prometheus Books

Hakesher Hamatemati: Hamatematika shell Hateva, Hateva shell Hamatematika, ve-Hazika La-evolutzia
(*הקשר המתמטי: המתמטיקה של הטבע, הטבע של המתמטיקה, והזיקה לאבולוציה*)

10 Kehilat Venezia St., Tel Aviv, Israel 6143401

Prometheus Books recognizes the following registered trademarks mentioned within the text: IBM®, NBA®; and Philadelphia 76ers®.

Cover images: spiral © Yang MingQi/Media Bakery; numbers © Chen Ping-Hung/Media Bakery; drawing in the upper left by Leonardo da Vinci, folio 518 recto, from the Biblioteca Ambrosiana in Milan, Italy

Jacket design by Jacqueline Nasso Cooke

Translated from the original Hebrew by Alan Hercberg

Unless otherwise specified, all images are by the author

Inquiries should be addressed to
Prometheus Books
59 John Glenn Drive
Amherst, New York 14228
VOICE: 716–691–0133 • FAX: 716–691–0137
WWW.PROMETHEUSBOOKS.COM

18 17 16 15 14   5 4 3 2 1

Library of Congress Cataloging-in-Publication Data

Artstein, Zvi, 1943– author.
[Hakesher hamatemati. English]
Mathematics and the real world : the remarkable role of evolution in the making of mathematics / by Zvi Artstein ; translated from Hebrew by Alan Hercberg.
pages cm
Includes bibliographical references and index.
ISBN 978-1-61614-091-5 (hardback) — ISBN 978-1-61614-546-0 (ebook)
1. Mathematics—Philosophy. I. Title.

QC6.A6913 2014
510.9—dc23

2014013908

Printed in the United States of America

To Yael

CONTENTS

PREFACE

There are many jokes about mathematicians. One of my favorites is about an engineer, an architect, and a mathematician who have been sentenced to be hanged. In the evening before the day set for the execution, the warden asks them for their last requests. The engineer asks to be allowed to present a new machine he has designed that can perform all household chores without any human intervention. The warden promises that the next day, before the hanging, he will have one hour in which to show his machine to the prison staff and his two fellow death-row inmates. The architect asks to be allowed to explain his new concept of residential accommodation, a modern house that keeps cool in the summer and warm in the winter, without expenditure on fuel. Again the warden promises that the next day, before the hanging, he will have one hour in which to present his idea to the prison staff and his two fellow death-row prisoners. The mathematician says that he has recently proved a mathematical theorem that will shake the foundations of mathematics, and he would like to reveal it in a lecture to an intelligent audience. The warden starts agreeing . . . and the engineer and the architect start shouting: "We want our execution to be brought forward to this evening!"

This joke appeals to me because it reflects the widespread attitude of the public to what can be expected from books and lectures on mathematics. We will give the reasons for this attitude later, and here we will just note that even in school we are exposed to indoctrination that causes us to relate differently to texts and lectures on mathematics than we do to other subjects. In school students are expected to solve mathematical exercises to show that they have understood the material. Other subjects such as history, literature, or even biology do not require such exercises. The impression that this creates is that without solving exercises there is no point in listening to mathematics. The development of intuitive understanding of a subject, without practicing what has been learned, is not accepted as understanding in the case of mathematics. That is so despite the fact that an intuitive grasp of a subject, without needing to put it into practice, is an acceptable objective in other scientific and general disciplines. This is misguided and misleading indoctrination that does an injustice to mathematics. Furthermore, that approach is alien to professional mathematicians too. Of course they must have a deep understanding of the topics they are researching, but an intuitive understanding of other mathematical subjects is sufficient. I will put forward an analogy that I would ask you to keep in mind as you read this book.

I love classical music and regularly attend concerts of the Israel Philharmonic Orchestra, and I greatly enjoy both live performances and recordings. I cannot read music, and I do not know the detailed history of music or the life stories of the different composers. I am confident that those who can read music or are familiar with the history of music enjoy what they hear in a way that is different from my enjoyment. I am not sure if they enjoy it more than I do because, for example, they may be conscious of any note played slightly inaccurately, whereas I would be totally oblivious of it. The experts understand the compositions on different levels from mine, but I enjoy the music immensely; perhaps not from the written notes, but from the tune. Not the trees, but the forest. There are hardly any "notes" in this book, nor trees, mainly a tune, mainly the forest. If one or a few notes appear here or there (at times using a different font, and preceded by a rule line), they can be skipped without breaking the thread of the text.

The different sections of the book are connected, but the ideas are presented in such a way that each section is self-contained and can be read independently of the others. The headings and titles of the sections and chapters indicate the central elements within them. It is advisable to start with chapter I, but then the reader can certainly go straight to the chapter on the mathematics of randomness or to the one on the mathematics of human behavior, or even jump to the last chapter on teaching mathematics.

Naturally, a book like this could not have been written without information, exchange of views, and help that I received from friends, colleagues, students, those who heard lectures on the topics covered in the book that I delivered in various forums, the translator and the editor, the publisher's team, and, of course, my family. There are too many people for me to be able to enumerate each one of them here. To all of them, my sincere thanks.

So what is it all about: The books deals with *the mathematics of nature*, *the nature of mathematics*, and their interrelationship. We will describe, by means of a historical review as well as from the aspect of current research, the link between mathematics and the physical world and the social world around us. The discussion in the book also relates to areas of science and society to which mathematics is relevant. We will therefore also present scientific facts and social situations described by mathematics. That presentation is not exhaustive or detailed, as we focus on the mathematical aspects of the various fields. The discussion will be accompanied by the constant presence of the question regarding the extent of the effect of the *evolution of the human race* on the development of mathematics and its applications. We will examine the claim that the manner in which the human brain was fashioned by millions of years of evolution affected humans' mathematical capabilities and the type of mathematics that is easy for humans to develop and understand. We will also show that, to a large extent, evolution is responsible for the difficulty we have in understanding certain other areas of mathematics. We will try to do all that with a minimum of musical notes but with much pleasing music.

Zvi Artstein
The Weizmann Institute of Science

CHAPTER I
EVOLUTION, MATHEMATICS, AND THE EVOLUTION OF MATHEMATICS

Could evolution have affected mathematics? • Can horses perform calculations? • Can rats count? • Do infants solve problems of addition and subtraction? • Which rectangles please us? • Why are clowns scary? • What color are the sheep in Ireland? • What number comes next in the sequence 4, 14, 23, 34, 42, 50, 59, . . . ? • Why square the circle? • How have optical illusions contributed to science?

1. EVOLUTION

The theory of evolution is attributed to Charles Darwin, but it was not Darwin who initiated the study of evolution. King Solomon stated, "There is nothing new under the sun," (Ecclesiastes 1:9), a philosophical statement alluding to the observation that the world is in a constant state of flux. At any given time we see the current situation around us, but we also follow changes that take place in our lifetime, and we are aware of changes occurring over periods of time that we are unable to observe directly. The evidence regarding changes that took place in the past often enables us to infer what caused those changes. That applies both to the physical world, such as rocks, flora, and fauna, and to society, including modes of behavior, fashion, literature, medical practices, and technology. The changes take place according to their own mechanism. Sometimes it is clear to us what survives, what is modified, and what becomes extinct, but it is not always easy to identify the mechanism.

Let us take as an example the Earth's surface. Some rocks exist for many years, while others are weathered and eroded by the wind almost as we watch. What causes the difference? Clearly it is the different rock textures that determine the differences in their ability to survive. Basalt will last, while limestone will crumble. There are no sand dunes on the tops of mountains because they would be blown away by the wind. We could say that the strong triumph, the fittest survive. We can deduce that being made of basalt is an advantage in the battle for survival on a mountain peak. That statement is trivial in the realm of rocks, and we do not usually examine rocks in terms of the competition for survival. But the conclusion that whatever is most suited to its environment survives is correct regarding rocks as well as human society. Historians, in discussing human history, try to understand why a particular society survived and another disappeared. Their conclusions generally refer to the advantages that the victors had over the vanquished. We can learn about the conditions under which a society or species developed from its specific characteristics. Likewise, from the conditions in which it developed, we can learn about the advantages that enabled it to win the battle for survival.

Darwin's great contribution to the theory of evolution was in identifying the mechanism by which different species of animals and plants changed and developed. Unlike Lamarck, who claimed that every species adapts to its environment and the characteristics it takes on are passed from each generation to the next, Darwin proposed a different mechanism responsible for the changes that every species undergoes. The mechanism consists of two main elements: mutation and selection. In the reproduction process, individuals undergo mutations that cause random and generally minor changes in their characteristics. The individuals with the best-suited characteristics reproduce at the fastest rate, and that constitutes the selection that results in successive generations of each species being better suited to the environmental conditions. The best-suited species among those competing for the same food resources are the ones that survive.

Charles Robert Darwin (1809–1882) was born in the town of Shrewsbury, England, to a well-established family. His father, Robert, was a wealthy physician. His grandfather, Erasmus Darwin, who died before Charles was born but whose writings were available to Charles, was a philosopher and a naturalist who favored the theory of evolution of the leading French naturalist Jean-Baptiste de Monet, Chevalier de Lamarck (1744–1829), known simply as Lamarck. Young Charles was exposed to scientific endeavors but was not a particularly industrious student. Instead of devoting his time to his studies, he preferred to observe nature and to collect various items, particularly beetles of different types. When he was twenty-three he was invited to join an expedition due to sail on a ship called the *Beagle* as its scientist; the main purpose of the expedition was to chart the shores of Australia and South America for the British Empire. His role was to collect and classify geological, zoological, and botanical specimens. In the course of the voyage Darwin noted that different but similar species could be found living in regions near each other, specifically on the Galapagos Islands. It was there that he conceived the model of evolution consisting of mutations and selection. It should be noted that Darwin, in developing his theory of the evolution of types of flora and fauna, was deeply influenced by the theory of the political philosopher Thomas Malthus (who lived half a century before him) on the demographic and economic development of human societies. Darwin's autobiography shows him to have been a modest and wise man; it is replete with statements indicating a deep understanding of evolution beyond the technique he developed. For example, in his reference to his elderly colleague Leonard Jenyns, with whom he had many discussions about nature, Darwin writes, "At first I disliked him from his somewhat grim and sarcastic expression, and it not often that a first impression is lost, but I was completely mistaken. . . ." The connection between "it is not often that a first impression is lost" and evolution will be discussed further on in the book.

Although Darwin shared his thoughts on evolution and the ample evidence he had found supporting his theory with his scientific colleagues, among them some of the best-known English scientists in those days, he refused to publish his findings. He agreed to publish his theory only after Alfred Wallace, a young naturalist researcher who had undertaken many voyages to South America and the Far East, submitted a paper for publication containing ideas similar to those of Darwin but with only a flimsy basis. Darwin's friends became aware of this and urged him to publish his book, *The Origin of Species*. As a result, in the first official presentation of the theory, Wallace's article and Darwin's theory appeared at the same time.

There were several reasons for Darwin's long hesitation before publishing his conclusions. Some derived from the possible conflict with the religious belief that different species exist because that is how they were created. Darwin's wife, Emma (née Wedgwood), was deeply religious, and Darwin did not wish to upset her. Another reason, however, no less important, for his hesitation to publish his mechanism of evolution was that despite the wealth of facts he had available supporting his theory of evolution, many aspects of the theory had not yet been demonstrated and lacked a scientific basis. In particular, Darwin could not offer a biological mechanism that would cause mutation. That mechanism was not discovered until the middle of the twentieth century, when the genes encoded in DNA molecules were revealed and were found to mutate randomly in the course of reproduction.

Mutation and selection of genes are the basis of the modern understanding of the process of evolution of plant and animal species. The genes carry the characteristics vital to the survival and development of each species. The ensemble of genes and the way they are expressed, at times as a reaction to the environment, define the characteristics of a species. Changes in the genes are responsible for changes in the species. Nevertheless, much can be learned about evolution without monitoring the changes in the genes themselves. By studying the conditions in which the species developed, survived, and was victorious in the evolutionary struggle, we can learn about the characteristics that are encoded in its genes and are passed on from generation to generation. The reverse of this claim is also correct: the characteristics observable at any given time enable us to learn about the conditions in which each species developed.

The following example shows how we can learn about the link between the conditions in which species developed and their characteristics today. I came across this example in the course of a trip I made to the Galapagos Islands a few years ago. It relates to the mating habits of birds.

The rock cormorant in the picture on the left cannot fly. It lives on cliffs close to the seashore that are exposed to strong winds. Its abilities to find twigs and to build a proper nest are vital to the survival of the species in these tough conditions. In its courtship, the male cormorant demonstrates to his potential mate his ability to gather twigs to build the nest they would share. The courted female will respond to the male's advances only if he can prove he has that capability. The middle picture is of the frigate bird. In its courting display, the male inflates his gular pouch until it becomes an enormous red balloon. He does this to show his intended mate the strength of his lungs and his ability to fly long distances to scoop fish out of the water. The picture on the right, the blue-footed booby, demonstrates entirely different qualities in its courtship display. The male of this species incubates the eggs and protects them by covering them with his large blue feet. He therefore tries to woo his potential mate by flaunting the size and shape of its feet, thus proving his ability to protect the eggs they will produce together against enemies and unsettled or rough weather.

These examples illustrate that the characteristics and behavioral patterns we can discern today indicate the characteristics that were of evolutionary importance, and those show us how each species survived the evolutionary struggle.

The essential characteristics that helped any particular population to win the battle for survival during the formation of the species are etched into its genes, and we may identify them as innate attributes. Cheetahs' speed, eagles' "eagle eyes" and cats' tree-climbing ability are all innate traits. A cheetah cub is born with the ability and basic instinct to run fast. It will need help from its parents to learn what it needs to fear, how to hunt, and even how to run more effectively, but the basic features of speed and hunting are carried in its genes. Similarly, the genes of a cat enable it to learn how to catch mice, and the innate characteristics of an eagle include its keen vision and ability to identify potential prey from a great height. Learning merely refines and improves the innate attributes. The attributes of each species enable us to learn about the conditions in which the species developed; similarly, knowing the conditions in which the species developed enables us to learn about the characteristics that evolved.

It is reasonable to assume that just as physical attributes of animal species are innate features etched into their genes, the same will apply to at least some mental attributes. Mental and social skills also play a role in the battle for survival, so that in these too, the selection process strengthens the features that help the species to overcome its rivals. Specifically, in the reproduction process mental attributes can also be changed and improved by mutation. In the following sections we will examine the mathematical capabilities of the human species from an evolutionary aspect. We will ask whether the abilities to understand and to use mathematics are the results of evolutionary development, or whether they may be by-products of a brain that developed to cope with other needs.

2. MATHEMATICAL ABILITY IN THE ANIMAL WORLD

If mathematical ability made a contribution in the evolutionary struggle that brought the human race to the position it currently occupies among the species, it may be assumed that other living beings would possess a certain degree of mathematical ability. But what does mathematical ability mean? Mathematics encompasses a broad range of topics and conceptual methods. The question to ask, therefore, is which of those mathematical features provide an evolutionary advantage? And the follow-up question is how can we identify these mathematical abilities in animals?

The most basic mathematical ability is counting. It is followed by the understanding of the concept of a number as an abstract object and the ability to perform simple arithmetic operations, such as addition and subtraction. We will start by discussing the existence of these simple elements in adult animals. A mother cat moves her kittens from place to place and generally does not forget a kitten or two, and when she has finished moving them, she does not usually go back again to check whether she has moved all of them. She may remember them individually, but it seems reasonable to state that the mother cat has a sense of quantity. The instinct of quantitative estimation clearly provides an evolutionary advantage, so we should not be surprised that adult animals possess that ability. But does that ability extend to the ability to count and to the possibility of performing arithmetical manipulations?

Before presenting several convincing examples showing that some animal species do have mathematical ability, a warning is in order. The results of experiments in general, and of animal experiments in particular, should be interpreted with great caution. A well-known illustration of this is the case of "Clever Hans." (More details and references concerning this story, and concerning the research mentioned later in this section, can be found in the monographs by Dehaene and by Devlin [2000] listed in the sources.) Toward the end of the nineteenth century a horse known as Clever Hans was exhibited on tour in Germany with its trainer, Wilhelm von Osten. The horse showed remarkable ability in adding, subtracting, finding the squares of numbers, simple division, and so on, all with a very high degree of success. The horse was wrong occasionally, but such errors occurred infrequently. The method by which the horse showed its abilities was that when an exercise was read out or written on a board, it would tap its hoof the number of times corresponding with the right answer. It was suspected that the act was simply a clever deception by which the trainer somehow or other managed to give the right answer to the horse. An official committee was appointed, headed by a psychologist named Carl Stumpf and whose members included the director of the Berlin Zoo. The committee checked, among other things, whether the horse could solve the problems if the trainer was not present, and it found that even then the horse could still give the correct answers. The conclusion was that some animals have a fairly advanced mathematical ability. Subsequently, more detailed examinations in 1907 by another psychologist, named Oscar Pfungst, showed that the horse did not know mathematics. The trainer was indeed reliable and honest, but the horse had learned to distinguish involuntary changes in his facial expressions and in the facial expressions of the audience when the trainer was not present. The horse understood from those facial expressions when it had reached the correct number of taps of its hoof. The presence of the trainer or an audience was essential. Pfungst found that if the trainer looked tense at a wrong answer, the horse answered according to the expression and not the correct answer. The research methods developed by Pfungst as a result of this case are now recognized as a breakthrough in psychological research.

Scientific experiments that were more soundly based have proven that some animals do indeed possess mathematical ability. The German zoologist Otto Koehler (1889–1974) proved as early as in the 1930s that some species of birds can identify a collection with a given number of elements. It is apparently not difficult to train a pigeon to choose every third seed when faced with a row of seeds. A squirrel can be trained so that when faced with boxes containing different quantities of nuts, it will choose the box with exactly five nuts. There is a limit to the numerical-identification ability of these animals. Koehler himself found that even the most capable animals could not identify collections with more than seven elements. The number appears in the literature also as a bound to the number of information units that a human brain can process. We will meet the number seven again later on in similar contexts. Still, these experiments demonstrate the mathematical ability to estimate quantity but do not yet prove an ability to count or to grasp the abstract concept of a number.

Adult crows are known to be able to count, within certain limitations. Food is placed near a building. The crow learns very quickly that it is dangerous to attempt to approach the food while someone is in the building. It cannot see into the building to check if anyone is inside or not, but it can see when someone enters or leaves it. The popular literature (without scientific checks, it must be said) reports situations in which several people enter the building one after the other. As long as they remain in the building, the crow keeps away. The people in the building then leave, one by one. With surprising accuracy the crow knows when all those it saw enter the building have left, and only then does it approach the food. Clearly there is a limit to crows' ability to be exact, just as there is a limit to humans' ability to keep track on exact large numbers. Crows managed to count up to five or six in this manner, with a high degree of accuracy. The ability to identify a collection with a given number of elements demonstrated by crows in this example and by other species is consistent with an evolutionary advantage.

The ability to count is clearly an advantage in the battle for survival, but its origin in the avian world is unclear. After all, how often in the evolution of crows did they encounter a situation in which they had to count the number of dangerous animals entering and leaving a building? Specifically, it is unclear whether this apparent counting is in fact counting in the mathematical sense. In other words, does the crow have the ability, whether conscious or not, to comprehend the number of the people entering the building, or does it simply remember who went in and who came out?

Monkeys were found to have a greater mathematical ability to count and compare. The following experiments were carried out by Guy Woodruff and David Premack of the University of Pennsylvania (their paper was published in 1981). A chimpanzee was shown a full glass and a half-full glass, and it was taught to choose the half-full glass every time. The same chimpanzee was then offered the choice of a whole apple or half an apple, and it chose the half apple. In other words, it generalized the mathematical principle from the glass to the apple. In a similar fashion, the chimpanzee was taught to demonstrate simple mathematical abilities, such as recognizing that the combination of half an apple and a quarter of an apple is three-quarters of an apple. In another experiment, two trays were placed before a chimpanzee. The first tray had two piles of pieces of chocolate, one pile with three pieces, and the other with four. The second tray had a pile of five pieces of chocolate and then a separate, single piece. In most cases, the chimpanzee chose the tray with the larger total number of pieces. This does not yet constitute proof that the chimpanzee understood the abstract concept of numbers or the addition of numbers, but it is evidence of mathematical abilities. This is not surprising, as such abilities constitute an evolutionary advantage.

Another experiment with animals proves that the concept of numbers in the abstract does exist to some degree among some, even among less-developed animals. The experiments were conducted by Russell Church and Warren Meck of Brown University (the research was published in 1984). It is not difficult to train rats so that when they hear two beeps, one after the other, they are given enough tasty food to satisfy them. Similarly, when they see two flashes of light, they can also safely eat the food. They were taught, however, that when they hear four beeps or see four light flashes, it is dangerous to eat the food, as they get an electric shock. The aural or visual signals, that is, the beeps or flashes, are received and processed in the brain via two different senses, hearing and sight. The rats reached a high level of reacting correctly, approaching the food if they heard two beeps or saw two flashes, and avoided doing so if they heard four beeps or saw four flashes. When the rats had been trained sufficiently, they heard two beeps that were immediately followed by two light flashes. How do you think they reacted? Did the rats consider the signals as a double invitation to eat the food, or did they interpret them as a four-signal warning to refrain? If they reacted according to the latter, it may be assumed that they recognized the number four as an independent concept, even though the signals received were of two different types. The answer: the rats clearly identified the number four and did not approach the food when they received four signals, even when they were received via different senses.

This experiment with rats still does not indicate arithmetic ability in these animals, nor does it give a definite proof that such abstract counting is an innate attribute, that is, a characteristic carried in their genes, as it may be the result of training made possible by the development of the brain for other purposes. It seems reasonable, however, that this ability is innate, mainly because of the evolutionary advantage given by the abilities to count and to recognize the concept of numbers. To be convinced beyond all doubt that a particular ability is innate, it should be identified in the animal when it is still very young. Such experiments with cubs and other animal young are obviously very difficult to perform. With human cubs, that is, babies, such experiments can be performed.

3. MATHEMATICAL ABILITY IN HUMANS

Before we present the evidence that mathematical ability is inborn in human beings, that is, embedded in their genes, we need to make two comments about the nature of the discussion. First, our use of the term *genes* from here on is a conceptual one and does not relate to any specific gene or set of genes. We will leave the identification of the genes responsible for mathematical ability to our biologist colleagues. For us, establishing the fact that it is innate is sufficient. Second, in the examples of animals and in the discussion in this section, we do not relate to the ability of any individuals. We do not ask whether the success in mathematics of a specific student is determined by his genes alone or whether it is due to environmental conditions or to his having good or less-good mathematics teachers. The discussion is concerned with the mathematical capabilities of the human race and the connection between that capability and the process of evolution, a process that has continued for millions of years, in the course of which the abilities under discussion were formed.

We will first address the simplest mathematical operations, that is, counting, addition, and subtraction. One of the basic principles of classical psychology is that babies are born with a brain that evolution has prepared for learning but that is initially void of all information. Babies learn about the world initially via observation and then via a combination of observation and experience. More-abstract learning appears later, with language development. This view was held and taught by no less than Sigmund Freud (1856–1939), the father of modern psychology. What he said related to knowledge in general and to mathematical ability. At first glance it does appear that regarding mathematical elements that description is correct. Only when they are three or four years old do children acquire the ability to count, and later on to add and subtract. At first they only recite what they have heard, one, two, three, and so on, without realizing that they can count. To illustrate, if given three balls, they may count one, two, three, four, five, counting the same ball more than once. Only at a later age do children begin to understand what counting is, and even later do they start performing simple arithmetic operations. A leading researcher and advocate of this approach was the famous psychologist Jean Piaget (1896–1980), who formulated a complete theory of cognitive development regarding the gradual acquisition of mathematical abilities from childhood to adulthood. (The reader can find elaborations on this issue and other research referred to later in this section in the monographs by Dehaene and Devlin [2000] listed in our sources.) In one of his experiments, Piaget showed children eight flowers, six roses and two chrysanthemums, and asked, "Are there more flowers or more roses?" A significant number of children answered roses. Piaget concluded that children have no intuition of set inclusion; in other words, the children had no understanding that with two sets, one of which includes the other—in our case the set of flowers includes the set of roses—the former is larger. In Piaget's time, it was believed that the relation between sets was the right basis for mathematics (a view that is currently becoming less and less accepted; more on this in the last chapter of the book). Accordingly, Piaget concluded that small children have no understanding of the connection between sizes of sets and of one set including another, let alone any ability to count or knowledge of simple arithmetic.

Nevertheless, the fact that the ability to count is not acquired until a child is several years old does not necessarily prove that the characteristic is not innate. In the observations mentioned above, including Piaget's experiments, counting and arithmetic were acquired together with the ability to communicate and to use a given language, generally the mother tongue. It is not surprising that communicating in a given language is not an innate attribute but a learned one. The ability to learn a language is an inborn characteristic, but acquiring the language itself takes several years. Before the child learns a language, his arithmetic abilities do not come into play, as seen in the above experiments. The obstacle is not the child's lack of ability to count but the fact that he has to answer questions that he does not comprehend, or he does not realize what the expected answer is, until he has had some more years of practice. It is easy to devise tests showing that understanding the question plays an important role in interpreting the results.

Children aged three to four years are shown four marbles and, nearby, four buttons; they are asked whether there are more marbles or more buttons. Most of them will answer that the number of marbles and buttons is the same. The buttons are then spread more widely, that is, spaced farther apart from each other, and again the question is asked, "Now, are there more marbles or more buttons?" Most of the young children will give the same answer, the number is the same. When this exercise is repeated with older children, aged five and six years, many of them answer that there are more buttons.

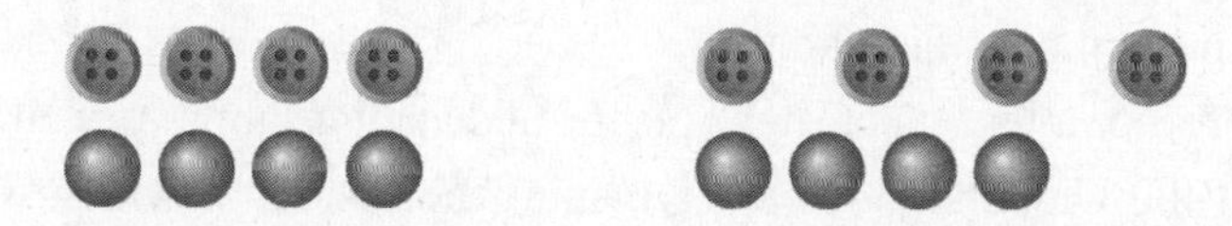

This does not indicate a drop in their mathematical aptitude. The correct explanation is that older children are not used to being asked the same trivial question more than once. They therefore conclude that the questioner must expect a different answer and assume that the question is about the distance between the items and not their number, and they answer accordingly.

There is clear evidence that very young infants relate to numbers and can even perform simple addition and subtraction. How can such cognitive capabilities be examined in babies only a few months old? Several parameters enable us to see whether a baby is excited or surprised. One is the length of time it looks at something. A baby can look at an object or a situation for a few seconds and then it will divert its gaze to something else. When it looks at something new or surprising—and for babies a few months old, new is also surprising—it holds its gaze on it for a longer period, for a few seconds more. A second parameter is the rate at which the baby sucks, say its pacifier. When it is excited or surprised, it sucks harder and more frequently.

An experiment undertaken by Ranka Bijeljac-Babic and colleagues in Paris (the results of which were published in 1991) showed that even newborn babies have a sense of numbers. They measured the intensity with which babies sucked while hearing meaningless words of three syllables, such as "defantok," "alovo," "kamkeman." At first, when the babies heard the words, they sucked harder until they became accustomed to the sounds, and then they reverted to normal sucking. Then two-syllable words were spoken, and this led to harder sucking again. This pattern repeated itself. Whenever the number of syllables in a series of words changed, the reaction of the babies changed too. In other words, even at such an early age, babies can recognize that word sounds consist of syllables and react to a change in the number of syllables between one series and another. The syllables in the "words" were chosen randomly to avoid their having any meaning or significance, so that the only explanation for the babies' reaction was the number of syllables.

A more complex experiment performed in the laboratory of Prentice Starkey of the University of Pennsylvania (the results of which were published in 1980) showed that distinguishing between different numbers is not restricted to one communication channel. Six-month-old babies were shown pairs of pictures with either two or three elements, say two in the picture on the left and three in the picture on the right. Different objects were shown each time, sometimes just geometric shapes, sometimes dots, and so on, and each time the colors were different; this was done to neutralize any possible effect of the content of the pictures. While the pictures were being shown the babies also heard notes or sounds, sometimes two and sometimes three, in random order, in order to cancel any possible effect of any structure in the order in which the notes were heard. When three notes were heard, the babies clearly preferred to look at the picture with the three images, and when they heard two notes, they turned their attention to the picture with two elements. They were exhibiting a counting operation or were at least comparing quantities perceived via two different senses, sight and hearing.

Another sophisticated experiment conducted by Karen Wynn of Yale University (results published in 1992) showed that babies have a natural sense of addition and subtraction. A screen was placed before babies of a few months, and they saw a figure going behind it. The screen was removed, and they saw the figure. Next, one figure went behind the screen, and then another figure followed. The screen was removed, and the babies saw the two figures. This was repeated several times until the babies became accustomed to what was happening. Then an arithmetically incorrect exercise was performed. One figure went behind the screen, and then a second figure. When the screen was removed, only one figure could be seen. To a highly significant degree, the several-months-old babies showed surprise. They expected two figures, and lo and behold, there was only one! The experiment was repeated with a number of variations to remove the possibility that the infants were simply used to the result of a particular exercise. It was highly significant that results that were arithmetically incorrect gained more of the babies' attention. Later, a similar experiment was conducted with adult rhesus monkeys. They showed signs of surprise when, for example, a banana was put inside a box, followed by a second one, and when the box was opened, there was only one banana inside.

These experiments were scrupulously and rigorously controlled, and it may be concluded from them that human beings' arithmetic abilities are genetic. Clearly these operations are performed in undeveloped brains and in no particular language, and there is thus no possibility for the baby to discuss the results with its parents or friends. When the child grows, it will have to learn how to express this mathematical ability in the everyday language it uses to talk to its parents. This learning is a process in itself. But simple arithmetic is innate in babies and is not a by-product of a brain that had been developed for completely different purposes. From this it may be deduced that simple arithmetic afforded an advantage in the evolutionary competition. This is not surprising. For those competing for food, the mathematical ability to distinguish large from small, the many from the few, and even addition and subtraction gives an evolutionary advantage. An individual with this ability will be better suited to a competitive environment than would other members of the same species with lower mathematical abilities.

How is this finding consistent with the finding that some primitive tribes, including some discovered recently in isolated locations, use only the numbers one, two, and three to describe their environment, and any larger quantities are referred to as "many"? If living beings such as birds or rats can differentiate between numbers greater than three, one would expect humans to be able to count better. The answer is simple: language developed much later among human beings in the process of evolution and placed greater emphasis on more important things than the less important. Those primitive tribes apparently are well aware of the difference between sets consisting of five or six objects, but their language is not rich enough to describe them because they had no need to devote terms to numbers greater than three. This does not contradict the fact that at the intuitive level their arithmetic capability is much higher. As a language develops, so does the ability to express and perform more-extensive arithmetic operations. Language developed relatively late in the general evolutionary process but is itself part of that process. The human brain is distinctive among living beings in its verbal communication abilities. Indirect evidence that arithmetic, counting, and the facility to perform addition and subtraction, for example, are the direct results of evolution and not just by-products of language can be found in recorded cases of people born with a malfunction of the brain that meant they could not count or perform addition or subtraction but whose other verbal capabilities were perfectly normal. Conversely, there are people with defective verbal capabilities who can perform arithmetic operations easily.

It should be noted that similar techniques of research into the evolutionary roots of mathematics can be used to discover abilities and features whose roots are evolutionary and that are unrelated to mathematics. Recently (in 2010) research by Karen Wynn of Yale, mentioned above, and her partner Paul Blum was published, showing that altruism and the aspiration for justice exist in babies a few months old, at an age when it is reasonable to assume they could not have absorbed these characteristics from the environment. This too is not surprising. The preference for a just distribution of resources is an attribute that helps a society survive the evolutionary struggle, and it is reasonable, therefore, that it is inherent at the genetic level.

4. MATHEMATICS THAT YIELDS AN EVOLUTIONARY ADVANTAGE

Mathematics has many aspects. The previous section showed that the ability to perform arithmetic calculations is the result of evolution. In this section we will indicate other branches of mathematical operations that, it may reasonably be assumed, provided an advantage in the evolutionary struggle. We will present evidence that those parts of mathematics were also incorporated in the genetic heritage. We may refer to this aspect of mathematics as *natural mathematics*. In the next section we will describe mathematical operations that are not natural, as they did not afford any evolutionary advantage in the hundreds of thousands of years during which the human genome was formed.

It is reasonable to assume that the ability to recognize geometrical elements gave an evolutionary advantage. As sources of food and water have typical geometric shapes, being able to recognize those shapes correctly constituted an advantage in the competition for sources of sustenance. But is there any evidence that, as a result of evolution, the recognition of geometric shapes is carried by the genes? We will soon turn our attention to such evidence but will first introduce what is known as the golden cut, or the golden-ratio rectangle.

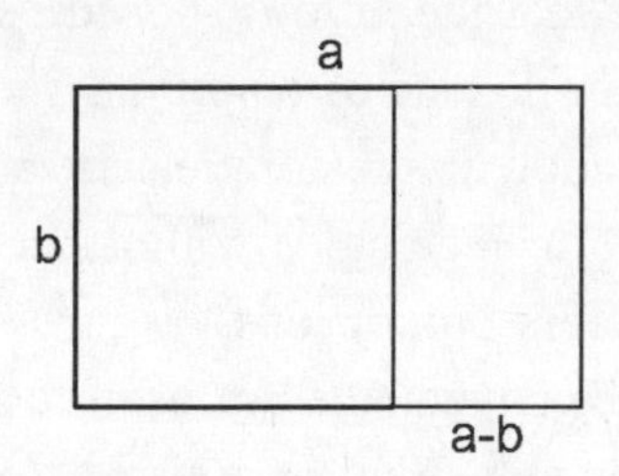

The golden-ratio rectangle is one in which the ratio between the longer side and the shorter side is such that if a square with sides the length of the shorter side of the rectangle is removed, the sides of the rectangle that remains will have the same ratio as the original one. We note, although this is not relevant to our tale, that it is not difficult to calculate the numerical value of the golden ratio (and skipping the calculation below will not impair the reader's understanding of what follows).

Denote the length of the rectangle by a, and the width by b. The relation required between the ratios is expressed by $\frac{a}{b} = \frac{b}{a-b}$. If we denote by x the desired ratio $\frac{a}{b}$, the unknown x satisfies the quadratic equation $x^2 - x = 1$, the solution of which is (recalling secondary-school mathematics) $\frac{1+\sqrt{5}}{2}$. This is the golden ratio, approximately 1.6180 in decimal numbers.

The golden ratio appears in many instances and processes in nature, and several of its attributes were known in ancient times. It has been identified in ancient architecture. For example, the dimensions of the Parthenon in Athens are amazingly close to those defined by the golden ratio. The ratio can also be discerned in Leonardo da Vinci's paintings; he also referred to it in his mathematical writings, although he did not state that he used it in his art.

The discovery of the golden ratio in various and sometimes-unexpected forms in nature resulted in the ancients attributing mystical properties to it, and they even referred to it as the *divine proportion*. For many years, a lively debate continued and is still continuing today among historians and artists on the question of whether builders and artists in ancient times made conscious use of the golden ratio in their architecture and art, or whether its frequent appearance is due to the fact that it is aesthetically pleasing. We will not join this open debate at this point but will just note that the ratio is indeed very pleasing to the eye. This has been proved in dozens of empirical studies, including studies showing that infants react with greater pleasure and calm to golden-ratio rectangles than they do to rectangles with other proportions, including those with greatly different proportions.

This needs to be explained. We are used to the fact that the pleasantness of a drawing or a painting or a shape to the adult eye is highly dependent on familiarity and education. For example, the attitude toward modern art initially was almost hostile, and it moderated over the years as the general public became more and more familiar with it. Babies have not had time to become familiar with any particular shape or form. What, then, is the origin of their preference for the golden ratio? The answer is simple: Evolution. An examination of the dimensions of the human head reveals that they are close to the golden ratio. Likewise, the proportions of sections of the human face, such as the ratio of the width and height of the eyes, the height and the width of the ears, and so on, are also close to the golden ratio. The evolutionary advantage to an infant who can recognize and is happy to discover a figure with those proportions is clear. Babies who are calm when they see their mother approaching, in contrast to exhibiting discomfort or even crying for help when they see a bird of prey nearby, have a greater chance of surviving. Hence the feeling of greater comfort when confronted with rectangular forms that have proportions similar to those of the human face rather than other forms is etched into the human genes. This has nothing to do with the golden ratio itself. In fact, research shows that babies also feel at ease with the shape of a hand, and the evolutionary reason for that is self-evident. Evolution rewards a baby who reacts with discomfort if held by a predator compared to its reaction when held by a human being. I would hazard a guess that if it were possible to perform similar experiments with birds, we would find that of all geometric forms, the most pleasing to a young chick would be an acute-angled triangle.

At this stage we may still wonder whether babies may have learned to feel comfortable with ratios similar to the golden ratio in the first weeks after their birth. The answer lies in the signs of discomfort and fear when they are faced with certain forms. Psychologists claim that about one-tenth of all children experience a primeval fear of clowns. Recently an occupation known as medical clowning has become widespread. It involves clowning activities meant to relax and help children requiring hospitalization. But cases have also been reported in which the activities of the medical clown only harmed the child, and the condition of the terrified children deteriorated when they saw the clown. This too is related to geometry and its evolutionary roots. The sight of a clown, with all his bright colors and the nonhuman proportion of his limbs and head, calls into play the same genes that make infants cry for their parents' help when they see a multicolored bird of prey or a tiger approaching. It is unreasonable to think that in the modern world children would "learn" to be afraid of clowns. These innate features are the inception of geometric recognition. (We will often refer below to the simple but illustrative metaphor of confrontation with a tiger.)

Another basic mathematical ability that almost certainly played a role in the evolutionary struggle is the ability to identify patterns. I am not familiar with controlled experiments that show that the tendency and ability to recognize patterns is ingrained in the genes, but imagine early man with a tiger stealthily creeping up on him in the grass, leaving a trail of flattened grass. The ability to identify the trail as a source of danger could be life-saving. Ability to recognize patterns is not restricted to visual patterns. Consider for instance patterns of sound. For most of us, hearing very few notes is enough to recognize a pattern and sometimes to identify an entire tune or melody. As recognizing patterns is an attribute that is helpful in the evolutionary struggle, those who had this ability had more offspring than those that lacked it. It is thus almost certain that the tendency to recognize patterns is passed on genetically. Less harm is caused by seeing a pattern where none exists than by failure to identify an existing pattern. Thus, the evolutionary tendency to identify patterns also results in identifying ostensible patterns, including patterns that do not exist. We can take the Bible code as an instance of incorrect identification. By constructing sentences consisting of only every *n*th word of the text in the Bible, it can apparently be shown that many of the events in modern life were allegedly foreseen in the Old Testament. Careful statistical tests proved that these patterns have no scientific reality. From the outset, however, the tendency to find patterns overcame scientific caution. In later sections we will come across other mental errors deriving from discovering patterns where they do not exist.

Much of mathematics, both in research and in the various stages of learning mathematics, focuses on the identification of patterns in sequences. Here are a few simple exercises. Continue the sequence:

$$2, 4, 6, 8, 10, \ldots$$

At a relatively early age, children will recognize the sequence of even numbers and will correctly give the next numbers in the sequence, 12 and 14. More knowledge is required to recognize the following sequence:

$$1, 4, 9, 16, 25, 36, \ldots,$$

but it is not difficult to see that the numbers in the sequence are the squares of the numbers 1, 2, 3, 4, 5, 6, so that the following numbers will be 49 and 64. We should point out and emphasize that these sequences do not necessarily continue as we have suggested. In other words, these extensions of the sequences do not derive from a logical necessity. Moreover, the answers are culture dependent. Here is an exercise attributed to the mathematician and historian Morris Kline. Continue the sequence:

4, 14, 23, 34, 42, 50, 59, . . .

The answer? 72. The numbers in the sequence are the numbers of the streets at which the Manhattan Subway C stops, and the next station is at 72nd Street. I would guess that if regular travelers on the New York subway were given this exercise, many would have given the answer 72. I have deliberately avoided saying that they would have given the right answer, because this is not a matter of right and wrong. The answer is right if that is what the questioner intended. It is easy to see, however, that the human race has the inborn intuition to continue series such as the above in a reasonable manner, and to understand what the questioner wants. (We will discuss again this exercise in the last chapter of this book.)

Clearly one must not exaggerate, and the story of the four-engine airplane flying from New York to London comes to mind. About an hour after takeoff, the pilot announces that one of the four engines has failed, but there is nothing to worry about. The other three are working as they should, and the flight would just take nine hours instead of the originally scheduled six. A short while later the pilot announces that a second engine had ceased functioning, but not to worry, the only effect was that the flight would now take twelve hours. A while later comes the third announcement, that the third engine is now out of action, so the flight time is now fifteen hours. At this point a passenger jumps up and asks, "Is there enough food and drink on board in case the fourth engine fails and the flight takes eighteen hours?" (It would be interesting to ask mathematics students to complete the sequence in the event that the fourth engine stopped working.)

Some continuations of sequences, even if there is no logical necessity, are directly connected with natural phenomena. Let us take, for example, the following sequence:

1, 1, 2, 3, 5, 8, 13, 21, . . .

Each number (from the third) in the sequence is the sum of the previous two numbers, so that the next two in the sequence would be 34 and 55, and so on. This is the Fibonacci sequence, named after the Italian mathematician Leonardo Fibonacci, or Leonardo of Pisa (1170–1250), whose book *Liber Abaci* (1202) included extensive development of the properties of this sequence. It reflects many aspects of development and growth in nature, as well as mathematical properties that are interesting in themselves. We describe one use of the sequence here.

Certain trees, including some types of mangrove, increase in number by a branch taking root in the ground and growing into a new trunk. A year has to pass, however, until a branch of a young mangrove can send out one of its branches from which a new tree will grow. Assume that a young mangrove is planted in the ground. After one year there will still be one mangrove tree, but after two years a branch of the first tree will also be growing, so there will be two mangroves. This is the beginning of the sequence 1, 1, 2. The next year, only the first tree can send out a branch to take root, so in the fourth year there will be three trees. The year after that, the two oldest mangroves will send out a branch each, so there will be a total of 2 + 3 = 5 trees growing, and we already have the sequence 1, 1, 2, 3, 5, and so on. Each year the number of new trunks is equal to the number of older trees (more than a year old), and the sequence describing their number of trees is the Fibonacci sequence. We will not expand the scope of this matter beyond the example quoted, but I will just add that if a number in the sequence is divided by the preceding number, the further along the sequence we go, the closer is the result to the golden proportion discussed above. This is another fact that convinced the ancients that they were observing a divine proportion or ratio. The fact that series whose extensions can be discovered intuitively are reflected in natural phenomena boosted the tendency to develop the ability to identify patterns throughout the generations.

We will summarize the observations in this and the previous section by stating that we can point to, and to some extent corroborate by means of experiments, mathematical abilities that throughout hundreds of thousands of years of evolution afforded an advantage in the evolutionary struggle for survival. The processes of mutation and selection by which evolution shaped the human race resulted in those abilities being etched into human genes.

5. MATHEMATICS WITH NO EVOLUTIONARY ADVANTAGE

In this section we will examine a number of aspects of mathematics that, apparently, are not carried by our genes because they did not provide an evolutionary advantage during the formation of the human species (other nonnatural aspects of mathematics will be discussed later on). The current discussion is speculative, but further on we will present evidence corroborating the observations made here. We emphasize once again that the lack of an evolutionary advantage we are referring to relates to a period in which the genes determining the human species were developing. That is why mathematics of the type we will discuss here is not natural to intuitive thinking. This does not mean that this aspect of mathematics is not important or useful. Just the opposite. This type of mathematical ability provides a great advantage in the later evolution of human societies, but the time that has elapsed since human societies developed is not long enough for these abilities to have been etched into their genes.

The language of mathematics makes much use of quantifiers, expressions such as "for every," or "there exists" that appear in mathematical propositions. For example, Pythagoras's famous theorem, which was proved as early as two thousand five hundred years ago, states that *for every* right-angled triangle, the sum of the squares on the two sides equals the square of the hypotenuse. The emphasis is on the quantifier "for every." Another useful claim states that every positive integer is the product of prime numbers. A recent famous example is Fermat's last theorem. The hypothesis that it was correct was formulated as early as the seventeenth century but was unproven until the proof by mathematician Andrew Wiles of Princeton University, which was not published until 1995. The theorem states that for every four natural numbers (i.e., positive integers) X, Y, Z

and n, if n is greater than or equal to 3, the sum $X^n + Y^n$ cannot equal Z^n. Throughout the thousands of years of development of modern mathematics, the proof that a particular property *always* holds was considered an achievement.

However, is it natural to examine whether a particular property *always* holds? When something occurs repeatedly under certain conditions, does it naturally give rise to the question whether it occurs *every* time those conditions hold? Not so. If experience shows that a tiger is a dangerous predator, the conclusion drawn is that if one meets a tiger one should flee or hide. Losing energy or time in abstract thought about whether that particular tiger always devours its prey, or whether every tiger is a dangerous predator, would not afford an evolutionary advantage.

Another concept often referred to in mathematics is the concept of infinity. The Greeks proved that there is an infinite number of prime numbers. Is the urge to prove this statement a natural one? On observing many elements, is it reasonable to ask whether there is an infinite number of them? Again, I think it is not. Imagine ancient man discovering that a certain region is teeming with tigers. Is it worthwhile for him to consider whether there is an infinite number of them, or would it be preferable for him to get as far away as possible from that area as quickly as possible? The question "Is there an infinite number of tigers?" and even the question "Are there many more tigers than the large and dangerous number that I have already seen?" are academic questions, which will only harm those who devote time and energy to them and hence will impair their chances of surviving in the evolutionary struggle.

Another type of claim developed by mathematics is expressed in the reference to facts that *cannot* exist. A statement such as "If A does not occur, then B will occur" is commonplace among teachers, students, and researchers of mathematics. We will come across many such examples further on. This way of thinking is also not natural. Activity of the human brain is based on association, on the recollection of things that happened. To base oneself on an event that did not take place may be possible and useful, but does not come easily or intuitively. When you enter a room, you look at what is in it and devote less thought to what is not there. We should repeat that we are not claiming that searching for an infinite number of mathematical elements, or proving that a certain property always holds, or relating to the negation of a possibility is an unworthy, unimportant, or uninteresting activity. What we are claiming is that those activities are not natural and that without a mathematical framework that suggests these possibilities, a reasonable person or an untrained student would not intuitively ask those questions.

Another attribute that is not innate in human nature is the need for rigor and precision. Mathematics is proud that a mathematical proof, provided it does not contain an error, is like an absolute truth. Mathematics therefore developed techniques of rigorous tests intended to lead to that absolute truth. Such an approach cannot have been derived from evolution. Genes do not direct humans to act rigorously to remove any possible doubt. The following anecdote illustrates this convincingly.

A mathematician, a physicist, and a biologist were sitting on a hill in Ireland and looking at the view. Two black sheep wander past them. The biologist says: "Look, the sheep in Ireland are black." The physicist corrects him: "There are black sheep in Ireland." "Absolutely not," says the mathematician, "In Ireland there are sheep that are black at least on one side."

Is the mathematician's claim, however rigorous and correct it may be, reasonable and useful in daily life? Of course not. In that sense, life is not mathematics. In life, even in ancient times, it is and was worthwhile and desirable to allow a lack of rigor, and even to allow errors, in order to achieve effectiveness. If a tiger's head can be seen above a bush, a man should not insist on being precise and saying that it has not been proven that the specific tiger has legs, but instead he had best distance himself from there as fast as he can.

We have claimed that the use of quantifiers and the interest in negatives or the reference to facts that cannot exist were not absorbed into the human brain during the evolutionary process and are not intuitive. Indirect evidence supporting this claim may be derived from studies that examined how many mathematical operations the human brain can perform consecutively. Calculations such as addition and subtraction can be performed one after the other almost without limit. A person can be asked to perform a long series of multiplications, additions, division, and so on, and if he manages to remember the order, for instance by discovering a pattern in it, he can internalize the instructions and develop intuition regarding the next operation. This does not apply to quantifiers and negations. "Every dog has a collar that is not green." That statement uses three concepts of logic: *every*; *has*; and *is not*. Studies have shown that even if someone can remember the order of the operations, the largest number of quantifiers that the brain can absorb is seven. Beyond that, even the most capable person cannot assess the outcome of the operation. It is interesting that the limit to the number of logical operations the human brain can absorb is seven, the same number as the maximum number of elements that animals can identify (see section 2 above). Other indirect evidence is provided by the existence of certain individuals, some of them autistic and some with Asperger's syndrome, who can perform complex arithmetic calculations with amazing speed and accuracy. However no individuals have been found who can similarly perform complex logical operations. The reason is apparently that the ability to perform arithmetic calculations exists in the brain naturally and is strengthened disproportionately in people whose limitations do not allow them to develop other abilities. Logic is not one of those extreme abilities.

Why is it important to identify mathematical abilities that are innate by virtue of evolution and to identify other attributes that are not innate? Humans think intuitively, associatively, and it is possible and easy to develop intuition based on natural abilities. Abilities contained in the genes are easier to develop, nurture, and use. It is harder to do that with abilities that are not natural to the human species. The recognition that there is a distinction between those two types of mathematical operations and understanding the source of that distinction are important to the understanding and utilization of human thought. In the sections that follow, we will see how these differences are significant to the development of mathematics, and in the last chapter of the book, we will discuss the implications of recognizing these differences for teaching mathematics.

6. MATHEMATICS IN EARLY CIVILIZATIONS

In this section we will review the mathematics that developed in the Babylonian, Assyrian, and Egyptian kingdoms. We will also look at the mathematics that developed independently and somewhat later in the Chinese dynasties. Although this survey does not cover the mathematics created in those realms exhaustively, it does correctly reflect the type of mathematics that developed. In particular, we will see that its development clearly traces what we have called the evolutionary advantage. These advantages of mathematics not only afforded humans an advantage over other living beings, but they also gave advantages to societies that developed mathematics over others that did not. The societies that ruled were those that developed the most up-to-date mathematics and that used it to establish and expand their power.

Reference to numbers and arithmetic existed prior to the Babylonian and Egyptian civilizations, but there is no direct evidence about where this mathematics existed or its level of development. Based on those remote tribes discovered in the last few centuries whose languages included only the numbers 1, 2, 3, many, we may assume that the mathematics these remote tribes used was minimal. In contrast, in 1960 human bones were discovered in the Belgian Congo that were dated to 20,000 BCE, and on them were signs that archaeologists and anthropologists believe express counting up to and beyond the number twenty. Thus we may conclude that when man lived and developed in small groups, was nomadic, and subsisted mainly by hunting, he used and even developed simple mathematics, which we referred to in previous sections as giving an evolutionary advantage.

The Babylonian kingdom was a mighty one, certainly for its time. Its origins date back to 4700 BCE. Its culture was based on Sumerian culture. Later, the Akkadian civilization became predominant, leading to cultural, economic, and social progress. The Akkadian contribution is attributed mainly to King Hammurabi, who ruled around 1750 BCE and who is famous mainly through the Code of Hammurabi that constituted the first-known comprehensive code of social behavior in the world. Around the year 1000 BCE, Assyrians started migrating from modern day Iran (Persia) and eventually dominated the Middle East until the Greek conquest under Alexander of Macedon, also known as Alexander the Great, in 330 BCE.

Our knowledge of Babylonian mathematics is based mainly on large numbers of potsherds discovered that served as the main means of written communication throughout the years of the kingdom. A particularly large collection was found on the site of the ancient Sumerian city of Nippur. Much of this was transferred to Yale University, and work on deciphering the writing has not yet been completed. Babylonian writing was in cuneiform, which had signs for numbers. The system used was based on the position of the digit, similar to the current decimal system of writing numbers, but for reasons that are not quite clear, the system of numbers was to the base of 60 (the decimal system was developed in India in about the sixth century CE and was introduced to the West by the Arabs in about the eighth century, but it was fully adopted by Europe only in the sixteenth century). The Babylonians did not have a symbol for the zero of current times. If we were to adopt the Babylonian system, the number 24 would signify twenty-four and also two hundred four. The reader would have to determine from the context what the writer intended to convey. In cases where the writer's intention was not clear, there would be a space to indicate the difference, but this occurred only in the last centuries of the kingdom. Thus, a space between the 2 and the 4 in the above example would indicate that the writer meant two hundred four. (In some potsherds, it was found that in places where we would currently write "0" there was a symbol that in texts served as a space or as a sign of separation. Some interpret this as the first use of the symbol for zero as a number.) Furthermore, the base 60 was not the only one used; sometimes the base 20 or 25 was used. In those cases too the reader had to decide from the context what base the writer was using. As far removed in time as we are from that practice, it seems to us that such usage was strange and must have made it difficult for the reader. We should be aware, however, that we act similarly in nonmathematical writing. Lack of clarity and even ambiguity are very common in both spoken and written language. The reader can generally understand what the writer means from the context. The reason for lack of clarity is evident. Precise formulations that would leave no room for misunderstanding would require great effort that would in general not be worth the benefit to be gained. Less precision is more efficient and hence is preferable in the evolutionary struggle. The Babylonians considered mathematical expressions as part of their language and did not think that they had to be more exact than the nonmathematical expressions.

Among the hundreds of thousands of potsherds found, many contain tables and calculations. The calculations include tables with sums of numbers and of squares, tables with interest on loans that may be taken, and even arithmetic exercises showing calculations of compound interest. Clearly these are our interpretations of what is recorded on the potsherds. The writing itself does not include explanations. We can generally assume that there was a commercial need for these calculations, but potsherds were also found with calculations whose purpose is unclear. One potsherd was inscribed with a calculation that in our current notation has of the form

$$1 + 2 + 2^2 + \ldots + 2^9 = 2^9 + (2^9 - 1) = 2^{10} - 1.$$

(The notation for raising numbers to different powers was not used until the sixteenth century, when René Descartes introduced it.) Other similar calculations involving powers were also found. It is clear that the Babylonians knew how to perform calculations involving numbers raised to different powers, but nowhere were explanations or formulae found for performing the calculation. Other potsherds contained calculations of areas and diagonals of rectangles, as well as calculations of the radius of a circle. In the terminology of today, one calculation gave the ratio of the circumference of a circle to its diameter as 3, and in another it was given as $3\frac{1}{8}$. These values are not far from the exact value of π found later, but there is no evidence that the Babylonians knew that the ratio of the circumference of a circle to its diameter was constant or that they tried to prove it. Babylonian mathematics lacks the element of proof, whether rigorous or not.

One of the better-known potsherds found at Nippur is known as Plimpton 322, the number taken from the catalog of potsherds in the collection in

Yale. The completion of the table is shown below; on the original potsherd, the left column is missing.

(120)	119	169
(3456)	3367	4825
(800)	4601	6649
(13500)	12709	18541
(72)	65	97
(360)	319	481
(2700)	2291	3541
(960)	799	1249
(600)	481	769
(6480)	4961	8161
(60)	45	75
(2400)	1679	2929
(240)	161	289
(2700)	1771	3229
(90)	56	106

The potsherd has been dated to 1800 BCE. It is not easy to decipher the writing, but it is generally accepted that apart from three mistakes, which can also be explained as the writer's errors, the potsherd shows the last two numbers of Pythagorean triples, that is, triples of positive integers that fulfill the equation $A^2 + B^2 = C^2$. The connection with Pythagoras's theorem (which we will discuss further on) is clear: the equation applies to the two sides and the hypotenuse of a right-angled triangle. We can see that the Babylonians who lived almost four thousand years ago managed to identify a pattern between numbers, to calculate, and to present Pythagorean triples with very large numbers, proving that they understood the geometric significance of the Pythagorean triples. Potsherds were also found with such exercises as the following: A rod is standing straight against a wall and its top is at a height of 13 cubits; the top slips down by one cubit. What is the distance of the bottom of the rod from the wall? The answer, that is, 5 cubits, is obtained by using the Pythagorean triple 5, 12, and 13. In all the exercises of this sort that were found, it seems that pre-calculated triples were used. There is no evidence of a formula or a method that the Babylonians used to perform these calculations, and no evidence exists that they expressed a hypothesis about the generality of the Pythagorean relation.

In addition to the lack of rigor in formulation and proof, it appears that the Babylonians were not strict about being exact in their calculations. In multiplication tables, mistakes are found that clearly derive from the fact that the writer did not consider precision in the answer to be important. An approximate result that would suffice for all practical purposes was sufficient. Mathematics was a practical tool and not an intrinsically theoretical subject.

The Chinese also developed and used mathematics that was quite advanced in their time. It developed later than Babylonian and Egyptian mathematics but without direct contact with those cultures. Our knowledge of Chinese mathematics is based on Indian writings of the first centuries CE that were copied and probably revised by the Arabs in a later period in the first millennium CE. Both the Indians and the Arabs were aware of the mathematics developed in Babylon and Egypt and later on in Greece, and this should be borne in mind when examining their interpretation of Chinese mathematics. We will consider only one element, and that is the relation between the sides of a right-angled triangle. Similar to the Babylonian texts, illustrated Chinese texts attributed to the twelfth century BCE show many exercises calculating lengths and areas based on the ratios in Pythagoras's theorem. For example, a wooden rod of length 6 cao (a Chinese measure) is up against a wall. The bottom of the rod is moved to 2 cao from the wall. How high will the top of the leaning rod be from the bottom of the wall? These textbooks show how to find the height at which the top of the leaning rod will touch the wall and give many concrete problems with given numbers. Although it is clear that the writers of the textbooks followed a general system in their writing, there are no indications that they tried to prove that their methods always worked or even that they tried to state their method in general terms.

The Egyptian kingdom dates back to about 4200 BCE, and it ruled under various dynasties until the Greek conquest around the fourth century BCE. There is no direct evidence about mathematics that developed in the early periods of the kingdom, but indirect evidence enables us to draw conclusions about the level of Egyptian mathematics. The construction of the pyramids, for example, required extensive knowledge of geometry and a very highly developed computational ability. The Great Pyramid of Giza, near Cairo, was built around 2560 BCE. Its base is a square, and if we divide the perimeter of the base by twice the height of the pyramid, the answer is remarkably close to π. It is unlikely that this proves that the designers of the pyramid left a hint that they knew what π was. The construction of the Abu Simbel temple in southern Egypt, on the banks of what is now Lake Nasser, definitely required advanced knowledge of engineering and astronomy. Once a year, in the afternoon hours, the rays of the sun illuminated the statue of King Ramses II. Many people are amazed by the enormous size of the pyramids and wonder how they were built with the resources available to the Egyptians of that time. I personally am not so taken by the huge size of the pyramids. A termite hill today is no less monumental, relative to the size of termites, and from an engineering point of view it is even more complex, as the termites take into account wind directions, the danger of flooding in that area, the need for ventilation in their tunnels, and so on. We understand how evolution developed those building abilities in termites. Due to the time that has passed, we understand less about the building methods of the pharaohs of Egypt, and therefore we admire the results. I have much greater admiration for the ability of the Egyptians to construct such a huge structure and to place the opening to face the sun in such a way that it shines on the statue of the king exactly once a year. Trial and error—the basis of evolutionary development—is not of much help in building temples with a king's statue exposed to the sun only once a year. It was the Egyptians' understanding of engineering and the calculations they could perform that brought them to such an impressive intellectual achievement.

Our direct knowledge of Egyptian mathematics is derived from the few papyri that have survived. These too contain a wealth of exercises. Among the better-known papyri is the Rhind papyrus, after the British antiquarian who discovered it in 1858, also called the Ahmes papyrus, after the person, apparently an early Egyptian teacher, who wrote it. The papyrus is housed in the British Museum in London. It contains many mathematical exercises, including additions and equations with several unknowns. The script used by the ancient Egyptians was hieroglyphics, which consists mainly of pictograms (or pictographs, pictures used for writing), which generally represented words but sometimes were used for syllables or letters. Hieroglyphics are usually carved into stone slabs. A simpler, more popular type of writing developed alongside hieroglyphics was the hieratic script, which was written in ink on papyri, and this is what was used for the exercises found on the Rhind papyrus. Hieratic script is written from right to left, like today's Hebrew and Arabic, and numbers are to the base 10, but the position of the digits was not important. Thus, for example, the symbol for ten was ∩, and _ represented four. Hence, the number twenty-eight could be written as _ _∩∩. There were special signs for simple fractions, but there was no sign for addition. When numbers had to be added, they were written side by side, and the reader was expected to understand from the context that they had to be added. We see that the Egyptians' relation to mathematical texts was similar to the Babylonians', that is, it was like their relation to language. In other words, there was no need for greater precision in mathematics than in normal written language.

A famous exercise from the Rhind papyrus shows 7 houses, 49 cats, 343 mice, 2,401 sacks of wheat, and 16,807 weights, with the answer given as 19,507. We note that the answer is the sum of $7 + 7^2 + 7^3 + 7^4 + 7^5$. We may conclude also that the Egyptians knew about adding numbers raised to different powers. We do not know how they did it. No general formula is shown for performing this addition or other exercises. The reader or student apparently had to learn how to solve other problems from the way the examples were solved. The correctness of the solution is not proved either.

The Egyptians' engineering capabilities also indicate their mathematical abilities in geometry. The papyri also contain exercises for calculating areas. The calculation of the area of a circle enables us to derive the value the Egyptians gave to the ratio π of the circumference of a circle to its diameter. In one of the exercises, the value is $\frac{16}{9}$ squared, which is approximately 3.16049. This is a fairly close approximation to the correct value. No proof is shown however, nor is there any evidence that the Egyptians knew or even assumed that the ratio of the circumference of a circle to its diameter is constant.

7. AND THEN CAME THE GREEKS

Greek mathematics that developed in what is called the classical period, between about 600 BCE and the rise to power of Alexander of Macedon (i.e., Alexander the Great) in the fourth century BCE, reflected dramatic changes in the approach to mathematics and in the methods of development, analysis, and uses of mathematics. The method formulated in that period served the Greeks themselves in the next few centuries and has remained, almost without change, the dominant system of mathematics until the present. Before we give a brief description of the developments, we should observe that after two thousand five hundred years of learning and using the method introduced by the Greeks, and generations during which we have become accustomed to such a system of analysis and debate, it is sometimes hard to appraise the significance of the dramatic turnaround that took place then. Today, the path paved by the Greeks seems natural and self-evident, but the new ideas are in contrast to what was expected from development in which evolutionary fundamentals were predominant. Greek mathematics constituted a sharp deviation from thousands of years of earlier mathematical activity and put forward an approach that largely contradicted the path dictated by healthy intuition. It is therefore understandable why the process of familiarization with the new ideas, and the development and incorporation of the method itself, took hundreds of years. This section deals with this aspect of the main developments. The next section will discuss what motivated the Greeks to instigate this revolution in mathematics.

In contrast to the Babylonian and Egyptian periods, we have no original writings from the classical Greek period. At that time writing was done with ink on papyrus, a method learned from the Egyptians, but the papyri did not survive. We learn about the development of mathematics in the classical period from comments found in much-later writings and from later versions of the ancient texts. Although those versions are copies of the original ancient texts, the practice at that time did not require the scribe to copy exactly, and he felt free to insert or omit texts, to correct errors (and to cause new errors), and so on, all in accordance with his understanding of the material. Even the most famous book on mathematics, Euclid's *Elements*, is known to us via versions copied hundreds of years after Euclid. Historical research of that period is based wherever possible on comparisons of later texts, copies of earlier texts made by various scribes. Although the picture we obtain is not a detailed one, it seems to be complete and reliable.

Thales of Miletus (in what is today Turkey) and his successors and disciples Anaximander and Anaximenes, also from Miletus, are credited with having started the reform. Information on Thales (640–546 BCE) comes from later sources. Plutarch, who lived in the first century CE, wrote of Thales that he was the first philosopher who was not a politician. Elsewhere it is written that Thales was the first to use his wisdom for practical purposes. It is not clear how we today should understand these statements, but it seems that Thales became very rich from trading. He traveled extensively in the ancient world, learned from the Babylonians and the Egyptians, and spent several years in Egypt. He became famous as a result of measuring the height of the pyramid in Giza. The method he used was to wait until the shadow of a rod was equal to its height, and then, using the similarity of the triangles (i.e., the properties of similar triangles) he claimed that the shadow of the pyramid was equal to its height. As he could measure the length of the shadow directly, Thales was able to arrive at the height of the pyramid. Based on this, he developed the geometry of similar triangles and used it to calculate the size of ships and their distance from the shore. He did not stop there. He *proved* that all triangles with the same length of base and the same two angles from the base are congruent. The Babylonians, Chinese, and Egyptians performed similar operations, but none of them found it necessary to formulate a general theory relating to geometric shapes or to *prove* that their method of calculation *always* gave the correct result.

Whether Thales was the first to introduce the concept of proof or whether it was attributed to him later, its revolutionary impact cannot be overstated. If you have been convinced that a certain assertion is correct, it would be a waste of time and resources to go back and prove it rigorously, especially if you try to prove that it always holds. Unproved propositions may be erroneous, but to exclude errors categorically requires an effort whose utility is not generally justified. The requirement for absolute proof would be an encumbrance in the evolutionary struggle. There was a reason that throughout thousands of years of the development of mathematics before Thales, mathematicians did not try to prove propositions they were convinced were correct. However, after Thales had taken the first step and Greek mathematicians in subsequent generations had followed his path, the concept of proof became a cornerstone of mathematics.

Other crucial milestones in the formulation of the new concepts in mathematics were the contributions by Pythagoras and his school. Pythagoras came from the island Samos, not far from the coast of Italy. According to the tradition, he was born, so it is believed, in 572 BCE, and it is generally thought that he was a pupil of Thales in Miletus. He went to study in Egypt, and when he returned to Samos he found a regime of tyrants and left for the town of Cortona in Italy, which was then under Greek rule, and founded the Order of Pythagoras. All sorts of mysteries have been attributed to that order, and it is difficult to separate truth from myth. It was involved in local politics in Cortona and considered itself part of the elite or upper stratum. It thus came into conflict with the democratic regime that came to power in the town, and, according to popular history, Pythagoras himself was murdered in 497 BCE. The members of the order dispersed, joining various seminaries in Greece, but they continued with their mathematical activities according to the Pythagorean tradition for about two hundred years. They customarily attributed every important theory or mathematical result to the founder of the order, so it is not clear what Pythagoras's own contributions were and what should be attributed to his followers.

One of Pythagoras's best-known contributions to mathematics is the theorem named after him: in a right-angled triangle, the sum of the squares of the sides equals the square of the hypotenuse. This is one of the most famous mathematical theorems, and until now, hundreds of different proofs of it have been published. Beyond the discovery of the general property in the theorem, Pythagoras's main contribution in this case was his *search* for a general property. As we saw above, the Babylonian's knew about Pythagorean triangles, that is, triangles with sides whose lengths were natural numbers that satisfied the theorem, and they made a list of such. The Chinese left written instructions on how to calculate the length of a side of a triangle if the lengths of the other two sides are known, and they gave many numerical examples and illustrations of various such triangles. Calculations left by the Egyptians show that they too knew of the relation of the lengths of the sides of a right triangle in many examples of specific triangles. It did not occur to any of them to even ask whether the property applied to *all* right triangles or to prove Pythagoras's theorem even for those triangles for which they had calculated the figures. They knew about the connection between the lengths of the sides, but they used it only in the context of specific calculations.

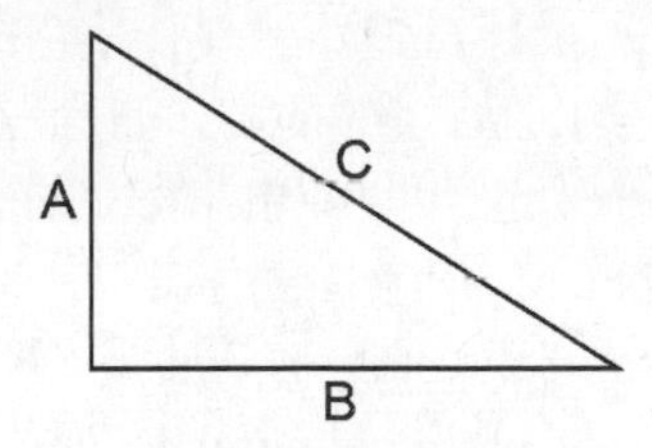

Moreover, Pythagoreans (and possibly Pythagoras himself) did not merely prove the relation between the sides of the triangle but looked for, and found, a formula according to which *all* Pythagorean triangles can be calculated.

The formula is (in the notation of today): for all two natural numbers u and v, such that u is bigger than v, define

$$A = 2uv,\ B = u^2 - v^2,\ C = u^2 + v^2.$$

A simple calculation shows that $A^2 + B^2 = C^2$, or, in other words, A, B, and C are the sides of a Pythagorean triangle. The Pythagoreans proved that all Pythagorean triangles are obtained in this manner (the claim that these constitute all the triangles appeared in Euclid's book, but without a proof).

Note the conceptual leap. The Babylonians and the Chinese compiled lists of many Pythagorean triangles; the Greeks found a proof that included all of them. The Babylonians made a great effort to discover Pythagorean triangles but did not think of trying to find a formula that would calculate all of them. Why, indeed, should anyone try to find *all* those triangles? What evolutionary advantage would be expressed by the desire to find all the numbers?

A major conceptual contribution of the Pythagoreans is related to the method of proof. In the next chapter we will discuss the connection between the Pythagoreans and their view of the world. At this stage we will note that they believed in the close link between numbers and geometry, and that the world consists of natural numbers and their ratios, that is, fractions, or in the language of the Greeks, sizes that can be expressed. To their great surprise, so it is told, they discovered that there are sizes that cannot be expressed, or in our terminology, irrational numbers. An example of such a number is the length of the diagonal of a square, the sides of which measure 1. According to one account, the Pythagoreans kept this revelation secret and threw their colleague Hippasus, who had leaked the existence of such numbers outside the Pythagorean school, into the sea. Another version of the story is that it was Hippasus himself who discovered the existence of irrational numbers and that he was thrown into the sea because of heresy, as his discovery destroyed the basis of belief in the structure of the world. One way or the other, what interests us is the following step-by-step proof of the hypothesis.

a. Consider a triangle consisting of two sides of a square with sides length 1 and its diagonal.
b. According to Pythagoras's theorem, the length of the hypotenuse is the square root of 2, denoted by $\sqrt{2}$.
c. Let us assume that the hypothesis is not true, that is, $\sqrt{2}$ it is a rational number; in other words, it can be written as a fraction or the ratio of two positive whole numbers, say $\frac{a}{b}$.
d. We can assume that either the numerator or the denominator is odd (because if they are both even, they can be divided by 2 until one of them is odd).
e. Now square the ratio $\frac{a}{b}$, and according to the hypothesis this is equal to 2, leading to the equation $a^2 = 2b^2$.
f. Hence a is an even number, which can therefore be written as $2c$.
g. Substituting $2c$ for a in the previous equation gives $4c^2 = 2b^2$.
h. Dividing both sides by 2 leads to the conclusion that b is also an even number.
i. But we specified that a and b are not both even. We have reached a contradiction, derived from the assumption that $\sqrt{2}$ is a rational number.
j. Conclusion: assuming that $\sqrt{2}$ is rational leads to a contradiction; hence $\sqrt{2}$ cannot be written as a fraction, and it is therefore an irrational number.

This type of proof is known as *reduction ad absurdum*. An argument that made use of contradiction was not only exceptionally innovative in its time, but it also runs counter to the natural way in which the human brain works. How could a claim that starts with the words "Assume that X is not . . ." be developed? The reader is invited to try to remember when and where, apart from in mathematics lessons, he or she intuitively used a thought process that assumed that something did not exist. Intuitive thought is founded on association, on links between current observation and the recognition of previous situations. A nonexistent event does not naturally arise as an association. After so many years of mathematical developments it is hard to assess how revolutionary this approach was. The real reason for the Pythagoreans hiding the discovery of irrational numbers may have been that they were not completely sure of the validity of proof via reduction ad absurdum. The uncertainty concerning such proofs reappeared in modern times, and we will discuss it in the chapter on the foundations of mathematics.

The Pythagoreans were familiar with prime numbers, that is, the natural numbers that can be divided without a remainder by only themselves and

1, and they studied them extensively. Apart from anything else, the Greeks proved that the number of prime numbers is infinite. The proof is simple.

---

a. First, note that every number can be expressed as the product of prime factors of the number.
b. Multiply n prime numbers and add 1 to the product. We get a number that we denote by M.
c. If M is a prime number, we have found one not among the n prime numbers that we had started with.
d. If M is not prime, consider a prime factor of M.
e. The prime factor can be divided into M without a remainder, hence the prime factor of M is different from each of the n prime numbers we started with as these, when divided into M, give a remainder 1.
f. Thus, in the second possibility (i.e., step d), we also found another prime number in addition to the n numbers that we multiplied.
g. We have thus shown that there are more than n prime numbers. But n is an arbitrary number, therefore the number of prime numbers is not finite and the proof is complete.

But why should anyone be interested in the question of whether the number of prime numbers is infinite? Where in evolution would the question whether there is an infinite number of any particular object be a meaningful question? Interest in the mathematical properties of prime numbers, including apparently useless properties, started with the Greeks, continued throughout generations of mathematicians, and is still today an important part of mathematical research. In current times uses of prime numbers have been discovered apart from the abstract mathematical interest, including commercial uses such as encoding, which we will discuss further on. For thousands of years the interest was purely mathematical. For the Greeks, however, it seems that the involvement in numbers was not simply motivated by curiosity but by the belief that thus they would better understand the world around them.

The next leap in the amazing change wrought by the Greeks in the development of mathematics is attributed to the Academy of Athens and its disciples, and in particular to its founder, Plato, his friend Eudoxus, and Plato's pupil Aristotle. The conceptual contribution of this group may be summarized as the formulation of the approach that bases mathematics on axioms and on logic as the essential tool in the system of deductive proof. As we will try to establish here and later in this book, these two contributions conflict with the natural intuition of human thought.

Plato (427–347 BCE) came from an aristocratic and influential family. He was a pupil of Socrates, who is considered the father of general and political Western philosophy. In his youth, Plato entertained political ambitions, but he abandoned them, perhaps because he saw what happened to Socrates, who was sentenced to death for his opposition to and criticism of the rulers of Athens. Plato traveled widely in the ancient world, visiting Egypt and the Greek colonies in Sicily, where he became acquainted with Egyptian mathematics and the Pythagoreans. On returning to Athens, he founded the first academy in the Western world. The academy had a decisive influence on contemporary science and philosophy. Plato was essentially a philosopher, and his interest in mathematics stemmed from his belief that the truth about the nature of science can be revealed only via mathematics. On the entrance of the academy he inscribed "Let none but geometers enter here." Plato went further and, in accordance with the philosophy he developed in other fields, claimed that mathematics, or mathematical results, have an independent existence in the world of ideas that are not necessarily related to the earthly reality that we experience in daily life. Specifically, we do not invent mathematical results; we discover them. The right way to do this is to formulate the assumptions, which we will call axioms, and to use them to draw mathematical truths from them using deductive logic. To this end the axioms should be simple and self-explanatory. The smaller the number of axioms, the better. In modern mathematics it is generally accepted (admittedly, not by all researchers) that the researcher can choose his axioms freely. It is doubtful whether Plato would agree. He believed that the axioms are the link between man and mathematical truth, and they must therefore be "correct."

Formulating axioms and examining the situation under prescribed assumptions is today an accepted method of analysis not only in mathematics but also in many other fields. We should be aware, however, that this method is contrary to natural human thought. It is difficult to understand how evolution gives an advantage to someone who says, "I cannot see a tiger nearby and so I assume there is no tiger in the area." What benefit will accrue for someone who ignores certain attributes just because he has not assumed them? Even mathematicians reared on the system based on axioms cannot limit their intuition to axioms. First, they solve the problem confronting them intuitively, or guess how to solve it, and only then do they check whether their solution was based solely on the axioms or also on additional assumptions or on properties not consistent with the axioms. In the latter case they must look for another solution.

In Plato's time great emphasis was placed on abstract mathematical problems such as squaring the circle, dividing an angle into three equal angles, or doubling a cube (i.e., calculating the edge of a cube whose volume is double that of a given cube), and all these using only a ruler and a compass. In other words, they were attempting to draw a square with the same area as a given circle, using only a ruler and a compass, and likewise with the other problems.

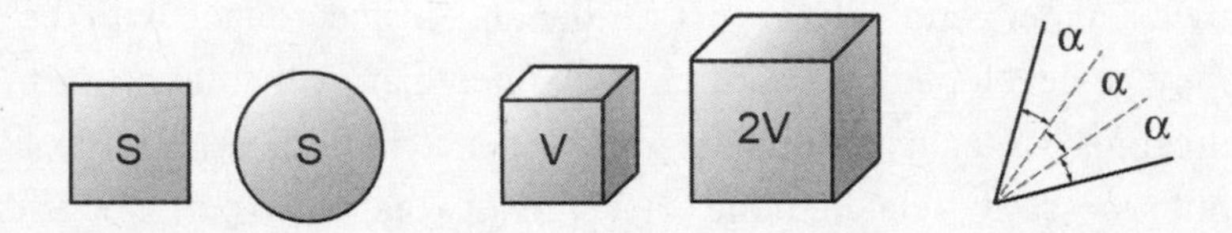

These problems were known before Plato. The problem of squaring the circle is attributed to the philosopher Anaxagoras, who thought about it while in prison, charged with impiety. The problems received even greater attention in the time of Plato, and this was at the same time as efforts were made to base mathematical proofs on as few assumptions as possible. The answer to the problems was that it is impossible to perform those tasks with just a ruler and a compass. The complete proof was not obtained until the nineteenth century. These and similar problems motivated research from the time of the Greeks until today.

The question arises, what made the Greeks interested in these questions? One story ascribes the problem of the doubling of the volume of a cube to a Greek ruler who was envious of his counterpart in a neighboring city and asked the builders of a mausoleum in that city to build one for him with double the volume. The story is not very convincing. No reasonable ruler would restrict his builders to the use of only a ruler and a compass. The origin of the idea not to use all the means available does not lie in the evolutionary struggle. Imagine ancient man fleeing from a tiger and thinking, "I wonder if I can run away on one leg only." Such individuals would not survive. Another version of the source of these problems is found in the writings of Plutarch. According to this story, the inhabitants of the city of Delos asked the local oracle for advice as to how to stop the constant disputes among themselves. The oracle's answer (not surprising, it may be said) was that they should double the volume of Apollo's altar in the city. The inhabitants asked Plato how to do this, and he, arguing that the oracle doubtless had proper mathematical intentions, interpreted the oracle's instructions to mean they should build with the use of just a ruler and a compass (also not surprising, as Plato wanted to promote his method). The point is that whichever version one prefers, questions that apparently do not relate to practical problems have occupied mathematicians since then.

Eudoxus (408–355 BCE) was born in the Greek city of Knidus on Cyprus and studied under the Pythagorean mathematician Archytas. Eudoxus traveled to Egypt to study and in 368 BCE joined Plato's academy in Athens. Eudoxus made many contributions to astronomy and mathematics. Here we will concentrate on only some of his significant contributions to the philosophy and practice of mathematics. Two of his innovations derived from his study of irrational numbers. Already in his time, many geometric dimensions were known that could not be expressed as a ratio of two integers (Plato had shown that the square roots of the prime numbers up to 17 were irrational numbers). Today we refer to both rational and irrational numbers as numbers, but in those days it was not clear in what sense irrational numbers were numbers. Eudoxus developed a mathematical theory that drew a distinction between numbers used to count individual elements (today referred to as natural numbers) and their ratios on the one hand, and numbers that measure geometrical lengths on the other. According to Eudoxus, mathematical operations in the two systems have different meanings. The interpretation of operations like addition, multiplication, and so on, on geometrical dimensions, is geometric. For example, multiplying $\sqrt{2}$ by $\sqrt{3}$ expresses the area of a rectangle with sides of lengths $\sqrt{2}$ and $\sqrt{3}$.

With regard to natural numbers, Eudoxus defined the ratio of two numbers, say n divided by m as the number of times m goes into n, and the product of two numbers, again say n and m, as the result of counting n elements m times. The result of this distinction was the separation of geometry from algebra or arithmetic, a separation that was bridged only in the seventeenth century by René Descartes. It is interesting to note that even today we use the geometric concept of "square" to express a number multiplied by itself. The need to distinguish between two types of numbers resulted in Eudoxus's using concepts that still today are considered the foundations of mathematics, that is, *definitions* and *axioms*. He defined rational numbers, a point, a line, length, and so on, and formulated several axioms precisely. This was apparently one of the first attempts to present precise definitions and axioms.

The need to give exact definitions is not a natural one. In discussions between humans, it is sufficient to reach a situation in which the participants know what is being discussed, and it is not necessary to waste time on exact definitions of the topic under discussion. To invest time and effort in defining what is a point, a length, or a plane, when everyone knows what these terms mean, seems superfluous. The process of making a definition, apparently, seemed unnecessary also during the thousands of years of mathematical development prior to the Greeks, but it did not seem so to the academicians of Athens, and mathematics inherited the practice from them and has preserved it until today.

Another contribution made by Eudoxus, essentially technical, is known as the method of exhaustion. This was an extension of a development by the Pythagoreans, who, after discovering that irrational numbers cannot be expressed by means of natural numbers, showed that they can be approximated by ratios of natural numbers. Eudoxus developed a system for calculating areas enclosed by general curves, such as a circle, by removing the areas within them, such as rectangles or other shapes whose areas are simple to calculate, until the total area to be calculated is "exhausted." Thus the area can be calculated by a close approximation. Eudoxus was very close to the concept of a limit, but he did not actually reach it. That was developed many years later by Archimedes, and is still used today as the basis of differential and integral calculus. Over and above the intrinsic technical contribution of the exhaustion method, the very presentation and formulation of the general method was itself an important contribution.

The third in the honor roll of the academy of Athens was Aristotle, a pupil of Plato. Aristotle was born in 384 BCE in a city close to Salonika, the capital of Macedonia, and, like Plato, came from an aristocratic family. His father was court physician to the Macedonian king Amyntas. At an early age Aristotle moved to Athens and became a pupil of Plato. Toward the end of Plato's life, or after his death (it is not clear enough which), Aristotle left Athens and founded the Royal Academy in Macedonia. His departure from Plato's academy was apparently the result of a difference of opinion over the scientific direction the academy was taking, and his decision may have been influenced by the fact that he was not appointed to be Plato's successor. He was also concerned about persecution by the Athenians, who considered him a Macedonian, then an enemy of Athens. He founded the Royal Academy of Macedonia, where he also taught Alexander the Great, who became the most prominent figure in the world of that time. Later Aristotle returned to Athens, where he founded his own academy, the Lyceum; its remains can be seen in Athens today. After Alexander's death, the Athenians accused Aristotle of supporting Macedonia, making him flee again. He returned to Macedonia, where he died in 322 BCE.

Aristotle's main contribution to the new mathematics was in the development of logic as a tool for analysis and for drawing conclusions. Syllogisms formulated by Aristotle still serve as the basis of logic. After such a long time, we are so familiar with these rules that their formulation seems simple, correct, and indisputable. We will claim that some of those rules, even if it is easy to agree with them via orderly thought, are not consistent with intuitive thought or with spoken language that developed naturally. We will deal with some of those rules now.

Following the generally accepted practice, we will use letters, say P or Q, to signify a claim or a statement. We will sometimes abbreviate. For example, "P implies Q" means "In every case, if P holds, then Q must hold" (in mathematics the terminology "In every case, if P is true, then Q must be true" is used). Similarly, we will sometimes write "P" when we mean that P holds (or is true), and we will sometimes write "not P" when we mean that P does not hold (namely, P is false).

The first rule is known as *modus ponens*, which is an example of an intuitive logical claim that may be reasonably assumed to be supported by reason as the result of evolution. The rule is as follows:

If P implies Q
and if P holds
then Q holds.

The following is such a claim: if it is raining, the sidewalk is wet; it is raining now, so the sidewalk is wet. The claim is an intuitive one because in the daily life of every living being, however inferior it may be, there are indications that this relation holds. The Pavlov effect is of the modus ponens type: the ringing of a bell implies that food has arrived. From this to mathematical inferences of the modus ponens type is not far, as we have seen with the animal and human examples cited above.

The next syllogism, the *modus tollens*, is different. It reads as follows:

If P implies Q
and if Q is false
then P is false.

The following is such a claim: if it is raining, the sidewalk is wet; the sidewalk is not wet, so it is not raining now. From a mathematical logic aspect the claim is correct, like the previous one. Yet it is much harder for the brain to absorb. The reason for the difficulty lies in the statement that an event is nonexistent. It is much easier for the brain to accept intuitively that an event occurs and to draw conclusions from it. To draw conclusions from an event that has not occurred is much harder. Many events do not take place, and evolution has not taught the brain to scan those nonexistent events and to draw conclusions from them.

Clearly we must differentiate between two types of negative that appear the same. "The sidewalk is not wet" is translated in the mind as "the sidewalk is dry," but it is easier to conclude that it is not raining from the claim that the sidewalk is dry than it is to draw the same conclusion (it is not raining) from the claim that the sidewalk is not wet. We can derive a recommendation from this for anyone speaking or trying to persuade someone of something: steer clear of arguments based on "not."

The difficulty in applying the syllogisms intuitively has been recognized by logicians, and they developed techniques to identify such errors, or syllogistic fallacies. Here is one example of the many that are available. A person declares the following:

No well-bred person reads tabloids.
I do not read tabloids.

He intends to imply that he is a cultured person. Many would agree, intuitively, at least. But the conclusion is not a logical consequence of his statements.

The connection between these two syllogisms is inherent in the following two rules, also formulated by Aristotle. The first is called the *law of excluded middle*:

Every proposition P either holds or does not hold, that is, P is either true or false.

The second is the *law of contradiction*:

A proposition cannot be both true and false, that is, P and not P cannot coexist.

The proof via reduction ad absurdum discussed above, and which, as we stated, is difficult to understand intuitively, is based on these rules. We want to prove P; we show that by assuming "not P" we arrive at a contradiction, so we can conclude that P holds. It sounds simple, and with a calm and well-ordered analysis the proposition is simple, but the use of this principle intuitively is by no means simple. What evolutionary advantage would a living being obtain from developing this principle intuitively? As will be seen below, the law of excluded middle will play a vital role in the study of the foundations of mathematics in the twentieth century.

Aristotle made other contributions to mathematics, physics, and philosophy, some of which we will describe later in this book. Here we will refer to his involvement in the concept of infinity. The concept of infinity did not appear in any earlier civilizations. When they referred to the term *infinity*, they meant a collection of elements too large to be counted or contained. The concept of infinity did however greatly intrigue the Greeks, especially in connection with the physical structure of the world and how long it had existed. Zeno, following his teacher, Parmenides, alludes to infinity in his paradoxes. We refer to the dichotomy paradox in section 51 and here only recall that the paradox states that a person who wishes to get somewhere would never reach his target, as he has to cover first half of the distance, then one-quarter of the distance, then one-eighth, and so on, in an infinite number of steps. Following this alleged paradox, Aristotle developed an elaborate theory of infinity that we do not describe here as it does not pertain to mathematics. Its contribution to mathematics was to distinguish between *potential infinity* and an infinite collection of elements. Potential infinity in Aristotle's terminology refers to finite collections of unlimited size, such as increasing finite collections of prime numbers. He drew a distinction between these and a set of infinite size. The latter type of infinity was not an acceptable mathematical concept, amenable to logical manipulations. Thus, according to him, the question "Are there more prime numbers or more even numbers?" is not a legitimate question. The study of these concepts was renewed by mathematicians in the nineteenth and twentieth centuries.

The contribution made by Greek mathematics in the classical period was summarized by Euclid. Not much is known about Euclid's life. He lived around 300 BCE and may have studied in the academy of Athens, but he carried out most of his work in Alexandria in Egypt. He was one of the founders of the famous academic center in that city. Although Euclid was active after the classical period, his main mathematical work (which consisted of thirteen volumes), known as Euclid's *Elements*, presented in an orderly and detailed fashion the mathematical knowledge developed in classical Greece. In addition, the *Elements* organized, established, and disseminated the novel approach developed at that time. Over and above the impressive collection of mathematical findings, the development was based on definitions, axioms, deductive proofs, and defined syllogisms. There are no copies left of the *Elements* that were actually written in Euclid's time. All the versions found (the earliest of which, fragments only, are from the first century CE, i.e., about four hundred years after Euclid; the oldest dated full version is from the ninth century CE) contain remarks, amendments, and additions by those copying the book. From a comparison of the different versions we can nonetheless conclude that Euclid himself organized and arranged the book according to the different subjects, gave the definitions, established his methods of proof, and laid the foundations of the whole approach. It is no wonder that the *Elements* became one of the most widely distributed books in the world of all time and that it has apparently been translated into more languages than any other, except for the Bible.

8. WHAT MOTIVATED THE GREEKS?

Why did the Greeks ask questions whose purposes were not clear and try to answer them via nonintuitive methods?

One reason suggested in the literature is a technical one. The Greeks found mistakes, inconsistencies, and contradictions in the calculations of the Babylonians and the Egyptians, and in order to resolve these they developed a more exact form of mathematics. I am not convinced that this was the reason. If you are uncertain about which of two calculations is more correct or you doubt the accuracy of a calculation, it is reasonable to assume that you would try to perform the calculation more accurately yourself and thus arrive at the right answer. In addition, the Greeks knew that in many areas the calculations of the Babylonians and the Egyptians were more accurate than their own.

A more plausible explanation relates to the political and economic situation in ancient Greece. Although it was a period when many wars were fought between cities and between small kingdoms, in general, a democratic aura prevailed, and political and social philosophies were highly developed. In an environment in which the study of philosophy is important, when there is no single ruler or government that requires instant achievements from its subjects, when there are no government-appointed committees that determine priorities for research, in an atmosphere when one can question and doubt everything and curiosity-driven study is a highly valued pursuit, in such an atmosphere enormous achievements can be made, even if it takes a very long time to derive benefit from them. To these considerations may be added the fact that the main contributors to developments in research came from established families, and they could study without being concerned with their livelihood and subsistence, a fact that clearly helped them to advance along unorthodox channels. These considerations explain how basic research developed, but they do not explain why it developed in nonintuitive directions and contrary to what evolution would have determined.

A documented explanation for the path followed by the Greeks derives from what are known as illusions. We will expand on this point because it will be relevant in the following chapters. The Greeks were familiar with geometrical or optical illusions, and they therefore tried to prove mathematical propositions without relying on appearances, in other words, relying only on axioms and logical deduction. We will describe two famous illusions from a later period.

The first is known as the Müller-Lyer illusion, named after the scientist who published it in 1889. The upper line in the diagram seems to be shorter than the lower one, despite the fact that they are of the same length (see the diagram). The usual explanation is that generally, in nature, we see a shape similar to the upper line when looking from the outside, for instance, looking closely at an edge of a three-dimensional cube, whereas we see the lower line as the more distant edge, looking into the cube. The brain uncontrollably corrects the signals that the eye receives, shortens the upper line and lengthens the lower one, in order to obtain the "right" length.

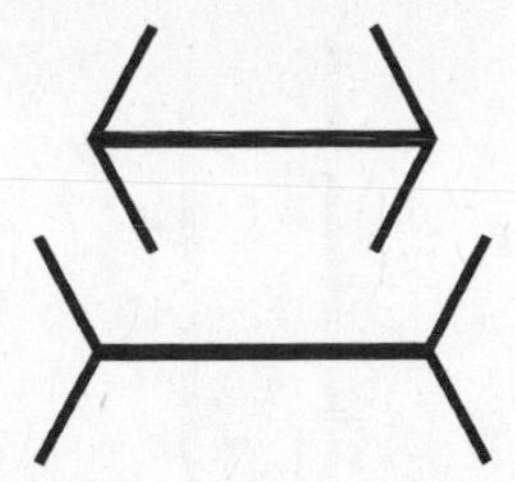

The answer lies in evolution. A correct interpretation of the signals gave an evolutionary advantage and therefore is embedded in the genes. Therefore the way the brain analyzes the information cannot be changed. By directing the eye to see only the horizontal lines we may be able to get the brain to see that the lines shown in a particular way, such as in the diagram, are equal, but we will not be able to do that with lines in situations where the brain interprets spontaneously. In any case, it would be inadvisable to change the way the brain interprets what the eye sees, because if we were to do so we would cause errors in the many situations in which the upper line is actually shorter than the lower one.

The second example is what is known as the Poggendorff illusion, published in 1860. To the untrained observer, the diagonal lines in the following figure do not seem to be sections of one broken straight line, but it can easily be shown that they are. In this instance too there is an explanation why the brain "misleads" us. The brain developed in such a way that it compares angles, not lines. The angles between the diagonals and the vertical lines make the brain create an illusion. Here the illusion does not derive from a correction the brain makes to the data it receives, in other words, a "software" correction; it derives from the "hardware" that the brain employs. The means by which the brain views geometry lead to such errors. In this case too the brain can be trained to avoid the error in specific cases, but it is not possible to carry out a repair that will prevent all such errors.

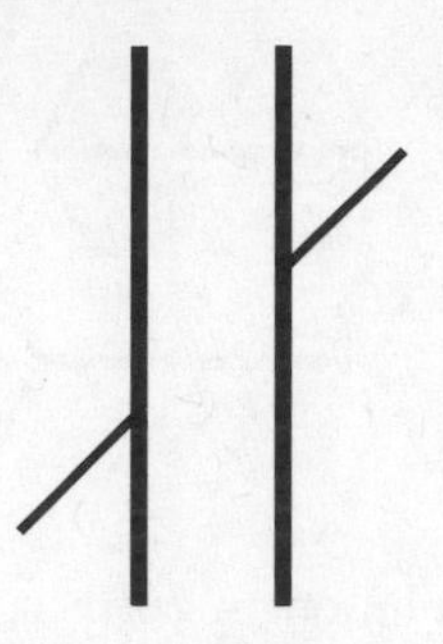

Illusions such as these occupied artists and engineers throughout the generations, and they used their knowledge to impress anyone interested in visual effects. My friend and mathematician Arrigo Cellina from Milan drew my attention to the apse in the San Satiro church, which is on Via Torino, close to the Piazza del Duomo (Cathedral Square), the main city square of Milan. The church was built in the fifteenth century. From the entrance one sees the nave, the pulpit, and, behind it, a large deep apse with a ceiling decorated with interesting paintings. On approaching the pulpit, however, one can see that the apse, its depth, and its dome are a very interesting optical illusion. Highly recommended.

The possibility of visual errors and illusions so captivated the Greeks that it drove them to extremes. In order not to rely on appearance or what one sees, for example in mathematical proofs about triangles, they would draw the triangles with sides that were not straight lines but curves. The purpose was to rely only on axioms, and as far as possible to avoid errors based on appearances or eyesight. Nevertheless, even the Greeks could not avoid having to relate to some extent to sketches and drawings. On the face of things, drawings are not relevant in cases of deductive proofs based only on axioms, but it seems that the brain cannot cope with abstract axioms without the help of metaphors or without reference to a model or previous experience. This characteristic of the brain, or perhaps this limitation, arises every time abstract mathematics is used to describe geometry or a natural phenomenon. In mathematical depictions of nature, as we will see in the next chapters, even though mathematics can stand alone and does not need a visual model, the brain does need a model or a metaphor to enable it to analyze and absorb the mathematics.

CHAPTER II
MATHEMATICS AND THE GREEKS' VIEW OF THE WORLD

Are there "holy" numbers? • What is the connection between numbers and musical notes? • Who discovered that the world is made of atoms? • Is mathematics discovered or created? • Why, in Aristotle's view, do we grow teeth? • Why is "up" considered improvement while "down" is seen as deterioration? • Do stars have to move in circles? • The Sun and the Earth, which moves around which? • What is preferable, precision or simplicity?

9. THE ORIGIN OF BASIC SCIENCE: ASKING QUESTIONS

In the course of the thousands of years of development of human society in the ancient Babylonian and Egyptian civilizations, those societies acquired extensive knowledge of mathematical calculations related to commerce, agriculture, and engineering on the one hand, and the prediction of celestial events on the other. The construction of the Abu Simbel temple is, as we have mentioned, a striking example. Its designers succeeded in planning a huge temple such that the rays of the Sun would illuminate the statue of King Ramses II just once a year. Other ancient cultures also had the ability to make calculations that take into account the movement of the heavenly bodies. For example, bearing in mind when it was built, the structure of Stonehenge on Salisbury Plain in England shows amazing harmony between the positions of the huge rocks and the times of the rising and setting of the Sun in the different seasons of the year. Years of observation of the movement of the stars led to detailed knowledge of the movements of the planets in the sky. The Egyptians and the Babylonians compiled detailed calendars that included measuring the length of the year (365 days, according to the Egyptians), and the relation between the lengths of the lunar months (based on the movements of the Moon) and the solar year (based on the Sun). They used these calendars to plan agricultural activity, among other things. Nevertheless, in all those many years, no attempt was made to discover the principles underlying those observations, that is, to look for rules that could explain the movements of the celestial bodies. Wherever attempts were made to explain the movements of the stars, these were confined to references to the gods responsible for the various heavenly bodies. Until the advent of the Greeks, developments in mathematics were in line with what we would expect from the principles of evolution. In other words, the mathematics that developed first was that which made a direct contribution in the evolutionary struggle between human beings and the other species. Later, mathematics developed that helped in the struggle between different societies competing for the same resources of land, food, energy, and the like. Basic questions are luxuries that the evolutionary struggle cannot allow. Until the Greeks, human beings were not free enough to ask basic questions.

The first evidence of an attempt to construct a physical description of the world is attributed to Thales and his successors in the city of Miletus. We met them in the previous chapter. Thales was also the first to claim that mathematics is the way to construct such a picture. The etymology of the word *mathematics* is the Greek word *μάθημα* (pronounced *mathema*), meaning lesson, understanding, learning. The word *mathematics* was also used to mean science itself (in Hebrew too, the word for science is derived from the word for knowledge). In addition to searching for the rules according to which the world operates, the Greeks also asked more philosophical questions, such as why are the various phenomena and properties of the cosmos as they are? What is the purpose of any particular law of nature? Questions of this sort are not natural questions from the evolutionary aspect. A species that starts to waste resources and energy searching for a purpose, a search that may yield results in the long term, lowers its chances of surviving in the short term. The debate on the essence and purpose of the laws of nature, teleology, started with the Greeks and continued throughout the generations and is still continuing today. For the Greeks, purpose meant a scientific reason. It is not clear what motivated Thales and the Greek philosophers who succeeded him to engage in these questions. The best explanation seems to be the academic freedom mentioned in the previous chapter. Despite the sharp change in the direction of mathematical development, away from the immediate evolutionary needs, we will see that evolution still had a major impact on the development of mathematics.

Thales himself did not come up with findings that helped understand the world (after many years Thales was credited with having foreseen the solar eclipse that occurred in the year 585 BCE. This claim is suspect, as the understanding and the mathematical tools available to him could not have enabled him to make such a prediction). However, the issues were raised by Thales and his school, issues such as what are the basic materials the world is formed from? What are the heavenly bodies made of? Where is the Earth situated? Is the Earth flat or a ball? Questions such as these launched the scientific era in human history. The principles laid down by Thales and his successors, such as basing explanations on as few assumptions as possible and looking for the simplest possible explanations, guided the development of science right through to our times.

10. THE FIRST MATHEMATICAL MODELS

The Pythagoreans were the first to formulate a model of the structure of the world. They believed that nature was based on natural numbers and looked for a reflection of these numbers in the structure of the world. They believed that the numbers from 1 to 4 had mystical, almost holy, significance and that the sum of those numbers, that is, 10, had special importance. One of the reasons underlying their belief was the direct link they saw between natural numbers and geometry. The numbers were presented as triangular numbers, square numbers, and so on, as follows:

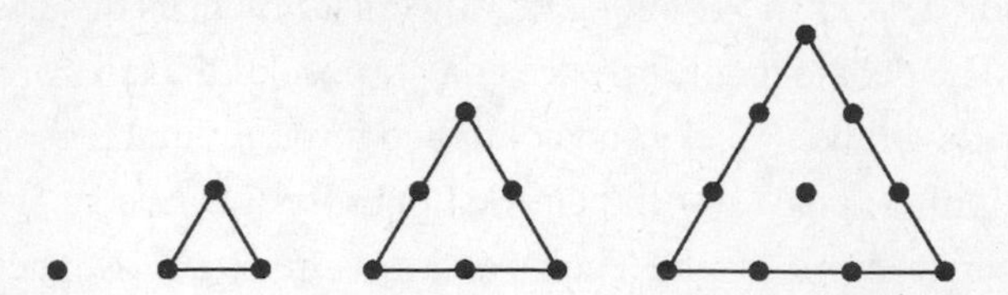

The triangular numbers were those that could be arranged on triangles: 1, 3, 6, 10, and so on, as the black dots in the illustration. The square numbers were those that can be arranged on squares: 1, 4, 9, 16, and so on, as the black dots in the next illustration.

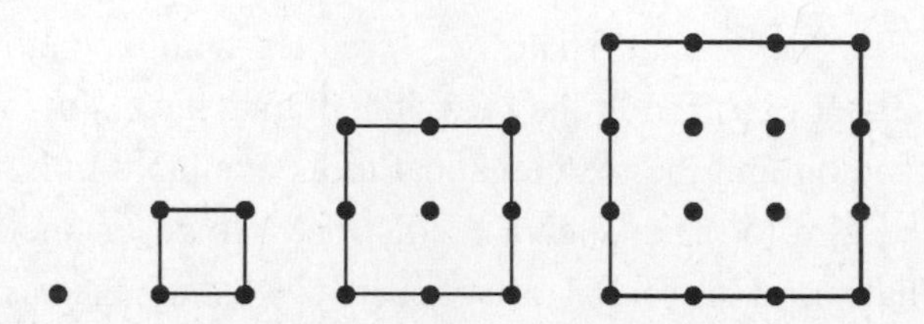

Similarly the pentagonal numbers are 1, 5, 12, 22, . . . and so on for the hexagonal numbers and beyond. As a consequence, the Greeks considered the numbers 1, 2, 3, and 4 as the dimensions of the world: a point, a line, a plane, and a space. Their sum is 10, and hence the importance, even the sacredness, of the number 10.

The Greeks used these illustrations to prove mathematical statements. For instance, it is easy to see from the diagram showing the triangular numbers that the nth triangular number in the series is the sum of the numbers from 1 to n, so that the fourth number, for example, would be $1 + 2 + 3 + 4 = 10$.

Another example of a geometric proof is the equality

$$1 + 3 + 5 + \ldots + (2n - 1) = n^2.$$

In schools today this equality is proven by induction (namely, checking it for $n = 1$ and verifying that the equality for $2n + 1$ follows from the equality for $2n - 1$). Although the geometric argument shown below is based on appearance, it is simpler, and it gives a visual explanation of why the equality holds.

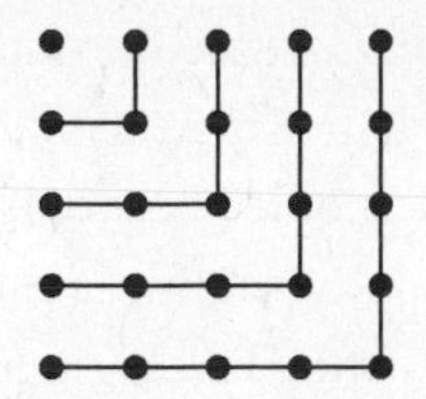

Another finding that strengthened the faith of the Pythagoreans that there was a very close link between numbers and the world was the link they found between numbers and music. By experimenting with stringed instruments they discovered that if the length of a string is halved, the note it produces is an octave higher than the original (even today, the physical mechanism that enables the ear to identify that it is the same note an octave higher is still not clear). Hence, the ratio 2:1 has physical significance; and, similarly, the interval of a fifth (in music), in the ratio of 3:2; and the perfect fourth, in the ratio of 4:3. These findings strengthened their belief that the numbers 1 to 4 had a significance beyond that of simple counting. Similarly, the arithmetic mean can be illustrated by the numbers 1, 2, and 3 (2 is the arithmetic mean of 1 and 3), and the geometric mean by the numbers 1, 2, and 4 (2 is the geometric mean of 1 and 4). Moreover, the harmonic mean can be demonstrated on the numbers 3, 4, and 6.

Reminder: The arithmetic mean of the numbers *a* and *b* is $\frac{a+b}{2}$. The geometric mean of the numbers a and *b* is $\sqrt{ab}$. The harmonic mean of a and *b* is *c* if $\frac{1}{c}$ is equal to $\frac{1}{2}(\frac{1}{a}+\frac{1}{b})$. The different notions have applications. For example, the geometric mean $\sqrt{ab}$ is the length of the side of the square whose area is the same as that of a rectangle with sides *a* and *b*. The harmonic mean is used when, say, the average speed that a given car travels from one city to another is *a* and the average speed during the way back is *b*. The average speed for the entire trip is the harmonic average of *a* and *b*.

The Pythagoreans were familiar with the various means and their link to geometry and mechanics. From this knowledge it was only a small step to the belief that the geometry according to which the world was constructed was made up of natural numbers. With the link between numbers and the world around us so close, it is reasonable to use numbers to look for a mathematical description of the world. As the Greeks considered the number 10 holy, they looked for a picture of the world consisting of ten elements. They knew of eight: the Earth, the Sun, the Moon, and the five planets known at that time. The Greeks also incorporated fire as a basic factor in the picture of the world and constructed an image of the world in which the heavenly bodies revolved around fire. So far we have nine elements. To complete the picture in accordance with their search for ten elements, the Greeks added another heavenly body, anti-Earth, and they positioned it on the other side of the fire element. That position explains why we cannot see the anti-Earth. The logic and method employed by the Pythagoreans to construct a model of this world may seem to be naive if not primitive. Yet we will see later that the scientific method employed by the Greeks is similar to the methods used by physicists throughout the generations and that are still used today, that is, the method of searching for a mathematical pattern that fits the known facts. As we have seen, to search for patterns is a characteristic embedded in human genes by evolution.

Although the Pythagoreans did not understand the nature of the planets, they did discern their irregular movement with respect to the other stars (in Greek, the word for planets and the word for wanderers come from the same root). They therefore assumed that planets are different in essence from the stars and focused on them in their picture of the universe. Some of the Greeks thought that the other stars were closer to the Earth than the planets were, the Sun and the Moon. It should be noted that the first Pythagorean model assumed that the nine heavenly bodies revolved around the fire; in other words, the model did not place the Earth in the center of the universe. The Pythagoreans also claimed that the Earth was a sphere, in contradiction to what our senses would lead us to think. Their reasoning is instructive. They stated that the Earth was perfect, and the most perfect geometric form was a sphere, and therefore the Earth is a sphere. Such reasoning contributed to the development of scientific thought throughout the ages, but at the same time, as we shall see, it constituted an obstacle to the development of science. It was not until the seventeenth century that science overcame the mental block resulting from the belief that the heavenly bodies moved in perfect circles.

The idea that nature is based on natural numbers and the understanding that the natural numbers are all made up of the number 1 added to itself a different number of times, led Leucippus in the fifth century BCE and his celebrated pupil Democritus (460–370 BCE) to claim that the world is made up of atoms. According to these two and their successors, atoms cannot be subdivided. These scientists disagreed over the question whether all atoms were the same or if there were different types of atoms, but they all agreed that the way the atoms are formed into more complex materials gives those materials their different properties, including their form, color, degree of hardness, and so on. They did not base this claim on only mathematical analogy but also gave other explanations. One was the principle of the preservation of matter, which they inherited from their predecessors from Miletus. If matter can be subdivided endlessly, at the end of the process it will disintegrate into particles without volume, and how can such particles reform into a real substance? It was only twentieth century mathematics that solved this dilemma; namely, it gave a mathematical framework in which a measurable length consisted of points, each of which was of zero length.

Another argument used by the atomists was the need to explain motion. If a substance is continuous and there is no space between one particle and the next, how can there be movement? They also claimed that the atoms, which our senses cannot perceive because of their very small size, are constantly in random and purposeless motion. A similar type of motion, Brownian motion (named after its discoverer, Robert Brown), relating to the irregular motion of minute particles suspended in a fluid, was discovered in the eighteenth century; it was later found that the movement of atoms is similar to Brownian motion. It was Albert Einstein, at the beginning of the twentieth century, who gave mathematical expression to the movement of atoms. In time, the Greek theory of atomism was recognized as being the forerunner of the atomistic approach to nature as it developed toward the end of the nineteenth century and the beginning of the twentieth. Stamps and banknotes issued by Greek governments in modern times carry an image of Democritus and the symbol of an atom. Nevertheless, the recognition of Leucippus and Democritus as the prophets of the modern atomistic approach ascribes to the Greeks greater merit than is warranted. The Greek atomists based themselves solely on philosophical study and mathematical analogies, with no supporting evidence. Aristotle rejected the claim of the Greek atomists for reasons we will discuss later, and his theory of the continuum was accepted by most of the Greek scientists and philosophers, although supporters of the theory of atomism were active until the first century CE. The atomic structure of the world reappeared and was accepted after thousands of years, but this time it was based on reliable physical evidence.

Another example of the construction of a pattern based on mathematical knowledge and adapting it to physical reality is the conformity of the perfect geometric solid shapes with basic components of nature. A geometric solid is called perfect if all its faces are identical in shape and size. A cube is an example of a perfect solid. Perfect solids were known thousands of years before the Greeks, but they were the only ones who tried to identify *all* of them. Later Greek writers attributed the discovery of perfect solids to Pythagoras, while others credited Theaetetus, a contemporary of Plato, with the discovery. In any event, in Plato's time it was known that there were only five perfect solids: the triangular pyramid, the cube, the octahedron, the dodecahedron, and the decahedron (see the diagram). On the one hand were five perfect solids, and on the other was the Greek's belief that nature had five basic elements: water, air, earth, fire, and the world. What could be more natural than to conclude that the perfect shapes reflect the main elements of nature? The matching is attributed to Plato himself: the pyramid is fire, the cube is earth, the octahedron reflects air, the dodecahedron reflects the world, and the decahedron, water.

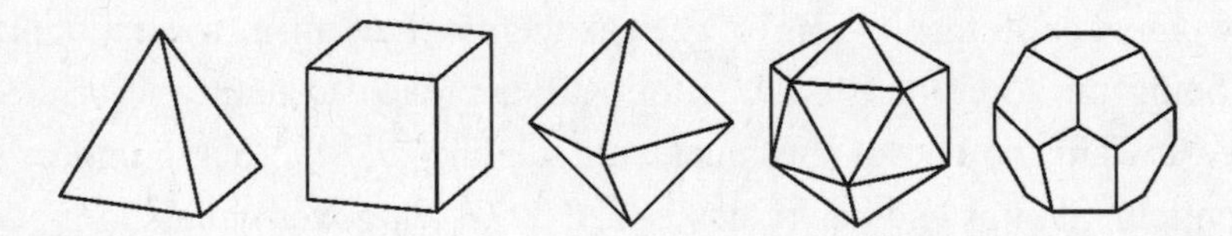

This matching today seems extremely naive, but it was in accordance with the mathematical and physical knowledge of that time. These perfect solids would play a role fifteen hundred years later, when Kepler, who knew that the world consisted of more elements than those Plato had enumerated, tried to find a function for the perfect solids in his description of the structure of the world.

11. PLATONISM VERSUS FORMALISM

The ever-closer link between mathematical findings and the depiction of nature required a philosophical study of the question "What is mathematics?" Similar questions about the essence of mathematics continue to be asked even today: Does mathematics stand on its own, or is it perforce bound up with the things it describes? What is the connection between mathematics and nature? Is mathematics created or discovered? Two of Greece's major classical philosophers, Plato and his student Aristotle, addressed these questions.

Their answers reveal differences between their views on the essence of mathematics, but they hardly differ in their ideas on how mathematics should be studied and used. Both start with axioms that are derived from nature but that are "self-evident," and for both, the next steps rely on the correct use of syllogisms and other logical tools. The differences between them are in their interpretation of the results and the role of those results in describing the world.

Plato's interpretation was firm: mathematics stands on its own. It exists in an abstract world, a world of ideas, irrespective of the people who discover it and of the phenomena it describes. The way to discover it is by using the power of logic and logical arguments. Axioms show the way to the starting point on the path of discovery (etymologically, the Greek word for *axiom* comes from the word meaning to think correctly). The axioms must be simple and obvious. We derive these simple axioms from the world around us, but care must be taken to choose only axioms that are irrefutably correct. When the axioms have been agreed upon, progress is made only through deduction, according to the rules of inference and logic, and this leads to the discovery of correct mathematics. Hence, according to Plato, mathematics is discovered, not created. The phenomena we observe in nature contain defects and are a contamination of the ideal mathematics. Nature is apt to mislead us, so that we must not rely on appearances but must derive the principles from nature and search for mathematics using the power of logic. In that way we will understand the principles underlying the natural phenomena. This approach is known as *Platonism*.

Aristotle's interpretation was different. He claimed that mathematics did not exist as an independent entity. Mathematics is simply the outcome of logical operations based on syllogisms, and the search starts with axioms. Mathematics is created, not discovered. The axioms and the mathematical results they yield have no intrinsic significance; they are formal results that have no purpose and no independent existence. Aristotle used the word *techne* for the creation of mathematics, the origin of the word *technology*. According to him, the importance of the formal operations comes to light immediately when one discovers "correct" axioms, that is, those that nature satisfies. Aristotle used the word *episteme* to describe the finding of the correct axioms, which later gave rise to the word *epistemology*, meaning the study and theory of knowledge, which provides the basis for the common sense that describes how the world operates. When the axioms are correct, the resulting mathematics derived from them constitutes a precise record of what is found in nature. Hence, that is the way to study nature, avoiding the pitfall that could be caused by false appearances. Mathematics does not exist in its own right but only as a formal procedure, the significance of which depends on its interpretation and how it describes nature. Aristotle's approach is today referred to as *formalism*.

In both interpretations the axioms are identified following observations. Aristotle, for instance, carried out extensive dissections of animals to learn about their anatomy. But once the axioms were adopted and mathematical results were derived, neither approach deemed it necessary to carry out experiments to compare mathematical results with what actually occurs in nature. Both of these great scholars and their Greek disciples actually opposed performing such experiments. They justified their opposition to experiments by claiming that appearances are affected by optical illusions and are likely to mislead, whereas logic is irrefutable. (It was not until thousands of years later that mental illusions were recognized and studied.) Thus, as soon as we find the self-evident axioms, the path forward using the power of logic is superior to progress based on appearances. This outlook held sway for thousands of years. It was not until the seventeenth century that the currently prevalent scientific practice, which demands experiments that corroborate the theory, became established. One possible reason for the fact that both Plato and Aristotle and most of the leading philosophers who succeeded them held extreme anti-experiment views is that they came from wealthy aristocratic families who considered physical work to be menial, inferior, and even contemptible.

The debate initiated by Plato and Aristotle on the essence of mathematics has continued for two thousand four hundred years and continues to this day. Scholarly articles supporting one view or the other, or proposing improvements to them, continue to appear in professional literature. The debate has no direct effect on mathematicians carrying out research or extending the boundaries of mathematics. Thus, so the story goes, it is quite normal for a mathematician, when asked if he or she is a Platonist or a formalist, to give one answer on weekdays and the opposite answer on the weekend. The reason is that if in the course of your work you discover a startling mathematical theorem or formula, you think of it as being an independent and significant entity, as Plato tells us. During the weekend, however, when you are asked to explain the essence of the entity you discovered, it is more convenient to avoid the discussion and to hide under the cover of formalism.

12. MODELS OF THE HEAVENLY BODIES

The Greeks inherited a wealth of astronomic measurements from the Babylonians and the Egyptians. These included information on the movement of the planets, the lengths of a year and a month, the cyclicality of the relation between the solar year and the lunar year, the effects of all these on agriculture, and so on. The Greeks made great efforts over many years to improve the measurements and to add new ones, and they developed the ability to forecast celestial events with the greatest precision. For example, whereas the Egyptians gave the length of the year as 365 days, the Greeks of the fifth century BCE determined it as 365 days, 6 hours, 18 minutes and 56 seconds, which is a deviation of just half an hour from the correct figure. In 130 BCE the Greeks achieved greater precision and set the length of the year at 365 days, 5 hours, 55 minutes and 12 seconds, a deviation of only six minutes and twenty-six seconds from the true figure. These measurements were achieved following the development of advanced mathematical methods of calculation and measurement. We will not expand on those methods here, as our focus is on how mathematics explains nature.

We will now review the conceptual and mathematical contributions made by the Greeks to the development of a model of the heavenly bodies, both in the classical period and thereafter. This development reached its peak in Ptolemy's (Claudius Ptolemaeus's) model. It should be mentioned again that there is no direct written evidence from the classical period, and the information available to us relating to that period is derived from remarks in much later writings, writings that reflected the authors' points of view. Therefore, we should not place too much credence on the precision or reliability of those descriptions. We will not be able to set out here the detailed development of the Greeks' picture of the heavenly bodies, but we will concentrate mainly on the conceptual development of the models.

The part of mathematics that the Greeks used to describe the movement of the heavenly bodies was geometry. From our standpoint today, two thousand five hundred years later, this role of geometry seems obvious. The heavenly bodies, including the planets, the Sun, and the Moon, move in space, and it seems natural to find out what geometric paths describe their movement. That, however, is an incorrect conclusion. The early Greeks had no knowledge of a physical space in which the heavenly bodies existed. The stars were spots of light in the heavens, and it was unclear what the Sun and the Moon were. Moreover, the distance of those objects, if indeed they were objects, from Earth was not known. At that level of understanding, to enlist geometry to describe the movement of the heavenly bodies was a bold pioneering step and by no means obvious. Geometry was known and well developed as a mathematical tool for describing earthly objects and as a useful device in measuring and building. The Greeks used the known earthly mathematics as a basis for cosmological research.

Plato advocated a view based on the Pythagorean outlook, according to which the heavenly bodies moved on the surfaces of spheres. Consistent with appearances, Plato claimed that the planets, the Sun, and the Moon circled the Earth, and those circles were all on one plane. He realized that this description did not cover the irregular movement of the planets and the Moon relative to the fixed stars in the sky. Specifically, according to his description, eclipses of the Moon should have occurred about once a month, but they did not. It was Plato who set the objective of improving the mathematical description of the movement of the heavenly bodies.

The challenge was taken up by Eudoxus, Plato's colleague in the Academy of Athens, whose contribution to mathematics was discussed in section 7. Eudoxus accepted Plato's assumptions that the Earth was static in its place and the heavenly bodies moved in circular orbits around it. Eudoxus proposed two significant innovations. First, he allowed the circles defining the movement of the planets, the Sun, and the Moon to be at an angle to each other, that is, on different planes. Second, he allowed each of the heavenly bodies to revolve on more than one sphere and the spheres to move in different directions and at different speeds.

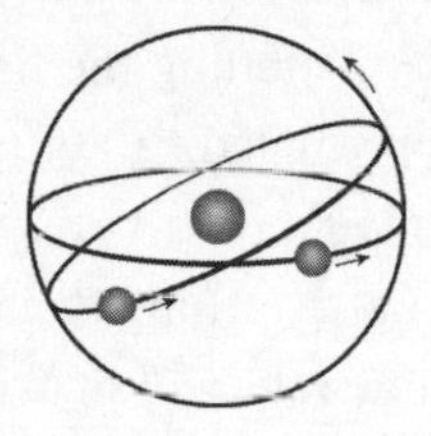

The flexibility afforded by these new ideas enabled Eudoxus, by means of sophisticated use of three-dimensional geometry, to define the number of spheres on which each planet was situated, its speed, and its direction so that most of the observations that were inconsistent with the previous model could now be explained. The movement of two planets, Venus and Mercury, still did not fit the model satisfactorily. It should be pointed out that the Greek astronomers did not stop at the geometric description of the paths of the heavenly bodies but struggled to calculate their orbits and to show how the models they proposed were consistent with the various observations and measurements they performed on the movement of the stars. The calculations required great mathematical ability combined with a deep understanding of geometry.

Heraclides of Pontus (in today's Turkey) proposed two far-reaching amendments to Eudoxus's model. Heraclides (388–310 BCE), an important philosopher in his own right, studied under Plato at the Academy of Athens and occasionally stood in for him as head of the academy when Plato was on his frequent journeys. The two amendments were as follows.

First, he claimed that Venus and Mercury did not orbit the Earth but revolved around the Sun, while the Sun itself orbited the Earth. This claim may be seen as the source of the epicycles model, a model attributed to Appolonius, whom we will meet further on. The second amendment was that the firmament and the planets do not revolve around the Earth at all, but the fact that the Earth revolves on its own axis makes us think that they do. Heraclides proposed the first amendment based on the fact that Mercury sometimes disappears behind the Sun. The assumption that Venus and Mercury orbited the Sun was more consistent with the observed movements of the planets than Eudoxus's model. Heraclides's argument in favor of the Earth revolving around its own axis was purely aesthetic. At that time, the size of the universe, including the firmament, relative to the size of the Earth, was already appreciated. It does not seem fitting, said Heraclides, that such a large sky should revolve around such a small body as the Earth, and he added that the Earth turning on its axis would explain equally well what we see. This argument based on aesthetics will recur again and again throughout all the years of the development of physical science. It is also consistent with the "self-evident" assumption, which Heraclides accepted, that the stars move in perfect circles.

We should mention that historians do not agree on the question of whether Heraclides himself actually proposed these amendments because there is no clear reference in Greek sources stating that he held those views. The attribution to Heraclides derives from the writings of Copernicus in the sixteenth century CE (see section 15). Nonetheless, whether he did or not, those ideas and others we shall refer to later on were heard in classical Greece but were not accepted by those who determined the mainstream in Greek science.

13. ON THE GREEK PERCEPTION OF SCIENCE

The chief architect of the mainstream Greek view of the model of the world was Aristotle. His rigorous philosophical approach, despite its many successes, also blocked original thought and led to the rejection of new ideas, ostensibly based on logic.

Aristotle's broad philosophical views fall outside our subject in this chapter, that is, the physical description of the world, and they encompass fields such as biology, zoology, and sociology. The philosophical approach of which Aristotle was a major proponent was that everything around us, including the laws of nature, has a purpose. Although the idea that the design of the world has a purpose arose many years before Aristotle and was presented by the school founded by Thales in Miletus, it was Aristotle who made the purpose underlying the laws of nature into a law of nature itself. The conclusions Aristotle drew from the principle of purposefulness (teleology) and how to use it to obtain a picture of the physical world had far-reaching implications.

The purpose as defined in Greek philosophy is nothing like the purpose that was and is still described by current monotheistic religions, that is, the need to satisfy the Creator. For Aristotle the purpose consisted of the search for sense and logic that were the basis of the physical laws. And as there is sense in formulating the laws of nature, according to Aristotle, they also satisfy the principles of beauty and aesthetics, and we must take that into account when searching for the laws of nature.

Aristotle based the principles of his theory on observations from nature. For example, he said, human beings grow teeth, the purpose of which is to chew food. Rain falls (or in Aristotle's words, the gods bring down rain), the purpose of which is to make grains grow, and similarly with other laws of nature. The purpose of the events that we see in nature enables us to discover the laws underlying the events, and vice versa, the purpose that we discover by virtue of our intelligence enables us to forecast those events. From a retrospective view it can be seen that although the cause and effect for Aristotle were totally different than those derived from the laws of evolution, the structure of Aristotelian thought is very similar to viewing nature via the evolutionary process. For Aristotle, the growth of teeth was a law of nature, and its purpose was to enable living beings to eat, while the evolutionary standpoint states that those living beings that grew teeth, enabling them to chew food, were the ones that developed. Likewise, Aristotle claimed that rain was a law of nature whose purpose was to enable grains to grow, whereas viewed from the evolutionary standpoint, grains developed in areas where there was rain. Either of these approaches helps us learn from the characteristics that we can observe how things developed. And vice versa, from the purpose, according to Aristotle, or from the conditions, according to evolution, we can predict the characteristics that we can see in nature.

Aristotle also expressed his views on the essence of space and motion in space. He saw bodies that fell to the ground, while other materials, such as flames or steam, rose. Whereas Plato tried to explain this by relating to the different weights of the various entities, Aristotle concluded that the purpose of pristine, pure elements such as steam or fire was to reach heaven, a pristine, pure place, while the purpose of materials such as ash or soil, which are clearly impure, was to reach the ground. His view was supported by the belief that the angels and gods, which are pure by definition, dwell in heaven. This led Aristotle to define a space that has directions, up and down. Upward is the direction toward the pure and the good, while downward is the spoiled and defective, toward the center of the Earth (even today, *upward* represents positive progress and *downward* signifies regression). This view of space answered the question about the antipodes, or specifically, what do up and down mean for people on opposite sides of the Earth. *Downward* for those on the other side of the Earth also means toward the center of the Earth, said Aristotle, hence they are not standing on their heads. This geometric view of the Earth as a sphere is of course consistent with ours. The purpose underlying Aristotle's directions up and down, relative to the Earth, firmly established the Earth as the center of the world for many years and predominated over other models, which in due course proved to be more correct.

With regard to the movement of various bodies, both earthly and heavenly, Aristotle observed that heavenly bodies moved in smooth, regular paths, a straight line or a circle, while earthly bodies moved along much more tortuous routes. The purpose that considered upward to represent purity while downward indicated inferiority led him to conclude that circular and straight paths were pure and other routes were defective. As it was geometry that served to describe the movement of the various bodies, Aristotle reasoned that the movements of heavenly bodies and earthly bodies were governed by different sets of rules. The gap between the different descriptions of earthly and heavenly motion persisted until the seventeenth century, when it was bridged by Newton. Aristotle also searched for a reason for the motion of the different bodies, and, by implication from the perception that the force exerted on a body causes it to move, he concluded that a force is responsible for all motion. Aristotle applied this conclusion to the heavenly bodies also. Hence, he claimed, the stars cannot move in a vacuum, and the world is filled with a substance he called *ether*. The ether also explained the source of the light and warmth of the Sun. The friction between the Sun and the ether creates heat. Aristotle went beyond that and rejected the model that described the world as consisting of atoms with a vacuum between them, as that would not allow for the exertion of the force required for motion. Hence, matter is continuous. The model of the atoms and the rejection of the assumption of the existence of the ether held until the nineteenth and twentieth centuries. Although many of the conclusions drawn from Aristotle's philosophy slowed the development of science, his philosophical principles nevertheless made an important contribution to scientific development.

14. MODELS OF THE HEAVENLY BODIES (CONT.)

Aristarchus of Samos (310–230 BCE) made a radical proposition about the motion of the heavenly bodies. He claimed that it was not the Earth but the Sun that was at the center of the universe and that the Earth and the other planets orbited the Sun. We know of the activity of Aristarchus from one essay that has survived in its entirety, in which he calculated the sizes of various astronomical bodies and distances, including the size of the Earth and the Sun, and the distance of the Earth from the Sun and the Moon. Archimedes, with whom Aristarchus was in contact, refers in his writings to the doctrine of Aristarchus. Aristarchus developed methods for calculating and measuring that resulted in assessments that were very advanced for their time but are very different from the facts as we know them today. For example, he assessed that the ratio of the distance of the Moon from the Earth to the distance of the Sun from the Earth was 1 to 19, whereas the correct ratio is 1 to 380. These and other measurements resulted in his proposing a heliocentric model (i.e., with the Sun at the center). The reason he gave was aestheticism. It was unreasonable to think that such a large body as the Sun would revolve around such a relatively small Earth. Aristarchus, who was affected by the views of the Pythagoreans, thought that the Sun was the fire at the center of the universe. He also realized that Eudoxus's model of spheres and its later developments can be more easily explained if it is assumed that the planets orbit the Sun and that the Earth also revolves on a circular path around it.

Although Aristarchus's model was widely known by the Greek astronomers, it was not generally accepted by them. There were several reasons for its rejection, some philosophical, and some scientific. The philosophical argument was the almost religious Aristotelian claim regarding the purity of the heavens as opposed to the impurity of Earth. One report states that Aristarchus himself was accused of heresy regarding those religious principles, but such accusations were not characteristic of the Greek environment that permitted pluralism of expression. The scientific objections to Aristarchus's model, however, were serious. One was that if the Earth moved around the Sun, then from different locations on its orbit the angles between the stars would be different, and that was not the case. This reason was quoted by Aristotle himself as an argument against the possibility that the Earth revolved around the Sun. It was not until many years later that the different angles between how we see the stars were revealed, differences that were too small for the Greeks to measure. Another argument against the centrality of the Sun was that if the Earth orbited the Sun, its speed would have to be so great that anything on the surface of the Earth would be propelled off it into space. We will address this claim when we discuss Ptolemy.

Further crucial progress in the calculation of cosmological data was made by Erastothenes (276–195 BCE). He was born in Cyrene, in what is today Libya, and carried out most of his scientific work in Alexandria, Egypt. He held the distinguished role of chief librarian of the widely known Great Library of Alexandria. Today's students know the name Erastothenes because of the "Sieve of Erastothenes," an (inefficient) way of finding all the prime numbers. He took it upon himself to calculate the dimensions of the Earth. He noticed that at noon, in the city of Syene (known today as Aswan), an upright pole does not cast a shadow, whereas at the same time in Alexandria, it does. By measuring the angle of the shadow in Alexandria and the distance between the two cities, which lie on the same longitude, he managed to estimate the dimensions of the Earth. The methods used by Aristarchus, Erastothenes, and their colleagues to perform their calculations are themselves very interesting, but beyond that, they provide evidence of the progress of practical mathematics in Greece, mathematics that served the Greek engineers and builders, who made remarkable achievements. Greek mathematics, fed by Egyptian and Babylonian ideas and achievements, constituted significant further development. Archimedes of Syracuse (287–212 BCE) was perhaps the best-known mathematician of that period for his mathematics-based engineering developments. We will allude to his contribution later.

Two far-reaching proposals regarding the model of the motion of celestial bodies were made by Apollonius of Perga (262–190 BCE). One was that the center of the circle on which a planet moved was not necessarily the center of the Earth. The other was that a planet can revolve along a perfect circle around a point that is itself revolving along a perfect circle around the Earth. He arrived at these views through his research into curves in a plane, particularly those created by orbiting in a small circle, known as an epicycle, the center of which moves along a larger circle, which the Greeks called a deferent. The curve thus created orbits around the deferent in a way very similar to the way the Moon revolves around the Sun. This motion is made up of two paths that sometimes move in opposite directions. Hence the motion is not constant but is sometimes forward and sometimes backward. This irregularity, combining progression and regression, is in some way similar to the irregularity of the paths of the planets, and Apollonius therefore proposed that the epicycles around the deferent were the paths of the planets. (See the diagram below, showing the Earth not at the center of the circle and the movement of the planet along the epicycle.) These two amendments to the previously held views enabled Apollonius, of course, after detailed calculations, to present a mathematical system of planetary motion that fitted well with the facts known at that time.

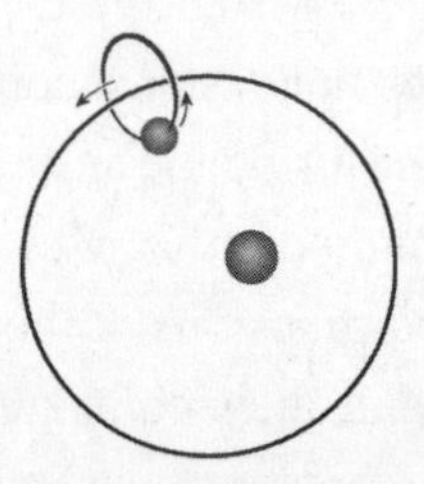

It is interesting that Apollonius, who is known for his contribution to geometry in general and who also carried out rigorous research into the properties of lines in a plane and the structure of three-dimensional bodies, was very familiar with ellipses. One of his better-known studies dealt with the structure of cones and the shape of their two-dimensional cross-sections. His research showed, for example, that the shape formed by a two-dimensional plane that intersects a cone could be a parabola, a hyperbola, or an ellipse (see the diagram below). Yet, although he was familiar with the shape of an ellipse, it did not occur to him to suggest that the paths followed by the planets were in fact ellipses. It took another fifteen hundred years to reach this conclusion.

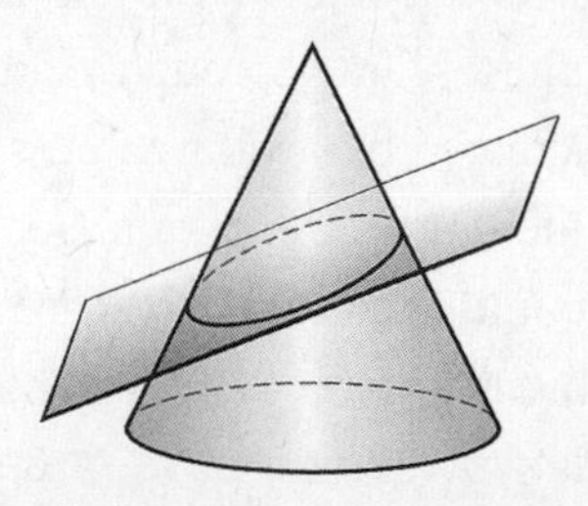

Hipparchus (190–120 BCE) took the mathematics of the model of the motion of celestial bodies to another level. Most of what we know of Hipparchus comes from the writings of Ptolemy, who called him the greatest Greek astronomer. Hipparchus did most of his scientific work on the island of Rhodes. Thirty-five years of astronomical observations that Hipparchus carried out on Rhodes combined with astronomical data from the Babylonians resulted in his constructing an astronomical model with previously unknown precision. For example, following his studies, the Greeks could forecast an eclipse of the Moon with an accuracy of within one hour. Hipparchus also discovered and measured the precession of the equinox (which we now know is caused by the movement of the plane of the Earth's orbit with regard to the position of the stars) and calculated the length of the cycle of this movement (about 2,600 years). His measurements greatly improved the data on the sizes of the heavenly bodies, the Sun, and the Moon, as well as their distance from the Earth, and data on the seasons, the times of the equinox, and so on. Hipparchus's major mathematical contribution was in the development of trigonometry. He defined trigonometric values, such as the sine and tangent of an angle, and found basic relations between them that helped him in his calculations. Following these definitions, he himself and his contemporaries constructed tables of what is today called the sine function (see the diagram), that is, the graph of the sine of every angle (the ratio of the line opposite the angle in a right triangle to the length of the hypotenuse), as well as other trigonometric functions, such as cosine and tangent. These functions fulfilled a major role in describing nature from the time of the Greeks to today.

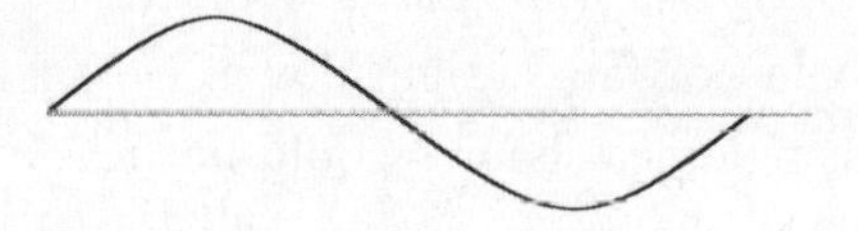

Hipparchus, and after him Ptolemy, adopted the geometric principles of Apollonius. Ptolemy was active in Alexandria from the year 90 CE to 169 CE, but these are approximations, as no biographies with details of his life have survived. His writings, however, have been preserved almost in their entirety, as they were copied and kept by the Arabs, and they provide much information on the model of the celestial bodies and Ptolemy's scientific approach. His main contribution was technical rather than conceptual. His basic starting point was the assumption that the heavenly bodies orbited the Earth, that their paths were epicycles whose centers were on a deferent that revolved around a point in space that was not necessarily the center of the Earth, and that the planes on which those deferents revolved were probably angled toward each other. With intensive and detailed computations, Ptolemy constructed a model, complete for his time, of the motion of the heavenly bodies. The model included seventy-two epicycles and larger cycles and represented a marked improvement on earlier models as it provided a superior fit of the observations to the model. The model gave far more accurate predictions than did the previous ones, and it remained the main model of the motion of celestial bodies for fifteen hundred years, longer than any other physical model in the history of science.

Ptolemy was aware of other ideas and approaches put forward by the Greek scientists over time and actually referred to them in his writings. He rejected Aristarchus's idea that the Earth revolved around the Sun with a quite rational claim based on calculations. Ptolemy's calculations showed that according to the heliocentric method with the Earth orbiting the Sun, the speed of the motion is incomprehensible and human thought cannot even imagine such a high speed. In particular he claimed that if the Earth revolved around the Sun, all animals and humans would fall off. He rejected with a similar argument the view still held at the time by Heraclides, namely, that the Earth revolved on its own axis. He added another argument to this, stating that if the Earth revolved on its own axis, a stone thrown upward would not fall straight down. He answered Heraclides's claim that huge cosmological bodies could not revolve around the relatively small Earth with the retort that the size of the bodies was not the determining factor, but their weight, and the Earth was very heavy, while the weight of the sky, the stars, and the ether in which they are situated is negligible. These are completely rational claims based on what our senses show us. Ptolemy acknowledged that Aristarchus's model, with the Earth revolving around the Sun, was simpler. He knew that there was contradiction between the simplicity of the mathematical theory on the one hand and what our senses perceive and the precision of the results on the other. In this situation, Plato would have preferred the simple mathematical theory. But Ptolemy's mathematical theory was so exact that it superseded the simplicity of other notions. This contrast between what our senses tell us and a simple but abstract model has accompanied science throughout the generations.

It should be stated that Ptolemy repeatedly said in his books that his model is just a mathematical description of nature. Specifically, he did not claim that his circles were a law of nature but that it was mathematics that best described nature. Moreover, he took the trouble to show that in some cases simpler models could yield more accurate results, but he was searching for a mathematical model that would incorporate all the phenomena. It was the Christian church that adopted Ptolemy's model as scientific truth and declared that God used mathematics to create the world.

CHAPTER III
MATHEMATICS AND THE VIEW OF THE WORLD IN EARLY MODERN TIMES

What causes the tides? • Why did Descartes say "I think, therefore I am"? • What is the Harmony of the World? • What effect does a new star have? • How can the third derivative help politicians get elected? • Why don't you need to be afraid of differential equations? • Why does the wall try to push back at us? • How was the mystery of the relation between the length of the string and the note it produces solved?

15. THE SUN REVERTS TO THE CENTER

The Renaissance period that started in the fifteenth century, also known as the early modern period, brought far-reaching developments in society, culture, and politics. These developments also encompassed a scientific revolution whose principles completely changed science and are as relevant today as they were then. Mathematics played a major role in that revolution.

Ptolemy's model of the heavenly bodies did not change much from when he formulated it in the second century CE until the sixteenth century. The model served as a sufficiently precise tool for predicting astronomical events, drawing up calendars, and so on. Ptolemy's model and the mathematics it was based on were studied in colleges and universities throughout the Middle East and Europe. The Arabs, who were at the forefront of scientific development at that time, enhanced the accuracy of the model by adding epicycles. At its peak the model incorporated seventy-seven deferents and epicycles along which the heavenly bodies moved in their orbits around the Earth. The complexity of the model, however, although it resulted in greater accuracy, also undermined Ptolemy's model.

Nicolaus Copernicus was born in 1473 in Thorn, Prussia, then part of the Great Polish Kingdom. He studied first in Cracow, which was a famous center of science, but most of his advanced studies he completed in Italy. There he became familiar with the Greek scientific literature, both the more classical and the writings of Ptolemy. He was a polymath. He knew several languages, including Latin and Greek, completed the study of law and medicine, and actually practiced medicine—all while studying mathematics and science, as well as astrology, then a highly respected profession. He then returned to Prussia and served as secretary, doctor, and astrologer to the bishop of Warmia, and then as economic administrator and advisor to the Warmia parliament. Copernicus devoted much time to astronomy, but it was not his sole occupation.

The idea of adopting and enhancing Aristarchus's heliocentric model came to him in the course of his studies in Italy. In his writings Copernicus refers to the effect of Aristarchus's model and the earlier ideas of the Pythagoreans. As early as in 1510 he wrote an essay on the principles of the model in which the Sun is at the center of the universe and the heavens and the planets, including the Earth, revolve around it. However, he circulated the essay to only a few colleagues. In the next few years he continued with the mathematical development of the model, including the completion of the astronomical measurements that he carried out himself over many years. His work was almost completed in 1533, but Copernicus still did not publish it. Nevertheless, news of it reached Europe, and requests for copies and encouragement for him to complete his work arrived from all over the continent. Scientists lectured widely on his theory, including in Rome before Pope Clement VII and some of his cardinals. They showed great interest in the findings and asked for copies of the work and the accompanying astronomical tables.

The Church, in general, did not object to the use of mathematics to describe nature and justified it by saying that it was self-evident that God had used mathematics in creating the world. Some sections of the Church, however, particularly the Protestants, objected to the description proposed by Copernicus. One of their objections was based on the biblical sentence (in Joshua), "Sun, stand still in Givon." If the Sun was stationary in any case, they claimed, there was no need to command it to stand still. Copernicus did not accept that criticism passively but answered firmly in a letter to Pope Paul III that someone who is ignorant in mathematics cannot judge a mathematical theory, and the holy scriptures teach us how to get to heaven but not how the heavens are constructed. Despite the assertiveness of his letter, the opposition of significant parts of the Church stopped Copernicus from publishing his findings sooner. The book containing his complete theory was submitted to the printers in 1543, and the first copy reached Copernicus when he lay on his deathbed, a few days before he died on May 24, 1543.

Copernicus's model adopted Ptolemy's mathematical method but amended the mathematics to be consistent with the idea that the Sun was at the center of the universe and that the planets revolved around it. Copernicus accepted the principle that dated back to the Pythagoreans, that celestial motion had to be along perfect circles, and therefore the planets revolve along circles around the Sun or on circles that move around a center that itself revolves on a circle around the Sun, that is, on an epicycle. Copernicus made sophisticated use of Ptolemy's mathematical systems, and his greatest achievement, in his own words, was to reach a level of accuracy similar to Ptolemy's but by using only thirty-four epicycles and deferents. Yet, to achieve accurate results, Copernicus located the Sun only close to the center of the deferent, and not at its center, in the same way as Apollonius the Greek had located the Earth in his geocentric model. The desire for simplicity was one of Copernicus's reasons for his faith in his model. One of the arguments he used was that God would not have chosen to use seventy-seven orbits when thirty-four were sufficient.

However, the quest for simplicity and aestheticism also had the effect of halting progress. Copernicus was convinced that for reasons of perfection and aestheticism, the orbits of the stars had to be circular. God could not and would not have created a world in which the paths of the stars were not perfect, that is to say, were not circles.

16. GIANTS' SHOULDERS

It was Newton who developed modern mathematics to describe nature. His response to the praise he received was, "I stood on the shoulders of giants." The circumstances in which he stated this do not prove the extent of his modesty or magnanimity, as we shall see further on, but the giants to whom he referred, Galileo Galilei, René Descartes, and Johannes Kepler, did make enormous contributions to mathematics and to understanding nature. In this section we will discuss the contribution of the first two to Newton's theory, and we will deal with Kepler in the next section.

Galileo Galilei was born in 1564 in Pisa, Italy, and died while under house arrest in his home in Florence in 1642. His family intended him to study religion and medicine, but he did not persevere in those fields and focused on the study of nature and mathematics, and he also exhibited a flair for commerce. He built a telescope following a Dutch invention that came to his attention and offered it to the city council of Venice for a generous annual stipend, to be used for early detection of enemy ships threatening the city. He showed the same flair and talent for science. He was the first to direct the telescope toward the heavens and the first to discover that the solar system consisted of more than the celestial bodies known to the Greeks. He discovered the four large moons of Jupiter, and his sharp political instincts led him to name them after members of the Medici family, the family of the Grand Duke of Tuscany, who later became Galileo's patron. He also used his telescope to study the surface of the Moon, realizing that the Moon had mountains and valleys, and he even calculated the height of the mountains. He also proved that the Moon was illuminated by light reflected from the Earth, and the directions of this reflected light deepened his faith in Copernicus's model. Galileo also tried to find additional proof supporting the heliocentric model. One of his "proofs" was an explanation of the ebb and flow of the tide. He claimed that they occurred as a result of the Sun drawing the water toward itself at the same time as the Earth is revolving around its own axis. This explanation is incorrect, because if it were correct, high tide would occur just once a day and not twice as actually happens. Galileo was so eager to confirm the heliocentric model that he found an "excuse" for the discrepancy in his explanation. He also rejected Kepler's explanation that the cause of the tides was the Moon. The full explanation was not revealed until the days of Newton, and it will be discussed in a later section. By using his telescope, Galileo also discovered the phases of the planet Venus, which varied from a full circle to a thin crescent that disappeared for a short time, similar to the way we see the Moon. This strengthened Galileo's opinion of the heliocentric universe. His findings made him famous all over the European continent, and he was known as a firm believer in Copernicus's model.

During many years of activity, Galileo was supported by many different groups, including the Church, but a combination of internal politics and ideology resulted in the Inquisition demanding that Galileo destroy his books and renounce his theory. His famous trial ended with his being sentenced to imprisonment, which was then changed to house arrest. In his trial he retracted his support of Copernicus. Years later the story spread that at the end of the trial he muttered to himself, "and yet it does move" (i.e., the Earth does move around the Sun).

The opposition of the Church to Galileo's discoveries was supported by, and some would say initiated by, the most prominent and influential scientists and philosophers of the time, who adhered to, taught, and believed in Aristotle's theory. They also rejected Galileo's actual discoveries. They claimed, for example, that what appeared to be the moons of Jupiter and the phases of Venus resulted from the use of lenses and did not exist in reality. They proved this by noting that what seemed to be the moons of Jupiter and the phases of Venus disappeared without the use of a telescope. Today their claim seems absurd. Lenses are used to enable us to see better. We must bear in mind, however, that the telescope itself, how it works, and its capabilities were then new and unknown to scientists. In a like manner, those philosophers did not accept the existence of the mountains that Galileo discovered on the Moon and explained that what seemed to be mountains on the Moon was a reflection of the mountains on Earth in a crystalline substance surrounding the Moon. Opposition of this kind continued throughout the years of Galileo's activity.

Galileo's main contribution to astronomy was in the field of observation; he did not contribute much in the area of theory. For the same reasons that guided astronomers from the time of the Greeks to Copernicus, Galileo accepted the assumption that celestial bodies moved in perfect circles and therefore vigorously opposed Kepler's model of elliptical orbits. At that time it was still believed that the laws of movement of celestial bodies were different than the laws of movement on Earth. Galileo therefore saw no discrepancy between the assumption that the stars moved in perfect circles and the actual movement of objects on Earth, which sometimes appears tortuous or crooked. The strictly controlled experiments that Galileo conducted on the movement of objects, and specifically on falling bodies, together with the mathematical explanations he used to describe that movement, constituted an essential element on which Newton's findings were based, but it was only as a result of Newton's findings that movement on Earth and that of heavenly bodies were first consolidated into one system. Galileo Galilei and the British philosopher Francis Bacon (1561–1626) were the principal formulators of the empirical method in modern science. The empirical method advocates learning from controlled experiments both to make new discoveries and to check, confirm, or refute scientific theories. This was a sharp deviation from the hitherto accepted episteme, or system of scientific knowledge, formulated by the Greeks that was apprehensive about biases and illusions inherent in observations, which they thought were likely to lead to incorrect theories.

The experiments that Galileo performed with moving and falling bodies were not innovative. Similar experiments had been carried out by others before him to discover the rules of the motion of bodies. Prominent among those earlier experimenters was the mathematician Niccolò Tartaglia (1499–1557), whose influence is mentioned in Galileo's writings; we will discuss some of his other contributions further on. The special element in Galileo's work was the almost-modern combination of theoretical mathematical arguments, which were then corroborated by experiments, and experiments that led to hypotheses about mathematical laws of nature. He dropped objects from the top of a tower (apparently the Leaning Tower of Pisa, although there is no clear-cut evidence of this) and found that the speed at which a body falls in a vacuum does not depend on its weight. In his first reports Galileo stated that the speed at which a heavy body falls is only slightly different from that of a lighter body, and he used that to show that Plato and Aristotle's theory that the speed at which a body falls is determined by its weight is not correct. Later, Galileo discovered that the differences in the speed of falling bodies in his original experiments were due to friction with air and that in a vacuum there is no difference in the speed at which the bodies fall. Experiments Galileo performed of rolling balls down a slope and measurements he took of the trajectories of stones projected parallel with the ground led him to conclude that one can distinguish between the movement in the accelerated fall of a dropped object and movement in a straight line. He also showed the numerical link he found between uniform motion and accelerated motion, such as:

Progress in time	1, 2, 3, 4, . . .
Progress in distance	1, 4, 6, 16, . . .

In other words, the distance is proportional to the square of the time. This relation between the numbers and their squares led Galileo to consider the concept of infinity, as we will see in section 59. The law of motion that Galileo derived from these measurements prevented him from discovering the general rule. He noted that the differences between the distances covered by a body between two equal and consecutive periods of time increase like the odd numbers, 3, 5, 7, 9, . . . Indeed, the equation $(n + 1)^2 - n^2 = 2n + 1$ proves easily that the differences between the squares of the integers are the odd numbers. Galileo thought it right to consider specifically the ordered manner of these differences as a law of nature. Thus he continued the Pythagorean tradition that explains nature via the relation between numbers. His experiments with a ball rolling down a slope confirmed these theoretical claims (the numerical confirmation was so exact that later researchers accused him of tampering with the empirical results to make them conform to the theory).

Galileo also found that the geometrical path of a ball thrown parallel with the ground is a parabola, a curve well known to the Greeks that had been studied in depth by Apollonius. The link with mathematics, in particular the fact that a parabola is the function that describes motion, was emphasized by Galileo. He also put forward an early version of the law of inertia, which states that in the absence of a force acting upon it, a body in motion will continue at the same uniform speed. Galileo thus repudiated Aristotle's approach that a force was required to start and maintain motion. Together with his empirical findings in astronomy and in the study of motion, Galileo's enormous contribution to the connection between science and mathematics—a contribution that served as a basis for Newton's laws—was his method of using observations to derive the relation between different elements such as time, speed, and acceleration. These precise mathematical connections, or functions, served Newton in his own research.

René Descartes disagreed with laws of nature in the form of the formulae proposed by Galileo. His objection was that they were derived from experiments and not from underlying physical principles. In this sense, Descartes was following the Aristotelian tradition. Descartes was born in 1596 to a wealthy family in La Haye, in the Loire valley in France. The town was later renamed Descartes. His family's prosperity meant that he was spared financial concerns throughout his life. A valet accompanied him wherever he went, even when Descartes volunteered as a soldier in the army of Duke Maximilian of Bavaria. This act reflects the spirit of adventure and wanderlust of the young Descartes, whose travels took him all over Europe. In Holland he met the mathematician Isaac Beeckman, who held fruitful discussions with him on philosophy and mathematics and who put to him mathematical questions that spurred his interest in mathematics and physics. Descartes spent many years in Holland but left suddenly to return to France to manage the family's affairs. In the same spirit of restiveness, he accepted an invitation from Queen Christina of Sweden to become her tutor and the chief scientist of her country. He became ill soon after arriving in Sweden and died in 1650.

Descartes's major contribution was in general philosophy. He is considered to be the father of modern philosophy, but we will not expand on that here. We will observe just that he tried to extend the logical approach to encompass all facets of life. He coined his famous saying "*Cogito, ergo sum*" ("I think, therefore I am") as a building block in the structure of logic used to reach philosophical statements. He realized that our senses indicate physical truths, like the sun rises, blood flows, and so on. But without the use of our mind, these "truths" do not make sense. He also claimed that the senses may fool us, hence we have to doubt every observation and check and recheck any logical step and conclusion that we arrive at. We have already stated that urging people to be constantly skeptical is urging them to behave in a way that is opposed to human nature, opposed to what evolution has prepared us for.

Descartes bequeathed to mathematics the consolidation of algebra and geometry. Since Eudoxus in classical Greece, algebra and geometry developed along parallel paths with very few meeting points. Algebra dealt with numbers and attempts to find numerical solutions to algebraic equations. A simple example that we encounter in school is seeking the solutions to the equation $2x + 5y = 8$, or the equation $x^2 + y^2 = 9$. Geometry tried to find the relation between various geometric shapes and their properties. Descartes demonstrated that geometric shapes can be described by algebraic equations. For instance, the solutions to the first equation we put down is a straight line in the (x,y)-plane, while the solutions to the second equation form a circle of radius 3. Descartes's contributions were not stated in the modern terms of a plane and its coordinates. An algebraic representation of a plane and its coordinates had not been created yet. But his ideas have led to the development of the system of coordinates for planes and spaces that we know today. The system of Cartesian coordinates taught today in all schools is a development based on Descartes's approach, and it is therefore named after him (his Latin name, which he used in many of his writings, was Renatus Cartesius). This mathematical tool enabled Newton to develop a new mathematics and formulate with it the laws of nature, as we will see later.

17. ELLIPSES VERSUS CIRCLES

An important step in presenting a picture of the world as we know it was made by Johannes Kepler. We will first state Kepler's three laws of planetary motion, which constitute his main contribution to the mathematical description of the world. We will then review the tortuous path by which he reached these understandings.

Kepler's first law: The path of the planets about the Sun is elliptical in shape, with the Sun being located at one focus of the ellipse.

Kepler's second law: The speed of each planet on its orbit is such that the line from its location on the ellipse to the Sun covers equal areas in equal periods of time.

Kepler's third law: The square of the time taken by a planet for a complete orbit around the Sun is proportional to the cube of its average distance from the Sun. This can be written as

$$T^2 = kD^3,$$

where T is the time of the planet to complete one round, D is the average distance of the planet from the Sun, and k is the same constant for all planets.

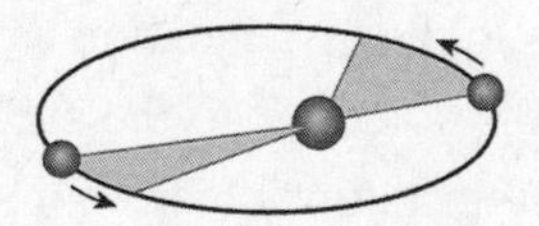

It is hard to overstate the mental difficulties Kepler overcame in presenting his three laws, and in particular the first one. The first law was formulated in accordance with the classical Greek approach describing the world with the help of geometry. At a remove of four hundred years, the transition from a circle to an ellipse seems simple. Today we actually consider a circle as a special instance of an ellipse in which the two foci converge. Kepler's first law, however, represented a contradiction to the deep-rooted tradition of two thousand years that as the firmament is perfect, the orbits of the planets must be perfect and, hence, circular. Kepler changed circles for ellipses and did not offer a reason or alternative purpose. His law was based on observations and measurements. At that time, those observations did not constitute clear proof of his claim, as the ellipses along which the planets moved were very close to circles, and the means for measurement in those days made it very difficult to differentiate between them. Kepler's ellipses did however enable the use of epicycles revolving around deferents to be abandoned, thereby offering a great simplification of the system.

Kepler's second law was also based on observations and measurements, without a reason or a purpose. This law also contradicted a two-thousand-year-old convention that was also based on the purpose of divine perfection, according to which the planets move at a uniform speed on their circular orbits, and the fact that we see the planets moving at varying speeds is the result of an optical illusion caused by their moving on epicycles whose center revolves on a deferent. Again, Kepler did not offer a reason or an explanation for the mathematical relation he found. He did not ignore divine wisdom, but he claimed that just as previously it had been thought that the Creator's wisdom had made the planets move at a uniform speed, it was the Creator's wisdom that made the planets depict equal angular areas in equal time intervals.

The second law compares a planet's speed at different locations on its path but does not refer to the actual speed itself. The third law does, and it too is based on observations. Kepler went as far as discovering the mathematical relation, which also is not precise, as it was not clear how to calculate the average distance of a planet from the Sun, but here again he did not explain or justify why his law was correct. The empirical confirmation of the law was possible only because the ellipses traced by the planets were almost circles, almost with fixed radii.

Kepler's laws constituted a crucial turning point in the relation between mathematics and the description of the solar system in at least two aspects. One was the abandonment of the unchallenged assumption that the purpose of the planets was to move along circular orbits. The second was the presentation of the laws of nature in the form of mathematical relations that were based purely on measurements and observations. These two aspects represented a sharp break with the general tradition of Greek thought, and in particular with Aristotelian thought, that had been the undisputed dominant system of thought until then. Kepler himself was greatly influenced by the Greek tradition, and the break with it caused him to do a lot of soul-searching, as we shall see.

Johannes Kepler was born in 1571 in the German town of Weil der Stadt, and he died in 1630. His family was a problematic one for various reasons: His father was a drunkard, violent toward his family and those around him, in trouble with the law, drawn to adventure, serving as a volunteer in the czar's army to fight against the rebel Protestants in Holland (although he himself was a Protestant). In doing so, he abandoned his wife, his first-born Johannes, and the latter's younger brother for long periods without arranging for their livelihood, and eventually he simply did not return from one of his journeys. Johannes was apparently born prematurely and was a sickly child who remained sickly all his life. His mother also had a reputation as a wild, unconventional, rebellious woman. Her mother had been hanged as a witch, and she herself almost met with the same fate. Johannes was not provided with any formal basic education. When he started learning in an orderly fashion at the University of Tübingen, he set out to study theology, but when he graduated at the age of twenty, he was offered a position as a teacher of mathematics and astronomy in Graz, in the Austro-Hungarian Empire. For reasons that are not very clear, he was recommended for the post by his teachers in Tübingen, although he had only scanty knowledge of mathematics and astronomy. It would seem that they wanted to get rid of Kepler, who was not a popular student. They may have wanted to remove him because he had publicly expressed support for Copernicus, although he did not have much knowledge of Copernicus's work. His support of Copernicus's views was due to the influence of classical Greek literature that he read, particularly Pythagorean, according to which the Sun was the source of heat in the world and was therefore at its center. Kepler accepted the position in Graz, among other things because he had no other economic alternatives, and in his new job he immersed himself in the study of astronomy and mathematics, which in the course of time resulted in his contribution via the findings we have described. This was one of the first instances of such a major contribution by a scientist who came from a disadvantaged section of the population. The great majority of successful scientists and philosophers prior to Kepler came from the well-established social strata.

From his youth, Kepler was fascinated by mathematics in general and specifically by astrology, and a large part of his income over the years derived from astrological tables he compiled. He became famous because of his successful astrological prediction of the Turkish invasion of Austria. It is not clear whether he himself actually believed in astrology. Some of his writings contain derisive references to it while, on the other hand, he expressed the view that there was a foundation for the belief that the planets have a direct effect on our lives and that astrology could be developed on a scientific basis.

His leanings toward the mystical together with his support of the Pythagorean and Platonic approach to nature led him to search for links between numbers and geometric shapes and astronomic structures. Kepler knew of Plato's attempt to relate the existence of five perfect geometric solids to five elements of nature. At an early stage of his career it occurred to Kepler that there was a link between those five perfect shapes and the six planets known since ancient times. That happened long before the idea of elliptical orbits came to him. The assumption was that the planets move along circles that are on spheres, that is, that envelope balls. Kepler's idea was that each such ball surrounds and supports a perfect geometric body and is also supported by a perfect body; in total five perfect bodies that separate the paths of the six planets. Complex calculations that took many years resulted in his proposal of a precise model of the orbits and the perfect bodies that separated them. In that model, Mercury was on the ball that was closest to the Sun. The ball is supported by the octagonal body, which is itself supported by Venus's ball, which is in turn supported by the icosahedron (the twenty-sided figure) that is supported by the Earth, which itself is supported by the dodecahedron (i.e., the twelve-sided figure), which is surrounded by the sphere of Mars, which is contained within the pyramid that is supported by Jupiter's ball. The latter is within the cube that is surrounded by Saturn's ball. Kepler took the trouble to construct a tin model of this structure and later succeeded in interesting his patron, the prince, to build a model of the universe out of pure silver, but the model was never completed.

Along with his attempts to construct and improve the accuracy of his geometric model, Kepler sought more, and more-accurate, data on the movement of the celestial bodies. These data were in the possession of the most famous astronomer of his time, Tycho Brahe. Brahe was a Danish nobleman who became famous through his observational proof that the star that "was born" suddenly in 1577 was located among the other "fixed," distant stars (which do not appear to move relative to each other), that is, it was not a lower body, such as a comet. At that time it was believed that the comets were closer to the Earth than was the Moon, and they were considered to be some sort of atmospheric disturbances. The appearance of the new star, followed by another new star that appeared in 1604 (these are known today as supernovae, stars that suddenly become far brighter than normal), were seminal events in astronomy in those days and constituted an important step toward undermining the Aristotelian approach that the firmament of the stars is fixed and unchanging. Ferdinand II, who became King of Denmark, built a sophisticated observatory for Tycho Brahe on a tower on a private island; there the Danish nobleman gathered a vast collection of accurate astronomical data, an effort unparalleled since the days of Ptolemy. Brahe's relationship with Prince Christian IV, Ferdinand's successor, was somewhat less cordial, and the astronomer departed for Prague with all his data, where, under the sponsorship of Emperor Rudolf II, he served as the imperial mathematician. Brahe did not believe in Copernicus's ideas but thought that the planets revolved around the Sun, which itself orbited the stationary Earth. Kepler, follower of Copernicus, set out for Prague to ask for Brahe's permission to use his data to support his theory of perfect bodies and to make it more accurate.

Kepler had another objective in mind in studying Brahe's data. The Pythagoreans had already noted that the relationships between musical harmonies are like those of natural numbers. Kepler, who was familiar with and influenced by this part of Greek mathematical writings, tried to match those harmonies with the orbits of the planets and the geometric bodies that separate them. In his book *Harmonices Mundi* (*Harmony of the World*), Kepler presents such an adaptation, in which Saturn and Jupiter are the basses, Mars is the tenor, the Earth and Venus are contraltos, and Mercury is the soprano. This description is accompanied by precise calculations with the support of astronomical readings that Kepler hoped to improve with the help of the information gathered by Tycho Brahe.

A few months after Kepler joined Brahe, the latter died and Kepler was appointed imperial mathematician. He continued with his calculations with the aid of Brahe's data, and after great effort and much soul-searching he arrived at the mathematical discoveries that we described at the beginning of this section. His strong loyalty to Copernicus's model aroused opposition from sections of the Church; in his later years he had to leave his post and return to Bavaria, where he died. Although he was convinced that the model he proposed was correct, he did not abandon his mystical notions about the harmonies of the heavens and the perfect bodies. In his last book he presented side by side his laws—Kepler's laws—that are still accepted today and his interpretation of the Pythagorean concepts of the celestial music in the hope that one day the two may be reconciled.

Kepler's laws did not provide an answer to the question that had arisen repeatedly regarding the heliocentric model: Why do people not fall off the Earth that is spinning so fast? The idea that the Earth somehow exerts a force that attracts people to it was raised in the time of the Greeks, but in itself the idea was not enough to provide a basis for the assertion. Without a quantitative theory—in other words, without a mathematical basis—the claim regarding the Earth's "attraction" lacked persuasive power. It is interesting to note that Kepler's answer to this difficulty indicates a thought process in terms of evolution. He did not seek a reason for the fact that people were not thrown off the Earth as a result of its rotation, but he claimed that a world in which people were thrown off Earth could not maintain humans. Hence, as humans do exist, the force of attraction also exists. With regard to the nature of the Earth's force, Kepler made do with the use of the term *magnetic connection* without any further explanation. It was Newton who invested mathematical content into the concept of gravity.

Viewed from the aspect of the motivation that drove them to their discoveries, the processes followed by Copernicus and Kepler were complete opposites. Copernicus rejected Ptolemy's model not because of its lack of precision but because of its lack of simplicity, and he preferred a less-accurate but more-aesthetic model. Kepler replaced circles with ellipses because the circles lacked accuracy, despite the fact that since the Greeks, ellipses were regarded as imperfect and thus unaesthetic. The dilemma of choosing between accuracy and simplicity has accompanied the development of science and continues even today.

The three giants, Galileo, Descartes, and Kepler, lived at the same time. Nothing is known of any meetings they may have had. Descartes supported Copernicus's astronomy but did not acknowledge that publicly, apparently because of concern that he would meet the same fate as Galileo. Descartes opposed Galileo's law of motion because it was based on observations and measurements and not on intellectual analysis. Kepler, as imperial mathematician, did not hesitate before complying with Galileo's request for a letter expressing support for his (Galileo's) findings. Galileo had no qualms about using Kepler's name and support for his findings, but he ignored Kepler's request for one of the telescopes he had built so that he, Kepler, could improve his data. Instead, Galileo opted to give the telescopes he had built to noblemen with no mathematical aptitude but with political influence. Kepler managed to obtain one of the telescopes through a mutual acquaintance. Galileo also opposed Kepler's law of ellipses and supported the Aristotelian axiom of planetary motion on perfect circles, while he himself had developed a different law of motion for earthly bodies, a law that was compatible with Kepler's law. Galileo did not mention Kepler in his scientific writings, presumably because he disdained his mystical Pythagorean arguments. Despite the lack of "harmony" between these three giants of scientific development, each made a crucial contribution to the imminent revolution in the relation between science and mathematics.

18. AND THEN CAME NEWTON

Isaac Newton developed mathematics that still today constitutes the basis for the description of nature. Beyond his technical achievement, which offered entirely new possibilities to describe the laws of nature with the help of mathematical relations, Newton's contribution represented something of an innovation also in the approach to the development of mathematics. The Greeks and their successors chose to advance known mathematical tools, principally geometry, which had developed naturally over many years for other purposes. Newton realized that there was nothing in the mathematics then available that was appropriate to his ideas, and he therefore formulated and developed special mathematics designed to describe nature.

Newton was born at Woolsthorpe Manor, Lincolnshire, England, on December 25, 1642, according to the English calendar, or January 7, 1643, according to the Gregorian calendar in use in most of Europe. (The Gregorian calendar was adopted by the Catholic countries in about 1582, and a little later by the Protestant and Orthodox countries. It was not adopted in England and its colonies until 1752.) Newton's father, a prosperous farmer, died shortly before Isaac's birth. Three years later, his mother remarried and went to live with her new husband, leaving Isaac at the farm in the hands of her mother. Isaac disliked his time at the farm and the way he was treated there, and this period in his life was apparently reflected in his somewhat-unsocial behavior and obsessively suspicious nature as long as he lived. When he was about nineteen, his mother returned to the farm after the death of her second husband and tried to turn him into a farmer. He hated the farm, however, and his mother was persuaded to send him to Trinity College, Cambridge. He did not shine as a student, but he immersed himself in the writings of Aristotle, Copernicus, Galileo, Descartes, and Kepler. In 1665 the Great Plague reached the outskirts of Cambridge, and the university closed for two years. Newton spent those two years at Woolsthorpe Manor, and there, alone, he developed his original ideas of the mathematics of nature. The university reopened, and Newton was accepted as a junior member of faculty and worked there with the well-known mathematician Sir Isaac Barrow. Barrow immediately realized Newton's capabilities and advanced and supported him, and in due course Newton was appointed to the distinguished chair position that Barrow had held. From there Newton's path to the universal acclaim he deserved should have been a smooth one, but it was full of pitfalls caused mainly by his querulous nature, his suspicious personality that prevented him from publishing his achievements, and his unsocial attitude toward his colleagues. We will mention some instances of this below. Newton eventually reached the pinnacle of scientific fame in England and Europe, was granted a title (Sir) later in life than might have been expected, given his achievements, and served as president of the Royal Society. He died in March 1727 (on the 20th according to the English calendar, and the 31st according to the Gregorian).

19. EVERYTHING YOU WANTED TO KNOW ABOUT INFINITESIMAL CALCULUS AND DIFFERENTIAL EQUATIONS

We will now present the concepts underlying the mathematics of differential and integral calculus (also known as infinitesimal calculus, namely, the calculus of small terms) and differential equations. No knowledge of or background in mathematics is required to understand these ideas, but it is worth investing a little effort and exercising a little patience to absorb the mathematical symbols that will appear here (alternatively, the reader can ignore the symbols and skip the technical part of the presentation).

In daily life we come across quantities that vary with time, for example, the position of a moving vehicle, the price of an article, or the temperature outside. In mathematics it is customary to notate the variable as a function, for example, $x(t)$, where the variable t relates to time, and for every t, that is, at time t, the value of $x(t)$ gives the position of the vehicle, the price of the article, or the outside temperature, as the case may be.

In addition to the function itself, we may be interested in the *rate of change* of the function. In the example of the moving vehicle, the rate of change in the vehicle's position measures its velocity. Where the function describes the price level in an economy, the rate of change measures inflation. The rate of change of a function is itself a function. That is because, for example, the speed of the vehicle can also change with time. In mathematics, the function that describes the rate of change is called the *derivative* of the original function and is denoted by x'. (Leibniz proposed giving the derivative the notation $\frac{dx}{dt}$ and that notation is still used today. It is important to note that this is purely notation and does not represent a division of one quantity by another.)

As the derivative is also a function, we can look at its rate of change as well. This would therefore be the derivative of a derivative, or in mathematical terminology, the *second derivative* of the original function, and it is written as x''. The second derivative can also be seen in terms of daily life. In the case of the moving vehicle, the first derivative gives the velocity; hence, the second derivative gives the rate of change of the velocity, that is, the rate of acceleration. In the example of prices, the first derivative gives the rate of inflation, and the second derivative gives the rate of change of inflation.

We could continue and define third, fourth, and higher derivatives. Sometimes these also can be interpreted in terms related to the meaning of the original function. The following is a quote from a speech by the American president Richard Nixon during the election campaign in 1972:

"The rate of increase of inflation is declining."

He was using the third derivative. If the function we are analyzing defines prices, then its derivative shows the rate of inflation, the second derivative is the rate of change of inflation, and the meaning of President Nixon's statement is that the third derivative of the price function is negative. That was the first time, wrote the mathematician Hugo Rossi in an article in 1996, that a president used the third derivative to get reelected.

We said that no knowledge of or skill in mathematical techniques is required to understand the concepts we have described. It is sufficient to grasp that the rate of change of a variable, such as speed, can itself change, and that is called the derivative. As we will soon see, these concepts, the derived function and higher derivatives, constitute the basis of the development of the connection between mathematics and nature. To enable the use of these concepts required the development of calculus, that is, efficient ways of calculating the rates of change, the derivatives, of functions. Vice versa, when the rate of change or the derivative is known, a method has to be found to calculate the original function. This function is called the *integral*. This branch of mathematics, calculus, was developed independently by Newton in England and by Leibniz in Germany. To understand it, knowledge of mathematical techniques was necessary, but the technical details of how to calculate derivatives and integrals are of concern mainly to researchers and those who use mathematics (and students!). To understand what is written in this section, those technical details are not at all important. We will give examples of such calculations further on. These were important for the corroboration of Newton's theory. We will just observe that the first sparks of calculus can be seen in the exhaustion method of Eudoxus and were significantly improved by Archimedes (see sections 7 and 10). Archimedes's improvement included the concept of the limit of geometric figures whose areas are easy to calculate. That limit is precisely the integral as defined by Newton and Leibniz. They took one step further, however, and generalized the method to calculate derivatives, a previously unknown mathematical concept. They raised the method to the level of mathematical rules by which derivatives and integrals can be calculated relatively simply. Both Newton and Leibniz applied the knowledge of their forebears, methods of calculating areas and slopes of tangents to specific instances, calculations that had been performed by many contemporary and earlier mathematicians, including famous figures such as Fermat, Descartes, Wallis, Barrow, and others. The contribution of Newton and Leibniz was to raise calculus from a collection of examples to the level of a general mathematical theory.

These parallel developments in England and Germany did not take place entirely peacefully, without discord between the two men, discord that enveloped the whole world of mathematics and included mutual allegations of stealing ideas and plagiarism. Newton was the first to start developing calculus, but because of his excessive mistrust, he did not publish his findings for many years, and when he did, they did not make for easy reading and understanding. Leibniz did publish his calculus in a much clearer fashion and in Europe was credited with being the discoverer of the system (as we noted, the notation above, still in current use, was that proposed by Leibniz). Newton, in contrast, created a mathematical concept that he called fluxion, which described the instantaneous change in a function. Newton went on the attack and accused Leibniz of stealing the idea. The truth is not completely clear. It appears that in the initial stages Leibniz came across a very early version of the concept and that was sufficient, for a genius such as he was, to develop the theory independently. In any event, Leibniz denied that his work was based on an early connection with Newton or Newton's work and even demanded that the Royal Society appoint a committee to investigate Newton's accusations. A committee was appointed, and after several months of investigation it declared in favor of Newton and did not absolve Leibniz of the accusation of plagiarizing the idea from Newton. It is not clear whether this finding was affected by the fact that at that time Newton was the president of the Royal Society.

This is an appropriate point to give a brief description of Gottfried Wilhelm von Leibniz, Newton's "competitor" in the development of calculus. Leibniz was born in 1646 in Leipzig, Saxony, to a family involved in issues of law and morality. From an early age Leibniz had access to the well-stocked library of his father, who was a professor of moral philosophy, close to what today is called social science. His father died when Gottfried was a young boy. The library introduced the young Leibniz to classical Greek literature, including Aristotle's ideas on morality and philosophy, which made a very deep impression on him. He was educated by his mother's family—her father was a professor of law. From his youth Leibniz was attracted to legal topics and completed a doctoral thesis on the subject when he was twenty, but he was not awarded the degree because he was found to lack experience of life and to be too young to hold such a title. He therefore left Leipzig and went to the University of Altdorf, where he quickly completed his doctorate in mathematics. He rejected an offer to remain as a professor at that university and in 1667 chose to work in the employ of Baron Christian von Boyneburg, who in the course of time became a close friend. In 1676 Leibniz moved to Hanover to serve the Duke of Hanover, and in due course he was appointed as the official historian of the Boyneburg family. Leibniz contributed greatly to mathematics in various fields, as we will see below, but he was not only a great scientist but also one of the greatest philosophers of all time. He died in Hanover in 1716.

Calculus has many uses in different fields, but the main reason for Newton's developing it was that it helped him solve differential equations—the tool he devised to express laws of nature—thus enabling him to use the laws of nature he had discovered. How to solve differential equations is a subject for mathematicians and those who use mathematics, but to understand what those equations are does not require a mathematical education, and certainly there is no reason to be deterred from continuing to read about them.

Since the dawn of the development of mathematics, mathematicians have tried to develop systems that would help them calculate an unknown quantity from the data available. Thus, Thales of Miletus calculated the height of the Great Pyramid in Egypt by means of the distance from the center of the pyramid and the angle at which the top of the pyramid was seen. Thus, the Babylonians calculated the length of the hypotenuse of *many* right-angled triangles using the known lengths of the other two sides. The Greeks had already formulated a general rule for calculating the length of the hypotenuse of such triangles, given the lengths of the other two sides, Pythogoras's theorem, and hence could calculate the length of the hypotenuse of *any* right-angled triangle. Presenting the connection between the unknown quantity and the data by means of an equation began only in the sixteenth century. The sign = to indicate equality was proposed by Robert Recorde, an Oxford University professor, in a book dated 1557. Since then we write equations, which generally describe situations in real life, and use mathematical techniques to solve them. The equations we are most familiar with relate to numbers. Thus the equation $a^2 + b^2 = x^2$ describes the calculation to obtain the length of the hypotenuse using Pythagoras's theorem, with x representing the length of the hypotenuse that we wish to obtain. Another well-known equation is the quadratic equation of the form $ax^2 + bx + c = 0$. In school we are taught how to find the value of the unknown x. Sometimes there is one solution, sometimes two, and sometimes none.

Newton studied a new type of equation that does not relate to unknown numbers but rather to unknown functions and their derivatives. Today these are referred to as *differential equations*. (The actual term came into use long after Newton.) The idea is that we write a mathematical equation that gives the relation between functions and their derivatives, and we do not look for solutions in the form of numbers but the functions that satisfy the equations. An example of such an equation is:

$$mx'' = -kx,$$

where the unknown is not a number, as we are used to, but a function $x(t)$. The equation can be described in words. It says that the second derivative multiplied by the number m is equal to the negative of the function itself multiplied by the number k (for the description of the equation, it is not important what k and m stand for). Sometimes additional conditions may exist that the solution must satisfy, for example, it may be specified that $x(0) = 0$. Mathematicians learn how to solve such equations. Sometimes there is just one solution, sometimes there are many, and sometimes, none. In day-to-day life the reader does not generally need to know the technique for solving differential equations, unless he or she is an engineer or a physicist using mathematics of this sort or is still a student.

20. NEWTON'S LAWS

Newton developed calculus for the purpose of giving a mathematical expression for motion in nature. In fact, his laws are formulated by differential equations. His was a remarkable achievement from various aspects. First, the contribution to physics itself. This was the first time that the same rules were applied to describe earthly and celestial motion. This conflicted with the Aristotelian approach, which was still followed in the seventeenth century. Even more impressive, however, was the use of the new mathematics of differential equations both to formulate the laws and to draw conclusions. We will briefly review the link between the new mathematics and the laws of nature.

Newton's first law, the law of inertia: Without a force being exerted upon it, a body will continue at a constant velocity.

Newton's second law: This is best known via the formula $F = ma$. In simple language, it says that the acceleration a of an object is proportional to the force F exerted upon it, and the proportion is determined by the mass m of the object.

Newton's third law: Every force has a reaction. A body that exerts a force on another body will have an equal and opposite force exerted on it by the second body.

The first law is actually derived from the second. Indeed, if the acceleration is zero, the speed is constant. Newton stated this as a law of its own because he saw it as a statement of the existence of "independent" motion, in contradiction to the Aristotelian doctrine that the cause of motion is a force exerted upon a body. Newton gave the appropriate credit to Galileo as the first person to conceive of the idea of inertia. When he was asked to explain the concept of speed itself—in other words, in relation to what, and how does one measure it—he answered that it was measured in relation to the stars that are fixed in the heavens.

The significance of the second law is mathematical, and so is the way to use it. The second law is a differential equation relating the force exerted on a body to the second derivative of the function of location or position, namely, the acceleration of the body.

The third law is technical and is very confusing. At a luncheon, I once met a senior physician, the head of a department in a well-known hospital; he told me that because of Newton's third law he had stopped studying physics and mathematics. "I decide to push against a wall," he said; "how can you say that the wall decides to push back at me?" This personification of the situation is indeed confusing, but it misses the purpose of the law, which is basically technical. To illustrate, consider this: A table is standing on the floor. The force of the Earth's gravity is acting upon it. Therefore, according to Newton's second law, the table ought to be falling toward the center of the Earth. It is not falling because of the opposition of the floor. That opposition is translated into the laws of motion in the statement that the floor is exerting a force on the table that is equal in size and opposite in direction to the force exerted on it by the weight of the table. Thus the sum of the forces acting on the table is zero and the table continues to stand, stationary. Another illustration of the third law is rowing. The oarsman pushes the water forward with the oars, in the direction he is facing, and the boat moves in the opposite direction. According to the second law, progress is made due to the force of reaction of the water to the movement of the oars.

Another central law of Newton's, which together with the second law enables much motion to be analyzed, is the law of gravity.

Newton's law of gravity: All two "point masses" (a point mass is a conceptualized zero-dimension mass) of mass M and m, respectively, at a distance r apart, attract each other with a force that is proportional to their masses and inversely proportional to the square of the distance between them. This can be written in the form of the equation

$$F = k\frac{mM}{r^2}$$

where F is the force, and the coefficient k is a constant.

We will state the obvious: Newton did not invent gravity (and the story of the apple is nothing more than a nice story that Newton himself apparently made up to keep troublesome inquirers away). Newton's contribution was the discovery of the *mathematical* relation underlying the law. The Babylonians, and after them the Greeks, had already claimed that the Earth exerts a force that pulls us toward it, and they had coined the word *magnetism* for that force. Formulae similar to that of Newton had been proposed before, and before he had published his law, three leading English scientists, Robert Hooke, Edmond Halley, and Christopher Wren, had discussed the question of what gravitational force would cause the planets to move on elliptical orbits. The three proposed a formula for the force that was similar to Newton's law, but beyond the proposal itself they did not know how to use it. Halley, a friend of Newton's who actually financed the publication of Newton's first book, asked Newton what orbit would be derived from such a formula for gravity and was happy to learn that Newton had already calculated that previously and had found that the orbit was an ellipse. Thus Newton showed that Kepler's first law can be derived from his, Newton's, laws, and shortly thereafter showed that all Kepler's laws can be derived from his own laws. It was the mathematical tools of infinitesimal calculus and differential equations that enabled Newton to formulate his laws of nature and to prove that Kepler's laws could be derived from the new laws.

The new mathematical tool did much more. Galileo described the path along which projected bodies fell on Earth by means of parabolas. Using differential equations, Newton proved that the parabolas resulted from his laws of motion. Here is the essence of the proof (which can be skipped without interrupting the flow of the text).

Newton showed that the derivative of the function $x(t) = \alpha t^n$, where α is a constant, is $\alpha n t^{n-1}$. In particular, if the second derivative has a fixed value g, its integral is gt, and the integral of the latter is $\frac{1}{2}gt^2$. This shows that as the Earth's gravitational pull g is constant over short distances; the parabolas that Galileo observed when he dropped bodies from the top of a tower fulfilled Newton's second law of motion.

Newton also examined the oscillations of a spring. The law of restoring force, according to which a spring oscillates, was formulated by Robert

Hooke, who was a bitter opponent of Newton's, and their harsh exchanges of words were well known in British academia. Hooke's law states that the force exerted by a spring is proportional to the spring's deviation from equilibrium. If this rule is applied in Newton's second law, the differential equation that appears in the last paragraph of the previous section, namely, $mx'' = -kx$, is obtained. The solution of the equation is also based on differential calculus. The essence of the solution is shown in the next paragraph (again, this can be omitted without interrupting the flow of the text).

Newton's and Leibniz's calculus showed that if $x(t) = \beta\sin(\alpha t)$, where α and β are constants, then the derived function is $\beta\alpha\cos(\alpha t)$, and the second derivative is given by $-\beta\alpha^2\cos(\alpha t)$. We will leave the full proof to mathematicians (and students), but a simple check shows that if we use the notation $\omega = \sqrt{\frac{k}{m}}$, then both the function $\beta\sin(\omega t)$ and the function $\beta\cos(\omega t)$ solve the differential equation $mx'' = -kx$. This shows that the oscillation function is a combination of the sine function and the cosine function.

In this way Newton proved that the empirical finding announced previously by Robert Hooke satisfies Newton's laws. This greatly impressed Hooke, Newton's opponent, and Hooke, despite the adversity between them, complimented Newton in public. Newton reacted with the oft-quoted "If I have seen further, it is by standing on the shoulders of giants." It is unclear whether the fact that Robert Hooke was short in stature was relevant to Newton's choice of words.

Newton's theory of motion was not readily or widely accepted right away, and many reservations were expressed about the findings. Two astronomical events that occurred many years after Newton's death removed all doubts. The first related to Newton's close colleague, the astronomer Edmond Halley, mentioned above. The paths of comets are affected by the gravity exerted by the larger planets and are therefore difficult to calculate, so much so that scientists expressed doubts as to whether comets also moved in accordance with Newton's laws of motion. Halley studied the trajectory of one comet, whose several appearances in the past had been documented, and he predicted that it would appear again at the end of 1785 or the beginning of 1786. Its appearance in December 1785 significantly reduced the doubts about the correctness of Newton's equations. Halley was honored by that comet being named after him. Its most recent appearance was in 1986, and it will be visible again in 2061.

The second event resulted from some inconsistencies between observations of the orbits of known planets and the predictions based on Newton's equations. This led to the discovery of a new planet being foretold, and eventually, in 1846, it, the planet Neptune, was discovered in space.

The laws of gravity also provided a satisfactory explanation for the ebb and flow of the tide. The explanation is that the Moon's gravitational pull has an element that is parallel to the surface of the Earth that has no effect on the height of the surface of the sea and an element that pulls toward the center of the Earth (see the diagram). That pull lowers the level of the water on both sides of the Earth so that the water level rises both on the side of the Earth nearest to the Moon and on the opposite, farthest, side (again, see the figure below). This also explains why high tide occurs twice daily, once when the sea is close to the Moon and once when it is on the opposite side of the Earth, farthest away from the Moon. The gravitational attraction of the Sun has the same effect, but because of its distance from the Earth the force is less strong. The ebb and flow of the tide are the results of the combined effect of these forces. There is a difference in the level of high tide at different phases of the lunar month, that is, at new moon and in the middle of the month, when the gravity of the Moon and the Sun reinforce each other, and the periods one-quarter and three-quarters through the Hebrew (lunar) month, when the Sun's gravitational pull reduces the effect of the Moon on high tide.

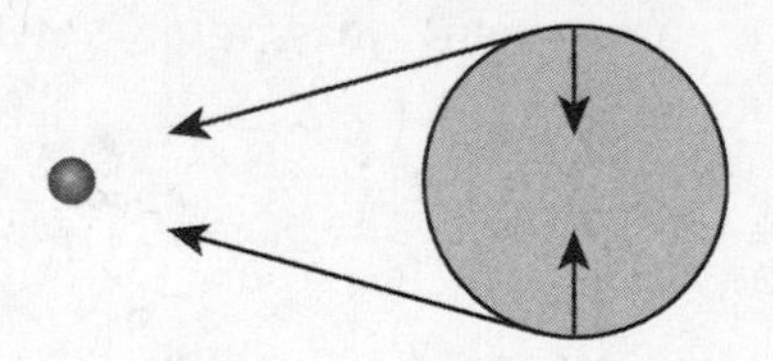

This explanation is very general, and in 1740 the French Royal Academy of Sciences offered a prize to anyone who would develop an exact detailed mathematical explanation. Such developments, based on

Newton's laws, were offered, separately, by Daniel Bernoulli, Leonhard Euler, Colin Maclaurin, and Antonine Cavalleri, all of whom were (jointly) awarded the prize. Sufficient numerical calculations describing the ebb and flow of the tides around the whole world were achieved only in the 1950s, at the Weizmann Institute of Science in Israel, with the appearance of electronic computers.

The mathematical formula that describes Newton's law of gravity omits one important link, and that is the question of the mechanism through which the attraction operates. Since Aristotle, the idea had become deeply entrenched that matter is continuous and that was what enabled force to be exerted by one body on another. Aristotle's opposition to the atomistic approach was based on, among other things, his puzzlement about how a force could be exerted despite the vacuum between the atoms. Newton's law of gravity gives a mathematical formula for the size of the force but does not refer to how the force is exerted. Newton did not ignore this question, and to answer it he adopted Aristotle's solution to the movement of heavenly bodies, taken up by Ptolemy, of the existence of the ether, that imperceptible material that fills the whole world. The great success of Newton's formula in predicting events and foreseeing orbits diverted research from studying what the ether was and trying to prove its existence.

Newton's contributions to science and mathematics far exceeded what we have described briefly above. They included fundamental contributions to optics, including the separation of white light into the colors of the spectrum, and in mathematics, for example, his binomial law, and so on. His great conceptual innovation in mathematics and science derived from his daring to create a new mathematics, specifically to describe phenomena in nature. After Newton, mathematicians did not hesitate to develop new branches of mathematics to describe nature better.

21. PURPOSE: THE PRINCIPLE OF LEAST ACTION

In his laws, Newton ignored the purpose that Aristotle's approach required. He made do with formulating mathematical laws through which, using mathematical operations, properties of the element being studied could be derived and predicted. A comparison of the prediction with the actual outcome is what corroborates a mathematical law of nature. When Newton was asked why his laws were formulated as they were, he answered that there was no doubt that God created a world that followed clear and simple mathematical laws. Tradition, however, and the approach that predominated in science until then were deep-rooted enough to make other scientists try to describe natural occurrences via basic principles rather than mere equations.

One route followed in the search for a purpose derived from an empirical law. Scientists had long known that the path of a beam of light passing from one medium to another is refracted at the point of transition. Willebrord van Roijen Snell (1580–1626), a Dutch physicist, discovered empirically and formulated what is known today as Snell's law of refraction. The law states that the sines of the angles of refraction are in the same ratio to each other as are the speeds of light in the different media (see the diagram below, in which the speeds of light in the two media are notated v_1 and v_2). The great French mathematician Pierre de Fermat (1601–1665) showed that Snell's law was equivalent to saying that the light beam passes from one point to another in a space such that it minimizes the time to get there. In other words, if we choose two points, A and B, on the beam, the point of refraction will be that which reduces the time the light takes to get from A to B to its minimum.

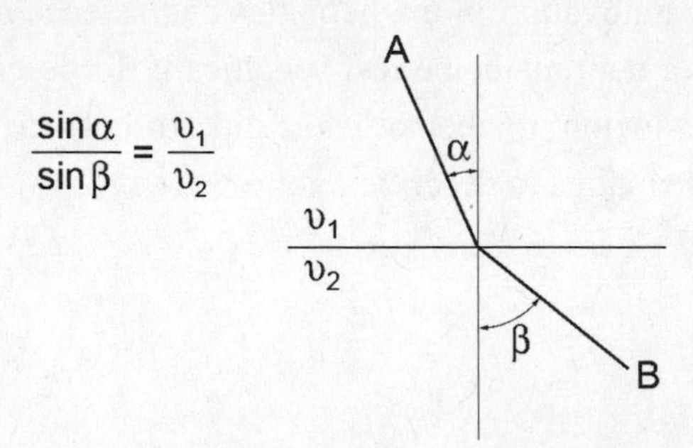

One way of interpreting Fermat's principle is to say that the light beam in effect solves a mathematical problem: it chooses a route that will get it as quickly as possible from one point to another. Fermat proved that the solution to the mathematical problem was given by Snell's formula. Obviously neither Fermat nor his successors claimed that the light beam had the purpose of reaching its destination as quickly as possible. They saw this property as a basic principle beyond a mathematical description, and thus, in the Greek tradition, the principle of the least time served as the purpose underlying the law of nature.

It is interesting to note that the Greeks themselves used a similar rationale. As is generally known, the angle of reflection of a beam of light striking a mirror equals the angle at which it strikes the mirror (or the angle of incidence equals the angle of reflection). Hero (or Heron) of Alexandria (10–70 CE) proved that the identity of the two angles leads to the conclusion that a beam of light from point A toward a mirror is reflected by the mirror and reaches a point B such that it moves from A to B (via the mirror) in the shortest possible time. Hero's reasoning, however, was inverted by Fermat. Hero considered it axiomatic that the angle of incidence equals the angle of reflection and proved that the light took the quickest route. Fermat proposed as an axiom, as a purpose, the principle of the minimum time and concluded that, in the case of the beam reflected from a mirror, the angles had to be equal.

From the time Newton's laws were published, many scientists tried to generalize Fermat's least time principle so that it would also apply to new laws. The most famous of those scientists were Leibniz, Euler, Lagrange, and Hamilton. The modern formulation of this type of principle, that is, the least action principle, is attributed to the latter. In this, the purpose is achieved by minimizing the integral along the path the body takes, integral of the physical quantity called the momentum of the system, that is, the multiple of the mass and the speed. This principle too is corroborated by experiments. Moreover, Newton's equations of motion can be derived from this principle. Hence, at least as far as mechanics are concerned, complete equivalence has been achieved between the direct mathematical description of the equations of motion and the description of the system by means of the purpose inherent in the least action principle. The reason for nature revealing such "efficiency" is a question with no clear answer.

Moreover, sometimes we try to attribute to nature an element of efficiency that it apparently does not have. For example, parts of lava fields have hardened into hexagonal shapes, as can be seen in the photograph below, taken in Iceland. The explanation offered by the tour guide and by scientists I have asked about this is that covering an area with hexagons is a solution to the minimum energy problem, which is rather difficult to formulate. The hexagonal structure of honeycombs is explained in a similar fashion. The bees try to "solve" the problem of constructing a surface of cells of a given magnitude with minimum length of walls. The intention is to minimize the amount of wax required. The ancient Greeks already conjectured that hexagons offer a solution to this minimization problem, called the honeycomb conjecture. Many tried to resolve the issue, but a complete proof of this mathematical fact was published in only 2001 by Thomas C. Hales of the University of Michigan, more recently of the University of Pittsburgh.

In regard to lava hexagons, I offer an alternative explanation. The support offered by one hexagon to its neighbors makes the hexagonal structure the most stable, the most able to withstand external forces of displacement. At the time of the formation of the lava fields, various weird and wonderful shapes were formed, covering different areas, such as squares and triangles. Those small areas covered by hexagons best survived the shocks and earthquakes that occurred there, which is why that is the shape we can see today, millions of years after its formation. That is also why the area covered by hexagons is only a small part of the total lava fields, with only smaller areas covered by other shapes that survived too. With regard to honeycombs, evolution may have taught bees to solve the honeycomb problem, but it may also be the case that the stability aspect also played a role; in other words, evolution selected the bees that constructed stabler honeycombs.

22. THE WAVE EQUATION

The laws of nature set out in the form of the relation between functions and their rates of change, that is, using differential equations, have, since Newton, become the major instrument used to understand nature through mathematics. Newton laid the foundation, and from his days until today scientists use his method and propose new equations to describe additional natural phenomena. If and when the experimental results corroborate the correctness of the equation, it is customary to name the equation after the scientist who proposed it. The following is a partial list of equations mentioned in the titles of lectures given recently in mathematical conferences in which I participated: Euler equation, Riccati equation, Navier-Stokes equation, Korteweg-de Vries equation, Burgers equation, Smoluchowski equation, Euler-Lagrange equation, Lyapunov equation, Bellman equation, Hamilton-Jacobi equation, Lotka-Volterra equation, Schrödinger equation, Kuramoto-Sivashinsky equation, Cucker-Smale equation, and so on. Each equation has its story and its use. Generally, the equations are formulated in terms of the relation between a function or a set of functions and their derivatives or integrals. The equations implement Newton's laws, sometimes in conjunction with other laws of nature, such as the law of conservation of energy or the law of conservation of matter. The names of the equations show that scientists have been using differential equations since Newton and continue to do so, and much remains to be done.

In this section we will present one equation that connects the distant past with the current era. This equation played a crucial role in the additional revolution in the relation between mathematics and nature, a revolution that we will discuss in the next chapter. The equation is not named after its formulator but after what it describes, namely, the wave equation. We are particularly interested in one specific aspect of it, and that is the string equation. No mathematical background is needed to understand what follows, but it does call for patience and tolerance for the mathematical symbols (and these can also be skipped without impairing the understanding of the text).

Undulations, or wavelike motion, can be seen in many of the materials around us, from the actual waves in the sea to the vibrations of string or membranes stretched to different extents. As we stated in the previous section, the equation that describes the motion of a spring or a pendulum was put forward as early as in Newton's days. In discussing the oscillation of a stretched piece of string or the movement of the waves in the sea, the situation is slightly more complex as the height of the wave changes from one place to another, and it also changes with time. That is, to describe a wave requires a function of time and place. The height of the wave may be notated by u, and for every location x and time t, the quantity $u(x,t)$ will represent the height of the wave at that location and at that time. We can relate to the rate of change of the height of the wave according to the time at a fixed location, and according to the location at a fixed point in time. The rate of change of the function according to location at a fixed time (imagine the profile of a wave at a given moment in time) is written in mathematics as $\partial_x u(x,t)$, and it is called the *partial derivative* by location (the notation ∂ was introduced to mathematics by the Marquis de Condorcet in 1770; some attribute it to the mathematician Adrien-Marie Legendre in 1786). Similarly, the expression $\partial_t u(x,t)$ indicates the rate of change of the function according to time (imagine the increase and decrease in the height of the wave at a given location). The second derivatives, that is to say the rate of change of the function of the rate of change, are written in mathematics as $\partial_{xx} u(x,t)$, and $\partial_{tt} u(x,t)$, respectively (see the diagram below of an advancing wave). The height of the wave varies with time and location. The law of its change will be given by a differential equation.

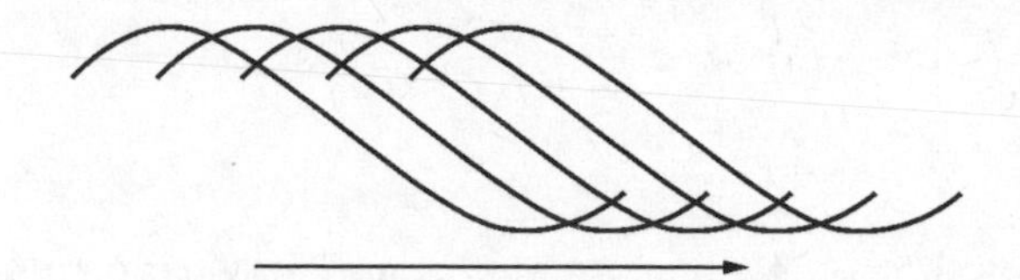

Early versions of differential equations describing waves were first suggested by Leonhard Euler (1707–1783) in 1734 and by Jean D'Alembert (1717–1783) in 1743. The breakthrough in the understanding of waves in general, and the vibrations of a taut string in particular, came in 1746 and 1748, when D'Alembert and then Euler published different solutions to the equation. It eventually transpired that the two solutions were in fact the same solution in different forms. For the sake of completeness we will give the equation itself:

$$\partial_{tt}u(x,t) = c^2\, \partial_{xx}u(x,t)$$

In ordinary language this says that the acceleration of the height of the wave according to time changes in proportion to the speed at which the rate of change of the height of the wave changes at the location where it is measured.

When the solutions to this equation were being studied, a surprising link was found with discoveries made by the Greeks, of course with no mathematical explanation, regarding the length of a piece of string and the note it produces when it is plucked. Without going into the technical details, we will just state that the relevant solutions of the equation describing the vibrations of a piece of string of length L, whose ends are fixed and unmoving, are a mixture of the sine and cosine functions that we have already encountered and that were known to the Greeks. Again for completeness, we set out a general form of solutions:

$$\sin\left(\frac{n\pi x}{L}\right)\left(\alpha_n \cos\left(\frac{n\pi ct}{L}\right) + \beta_n \sin\left(\frac{n\pi ct}{L}\right)\right)$$

There is an infinite number of solutions, as for every natural number n (and for all coefficients α_n and β_n) the formula provides a particular solution. It can be proved mathematically that every solution of the equation will be the sum of such particular solutions. The formula itself will not interest readers for whom mathematics and its use is not their profession, but two conclusions drawn from it are related directly both to the distant past and to future developments as follows.

Note that the length L of the string appears in the above expressions only in the denominator. The practical significance of this mathematical fact is that when the length of the string is halved, the frequency of the vibrations (the speed at which the sine and cosine functions change) is doubled. Thus, with a delay of some two thousand years, mathematics solved the mystery of the source of the finding dating back to Pythagoras, that when the length of the string is shortened by a half, the note it produces rises by exactly a full octave. The string equation connects a full octave, which the ear can identify naturally, with doubling the speed of vibration of the string. In a similar fashion, it is possible to obtain other findings from Pythagoras's time regarding the pitch of a note and the speed of vibration of the string.

Moreover, the equation shows that the vibration of the string consists of the sum of the vibrations shown in the formula for every value of n equal to 1, 2, 3, and so on. These vibrations have a "clean" frequency given by $\frac{nct}{L}$ for natural n. These frequencies are called *natural*, or *characteristic*, frequencies of the string. They will play a central role in the modern description of nature, which we will present in the next chapter.

23. ON THE PERCEPTION OF SCIENCE IN MODERN TIMES

Following the huge strides taken in the development and progress in understanding the world, a modern philosophical approach took shape with regard to how science operates and how nature is understood. We will summarize that philosophy very briefly. First, mathematics occupied its rightful position in the center of the stage. No scientific finding was considered understandable unless it was supported by mathematics that described it both qualitatively and quantitatively. Second, the requirement that mathematical principles explaining the findings must reflect some sort of purpose was abandoned. The very fact that a law of nature can be described by a simple and elegant mathematical description constitutes a purpose. The description can be in the form of an equation that nature "solves," or in the form of a "purpose," for example, to minimize a certain quantity, such as minimizing the time a process takes or minimizing effort. Newton's contribution showed that a special, new mathematics can be developed that is appropriate for the description and analysis of natural phenomena that can be measured, and he thus cleared the path for the search of new mathematical systems to describe the laws of nature.

The empirical approach developed by Galileo Galilei in Italy and Francis Bacon in England (according to which the mathematical developments must be initiated based on experiments) has been adopted, but the predictions provided by mathematics would not be acceptable unless backed up by the results of experiments. In all aspects of the essence of mathematics and the link between mathematics and nature, despite the fact that the Aristotelian search for a purpose as the basis for all natural phenomena was abandoned, the new philosophy adopted Aristotle's formalistic attitude to mathematics (see section 11). The philosophy that developed in the early modern age still holds today, with certain refinements that we will discuss in the next chapter.

CHAPTER IV
MATHEMATICS AND THE MODERN VIEW OF THE WORLD

What can an electric current be useful for? • Is the universe paved with gears? • Are mathematics and physics one and the same? • How did the axiom of parallels affect the theory of relativity? • Who discovered the formula $E = mc^2$? • How can you bend light beams? • Are we actually waves? • How do elementary particles group together? • Are humans made of strings? • How many dimensions do we live in?

24. ELECTRICITY AND MAGNETISM

In the middle of the nineteenth century there was a dramatic development in the use of mathematics to describe the world. It came in the wake of results of experiments in electricity and magnetism. The mathematical explanations of these discoveries led both to further surprising revelations and to a revolution in the approach to the mathematical description of nature. In a certain sense, mathematics that describes physics became physics itself. In this section we will briefly review the experimental discoveries that resulted in that revolution.

Static electricity and magnetism were known in the times of the ancient Greeks and ancient Chinese, and possibly even earlier. Thales of Miletus knew that when amber is rubbed with a cloth, the amber attracts light objects. Today we understand that the rubbing generates static electricity that causes the attraction. The word *electricity* comes from the Greek word for amber. Magnetism was also a known phenomenon, and the word *magnet* was taken from the town Magnesia in Turkey, part of Asia Minor, which was then under the Greeks. The Greeks knew that an iron bar suspended from a cord settles in a north-south direction. Compasses based on that property were already in use in the eleventh century. In the spirit of the Greek tradition, however, no experiments were performed to study those phenomena. Throughout ancient times it was thought that magnetism and electricity were totally unrelated.

In the sixteenth century, following the scientific revolution of the modern era led by Galileo, Francis Bacon, and their contemporaries, scientists began performing controlled experiments to study and understand different natural phenomena, including magnetism and static electricity. Among the pioneers in this field was the British physicist William Gilbert (1540–1603), who carried out controlled experiments and was the first to discover that magnets have two poles, north and south. Like poles repel each other, while unlike poles attract each other. Gilbert also found that there were two types of static electricity, which also repel or attract each other like magnets. Yet he did not realize the connection between static electricity and magnetism. More than a hundred years passed, and in the light of Newton's success in formulating the laws of gravity and its uses, scientists tried to find a quantitative expression for magnetic forces. The French physicist Charles-Augustin de Coulomb (1726–1806), after whom the unit of electrical charge (coulomb) is named, discovered that the power of attraction between two magnets and the repulsive force of electrical charges act in a similar fashion to the force of gravity; in other words, the force is proportional to the size of the charge and reduces in proportion to the square of the distance. The mathematical expression was of a familiar form, and hence the law was accepted relatively easily. Moreover, an understanding started crystallizing that perhaps something of deeper significance was taking place, and that was the uniformity of the mathematical forms that describe nature. Further progress in understanding the essence of electricity was made by the Italian physicist Luigi Galvani (1737–1798), who showed that static electricity can cause a mechanical action. Among other things, he connected static electricity to frogs' legs and found that it made the legs jump. This effect was given the name *gal-*

vanism, and still today students carry out those experiments in school. The Italian count Alessandro Volta (1745–1827), whose name is used for the unit of electric potential (the volt), showed that if one connects material with static electricity to material without static electricity by means of a metal bar, an electrical current is generated. He also showed how chemical processes can create static electricity and used that to build a primitive electric battery, the principle of which is used still today in the battery industry.

Until the beginning of the nineteenth century nothing was known of a physical connection between electricity and magnetism. The first such connection was brought to light in 1819 by the Danish physicist Hans Oersted (1777–1851). His discovery, apparently serendipitous, was that the needle of a compass changes direction when in the vicinity of an electric current. In other words, the current emits a force around it that affects the magnet. In about 1831, Joseph Henry (1797–1878) in the United States and Michael Faraday (1791–1867), one of England's leading physicists, discovered the second side of the connection between electricity and magnetism. They showed (independently of each other) that when a metal wire is passed close to a magnet, an electric current is produced in the metal. As an aside, we should add that Faraday, who devoted much effort to making science accessible to the public, was famous enough to merit a visit to his laboratory by King William IV. The king saw the experiment and asked, "Professor Faraday, of what use can this discovery be?" Faraday answered, "I don't know, but you will certainly be able collect a lot of tax on the results of this research."

Faraday's experiments were thorough and extensive. He formulated the quantitative relationship among the strength of the electric current and the speed at which the metal wire moved and the distance from the magnet. It was no great surprise to discover that when the wire moved repeatedly in circles near the magnet, the strength of the current generated acted like the function known to and used by the Greeks to describe the movement of the celestial bodies, and which had been used for about a hundred years to describe the movement of a pendulum, that is, the sine function. The nature of the link between the magnet and the wire was, however, not clear.

To understand the connection, Faraday used a concept that he developed himself, that of a magnetic field. No one knew what exactly the magnetic field was or how it worked, but the force it exerted could be measured, so it was not difficult to accept its existence.

The question was still unanswered as to the medium through which the magnetic force moved to create the current in the wire and then from the electric current in the wire to the movement of the magnet of the compass. Faraday suggested that ether was the medium through which the force was transferred, that is, the same material that fills every space around us, which the Greeks had used to explain the movement of the heavenly bodies and which was used later, in Newton's days, in relation to gravity. The formulae that Faraday developed were used to describe those quantities that could be measured and about which direct evidence could be obtained via the senses, extending Newton's approach. The concept of the magnetic field was more abstract but could be accepted because its activity took place in that very same ether. The explanation was that the magnet caused some kind of transformation in the ether that exerted a force on the electric particle, similar to the transformation that gravity apparently exerted in the ether.

The descriptions and measurements of the electrical and magnetic effects in Faraday's time yielded much information about the connection between electricity and magnetism, but the knowledge did not go beyond the measurement of the actual forces that could be measured and a quantitative description of the connection between them. The magnetic field that operated via that elusive material called the ether served to provide a mechanical explanation for the action of the forces, but that was not a mathematical explanation in the spirit of the modern era. There were no mathematical equations that the electrical and magnetic effects satisfied, nor was any purpose defined along the lines of, say, the least action principle that explained those effects.

25. AND THEN CAME MAXWELL

James Clerk Maxwell was born in 1832 in Edinburgh, Scotland, to a well-established yet not particularly wealthy family. His father, John Clerk, was a fairly successful lawyer who inherited from childless relatives a country estate in the area of Glenlair, not far from the capital, Edinburgh, on the condition that he adopted their family name, Maxwell. This he did, and he moved to the estate with his wife, Francis, and his firstborn son, James. James's mother died when he was eight years old, and his first years of schooling were at home with a young private tutor. It was only when James reached the age of ten that relatives persuaded his father to send him to a normal school in Edinburgh. In school he was found to be an outstanding student, and he continued his studies at Edinburgh University. There too he excelled, and in the course of his studies he moved to Cambridge University in England, where he completed his studies cum laude, second among all the graduates. Heading the list was his friend Edward Routh (1831–1907), who also became a distinguished mathematician and actually researched areas in which Maxwell had laid the foundations. While at Cambridge, Maxwell wrote several important papers that earned him a fellowship at Trinity College in Cambridge. His father's death brought Maxwell back to Scotland at the age of twenty-five, and he held the position of chair of natural philosophy at Marischal College, Aberdeen. There he wrote an important paper on the stability of Saturn's rings, a paper that earned him the prestigious Adams Prize. There he also married his sweetheart, Katherine Mary Dewar, the daughter of the principal of Marischal College. Neither his family connection nor his scientific fame helped him when the college merged two departments and a professor from the other department was chosen to stay on. Maxwell had to leave, and he returned to England, to a post at King's College, London.

His achievements in London were astounding. There he completed his work on colors, a subject that had absorbed and fascinated him for many years (we will expand on this below), and, his crowning glory, he published his study on electricity and magnetism. His success and fame, among other things his appointment to the Royal Society, did not change his modest, family-oriented personality. After suffering from an infection resulting from a cut incurred when he fell from a horse, in 1865 Maxwell chose to return to Scotland and lived on his estate in Glenlair. He continued with his academic work and published two seminal papers, one on the structure of stabilizers and the other on what is today called statistical mechanics; on these too we will expand below. Maxwell did not succeed in obtaining a position at St. Andrews, and in 1871 he decided to move back to England to head the Cavendish Laboratory of Experimental Physics in Cambridge. He continued his theoretical work and published several important papers and books. He died in Cambridge in 1879.

Maxwell's greatest achievement was his contribution to electricity and magnetism, but it should be noted that other research he carried out led to major changes in other areas of science as well. The composition of the colors of the spectrum was a subject that occupied Maxwell for many years. Newton showed how the elements of white light can be revealed, that is, all the colors of the spectrum, by passing the light through a prism. After Newton, many attempted to understand the links between the different colors and the rules according to which, by mixing various colors, we create new ones. It was Maxwell who discovered and proposed the structure known by the initials he used, RGB (i.e., the three primary colors, red, green, and blue), which when mixed in the correct proportions can produce all the colors the human eye can perceive. Even today this system is used in our daily lives, including television broadcasts and printing of pictures. Maxwell was also the first, in 1861, while at King's College, London, to produce a color picture. Another area of his work laid the foundation of the area of physics called statistical mechanics, the mathematical laws according to which gases move. Maxwell's ideas were later developed further by the Austrian mathematician Ludwig Boltzmann (1844–1906) at the University of Vienna. Today the equations defining the movement of gases are called the Maxwell-Boltzmann equations, or Boltzmann equations. Yet another subject in which Maxwell set the foundations was the theory of stability in control systems; he wrote this in an article in 1868 as a solution to an engineering problem that engineers of steamships had encountered, and asked for his help (we will review this fascinating topic in section 64, in the context of a discussion of applied mathematics).

The search for the laws of physics underlying electricity and magnetism occupied Maxwell and many of his colleagues for several years. After working intensively, Maxwell constructed a system of differential equations whose solutions described all the observable effects that could be measured at that time. The mathematics he used to formulate the equations was based on the calculus proposed by Newton and the further developments of the system, especially partial differential equations, such as the wave equation discussed in section 22. The equations developed by Maxwell related to the same magnetic and electrical fields that Faraday had suggested and proposed relations between them, from which the effects and the different forces observed in the laboratory could be derived. The equations, however, like Newton's law of gravity, did not include an explanation of how the electric and magnetic forces pass through space from the source of the force to the object on which it is exerted. Moreover, unlike the case with the law of gravity, the ether could not provide an explanation for the transition of the electrical and magnetic forces. The reason is "technical" (and can be skipped). The direction of the magnetic field created by the current is perpendicular to the direction of the current, while the force exerted on the current by the magnetic field is perpendicular to the magnetic field and varies according to whether the field gets stronger or weaker (readers who have studied electricity and magnetism will certainly remember the left-hand and right-hand rules to describe the directions of these forces). Such a combination is impossible if the force passes through mechanical stress in any type of material.

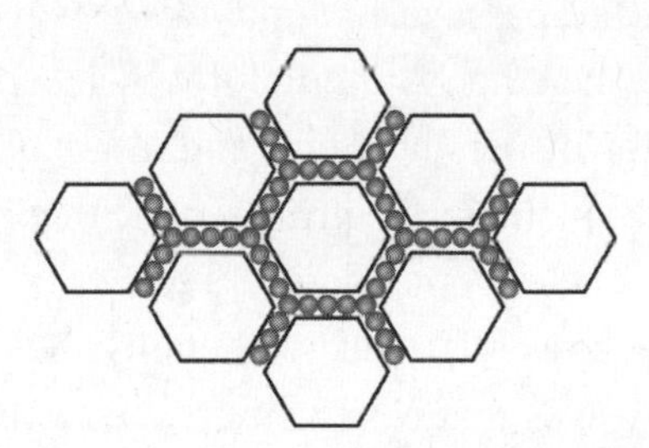

At this stage Maxwell proposed a mechanical model through which the electrical and magnetic force would pass. And indeed, after working on this for more than a year, Maxwell succeeded in constructing a theoretical system of gears and bearings of different sizes and directions (see the diagram) whose revolutions, if the space was covered by them, would explain the directions of the transfer of the electric and magnetic force as they were then known. Examining this model of bearings, Maxwell noticed that the same bearings and gears would probably provide a new type of solution to his equations, namely, waves in the magnetic fields themselves. He called these electromagnetic waves.

Maxwell published his gear model as one according to which the magnetic and electrical forces could operate, but he did not claim that a material with the same properties as the gears in fact existed. Yet, he published his hypothesis that the new solutions he found to the equations, namely, electromagnetic waves, exist in reality and added another bold hypothesis that light itself is an electromagnetic wave.

The fact that light has the properties of a wave had been discovered long before by Christiaan Hyugens (whom we will meet again further on). Since the days of Newton there had been a debate as to whether light consisted of particles or had wavelike properties. Newton was of the opinion that light was made up of particles, but many others championed the wave model. The wave model was based mainly on the results of work by the French physicist Augustin-Jean Fresnel (1788–1827), who gave a precise explanation, with the aid of the wave model, of light diffraction through the creation of a mosaic of light and bands of shadow when light is passed through narrow slits. Fresnel and others showed that diffraction can be clearly explained as the passage of light like a wave through that same intangible substance, the ether. Although it was acceptable to consider the ether as a medium for dispersing light, and although he himself showed that the ether cannot explain his electromagnetic theory, Maxwell ventured to claim that light is an electromagnetic wave.

Maxwell's conjectures were received coolly. Nobody had felt or seen the electromagnetic waves that Maxwell claimed existed, nor had anyone seen their activity. The similarity between light and the electromagnetic wave was perceived simply as a coincidence. The dispersion of light could be explained with the help of the ether, but the use of the ether contradicted the assumption that light satisfied Maxwell's equations. He himself proposed his gears as a metaphor and did not claim that some kind of gears actually pave space. Even his supporters among his professional colleagues tried to encourage him to find alternative explanations for the way in which magnetic and electrical forces are interwoven. In the approach that had prevailed since the Greeks, through Newton, and on until the nineteenth century, acceptable explanations for the application of a force had to indicate the means through which it was transmitted.

Maxwell then offered an improved formulation of his equations describing the connection between magnetic and electrical forces that was more complete and symmetrical than his earlier one. The development of the equations depended on only basic rules of dynamics, such as the conservation of energy, and totally ignored the mechanism through which the force was transmitted or the energy conserved. The only justification for the existence of the waves was the fact that they provided a solution to the equations, solutions that have wavelike characteristics. Unlike waves in the sea or sound waves, however, they are not necessarily the result of motion in any medium. The equation defines them as waves. Maxwell persisted in his view that light was an electromagnetic phenomenon. (The formulation in use today is a further simplification and improvement developed later by Oliver Heaviside, 1850–1925.)

This was a revolutionary approach to understanding nature and to the link between mathematics and nature. The Greeks employed the mathematics already known to them to describe natural occurrences accurately and logically, mainly to avoid errors and optical illusions. Newton and his successors, including D'Alembert and Euler whom we came across in connection with the wave equation, used new mathematics to describe quantities and effects that humans can perceive or feel and about which they can even develop intuition, for example the oscillations of a string. Maxwell's improved equations describe electromagnetic waves without reference to the physical nature of the wave. Maxwell's was a twofold innovation. First, the equations ignored how the factors they described operated, that is, the physics of the phenomenon. Second, no one had seen, heard, or felt the result predicted by the equations, the electromagnetic waves, and in particular there was no intuition regarding how they operated.

The same problem of lack of intuition explaining a remote effect apparently exists with regard to gravity and Newton's laws. The important word here is *apparently*. The existence of the ether, that unseen, intangible medium that can transfer a force, explained the working of gravity. It was only in the wake of Maxwell's revolution that it became accepted that gravity also acts without a material medium. The predictions provided by Newton's equations, for example about the existence of the planet Neptune, were predictions about bodies of a known type with a firmly established intuition about their movement in nature. Maxwell's equations described physical phenomena using only mathematics. Maxwell in effect changed the way in which the world is described by mathematics. He forwent the physical explanation based on known physical quantities. He published his equations, as if declaring this is physics, the physics is inherent in mathematics. Although we cannot see or perceive the elements appearing in the mathematical equations in any other way, they are there, in nature. We can measure the effects of these elements on other bodies, measure the electricity they generate, utilize the force they exert, and so on, without being able to relate directly to their physical entity.

This conceptual revolution was not welcomed with open arms. Michael Faraday, then one of Britain's most eminent physicists, wrote to Maxwell in relatively blunt language (in this author's words, not verbatim): Would the mathematicians kindly translate the mathematical hieroglyphics in which their conclusions are written into comprehensible intuitive language so that the physicists might understand? As far as we know, Maxwell did not respond. In any case, the answer would have been negative. Physics is inherent in the mathematics that describes it.

Beyond the revolution in the approach, the difficulty in adopting Maxwell's theory lay in the fact that at that time all the elements that could be measured and that substantiated the correctness of the equations could also be explained using the physical mathematical relations that were known and accepted. Although there was not one comprehensive equation incorporating all the phenomena, the overall equation that Maxwell proposed had no physical basis in terms of those days, and it predicted effects that no one had seen previously. It took twenty-five years from the publication of Maxwell's work, and eight years after his death, until Heinrich Hertz (1857–1894), a German physicist from the University of Bonn, succeeded in 1887 in producing an electromagnetic wave in the laboratory. There is no need to detail the importance of this physical discovery. Soon afterward, Maxwell's claim that light is an electromagnetic wave with a frequency we can perceive, that is, see, was confirmed. Its uses include radio transmissions, television, cellular telephones, microwave ovens, x-rays, and many others, all based on electromagnetic waves.

We should note that although Maxwell presented his equations without reference to the ether, he did not claim that the ether did not exist. The existence of the ether was still "required," for example as a medium for Newton's gravitational forces. Maxwell claimed only that one could relate to how electromagnetic waves propagated by means of the mathematical equation they satisfied, without explaining the essence of the waves.

Maxwell's contribution was comparable to that of the Greeks in adopting mathematics to describe nature, and to that of Newton, who had the courage to develop a new mathematics to do so. Maxwell effectively changed the paradigm of the mathematical description of nature. No longer would a quantitative formulation of entities that the senses perceive directly or via a medium be the only way to proceed. Rather, the mathematical handling of abstract quantities, whose existence is justified simply by the fact that they explain effects that we can measure, became fully acceptable. Maxwell's contribution also changed the definition of a physicist. Today, if in the course of a conversation or a lecture I refer to Maxwell as a mathematician, some of my colleagues correct me, claiming that he was a physicist. Yet Faraday, one of the greatest physicists living in the time of Maxwell, refers to him as a mathematician who ought to translate his claims into the language of physics. Instead of doing that, Maxwell altered the definition of physics. Heinrich Hertz said, "Maxwell's theory is Maxwell's equation," a statement that gives the essence of modern physics. Physics is inherent in the mathematics that describes it.

26. DISCREPANCY BETWEEN MAXWELL'S THEORY AND NEWTON'S THEORY

Maxwell's equations were amazing in their predictions of the discoveries that came later and fitted the observations and measurements performed in relation to those discoveries. Likewise, Newton's original equations and other equations that were derived using the mathematical instruments that he developed fitted physical reality perfectly. Yet there was a discrepancy between their equations, a discrepancy that threw doubt on whether both described the same physical world. We will describe that discrepancy.

Newton's second law, $F = ma$, states that the acceleration a of an object is proportional to the force F exerted upon it. The law does not relate to the speed of the object while the force is being exerted on it, but to its acceleration, that is, the change in its speed. Neither does the law of gravity relate to the speed of the object. The gravitational force acting on a body does not vary if it is moving or stationary. This non-dependence on the speed of the object is a blessing to mathematicians and physicists, and of course to engineers, because when measuring a change in speed as a result of the application of a force, there is no difference whether it is measured in relation to the Earth or in relation to another system of coordinates, such as a moving train. Newton presumed that the world had an absolute system of coordinates that can be used to measure the speed of every moving object. Although we cannot accurately identify this system, we are fortunate that it is not of any importance, since the laws of motion do not change if the speeds are measured with reference to some speed that is constant relative to the absolute system. Such systems are called inertial systems. The laws of motion do not change if one moves from one inertial system to another.

This property is lacking in Maxwell's equations. A transition from one system of coordinates to another, even if the second is moving at a constant speed relative to the first, alters the equation. We did not state Maxwell's equation explicitly, but it is not necessary to know its details to understand why a move from one system to another changes it. Maxwell's equation describing electromagnetic waves makes use of the speed of the wave.

When the speed appears explicitly in the equation, a change in the coordinates according to which the speed is measured changes the equation. In other words, Maxwell's equations are inconsistent with the invariance with respect to the inertial system in which Newton's equations are formulated.

The state of physics at that time can be summarized as follows: Newton's equations, which relate to a geometry that is consistent with our senses and whose relation has been accepted for hundreds and even thousands of years, accurately described many physical aspects of motion, including the propagation of waves in an earthly medium such as air or water. The movement described by the equations fits our intuition regarding the movement of objects, intuition that was formed in the course of the development of the human species. Maxwell's equations, on the other hand, predicted the existence of waves without indicating the medium in which those waves move, and their formulation is inconsistent with the known and familiar geometry. Those equations, however, were also astonishingly effective in predicting the physical effects relevant to them.

How do we proceed from here? One possibility is not to try to reconcile the two approaches. No discrepancy has been found between the two sets of equations. They describe different physical effects, and nobody can guarantee that there is one mathematical theory that covers the different effects in physics. A second possibility is to try to change one of the systems of equations for another system whose structure is consistent with that of the second. Many attempts were indeed made to replace Maxwell's relatively new equations with others in order to remove the discrepancy; none succeeded. Then Einstein came along and proposed a surprising third solution: he changed the geometric depiction of the world, that is to say he suggested a description of the geometry of the world that differs from what we feel it is but that reconciles the two systems of equations. Einstein's contribution was a huge breakthrough. It too came about following findings in physics, which we will discuss in section 28, and research in mathematics over many years into the geometry of the world. The salient aspects of that mathematical research are described in the next section.

27. THE GEOMETRY OF THE WORLD

To describe Einstein's theory of relativity with the correct perspective we must go back more than two millennia. We will repeat what axioms meant to the Greeks. They were basic working assumptions from which, by using the power of logic, they could develop mathematics further. The axioms themselves were obvious facts, physical truths, or ideal mathematical truths that did not need to be explained or substantiated. The geometry of the space in which we live could of course be examined mathematically, and Euclid summarized the mathematics of geometry in his book *Elements*, where he presented the axioms of geometric space that had been formulated in classical Greece. Euclid formulated ten axioms and postulates (modern formulations show fourteen), including "self-evident" axioms such as, given two points, there is a straight line between them, and that a point in a plane and a given length from it define a circle with the point at its center and the given distance as its radius. Euclid formulated his fifth axiom, which in time became the parallel-lines postulate, as follows (see the upper part of the diagram below).

> *If a line transversing two lines forms on one side of it two angles whose sum is less than that of two right angles (i.e., less than 180 degrees), the extensions of the two lines will meet on that side.*

The formulation of the parallel-lines postulate accepted today states (see the lower part of the diagram):

> *Given a line and a point not on it, there is only one line that passes through that point that is parallel with the first line; parallel means that the extensions of the lines will never meet.*

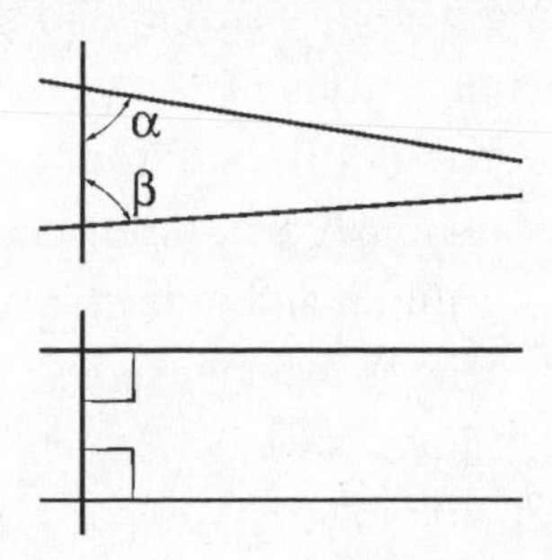

Euclid himself made no reference to parallel lines, that is, straight lines that never meet. The reason was, apparently, the desire to avoid the assumption that there are lines that are infinite, having no end. As we wrote in section 7, Aristotle distinguished between infinity and potential infinity. He claimed that potential infinity existed so that, for example, we can draw lines as long as we want, but he denied the existence of infinity, that is, he denied that an infinite line exists. Euclid adopted that view in his books and avoided presenting and dealing with infinite quantities.

His fifth axiom caused unease even in Euclid's lifetime. The critics claimed that it was not obvious nor apparent from observing nature, and therefore it was not suitable to be included as an axiom. Attempts were made, therefore, to prove that property using other, self-evident axioms, but those attempts failed. Interest in this question continued for hundreds of years. Particularly noteworthy were the contribution and discussions of Omar Khayyám (1048–1131) and his school in Persia. It was there that Euclid's original fifth axiom was replaced by the parallel-lines postulate, which is accepted still today. There too it was proven that if one assumes the existence of infinite lines, the two versions are the same.

An interesting development that illustrates the complexity of the connection between mathematics and nature was made by the Italian mathematician Geraolamo Saccheri (1667–1733) of the University of Pavia. We will give the details of the case to show the difficulties that logical arguments are likely to cause. Saccheri used proof by contradiction, that is, he chose to assume that the parallel-lines axiom was not correct and tried to obtain a contradiction between that assumption and the other statements that could be proven using the other axioms. In that way he would show that there was a contradiction deriving from the assumption that the axiom was incorrect so that the parallel-lines axiom is derived from the other axioms. We have shown previously that proof via reduction ad absurdum, or by contradiction, is not natural, and indeed it is difficult even to follow the steps that guided Saccheri, as we now show.

The parallel-lines axiom can be divided into two claims. The first is that there is a line parallel with the first line, and the second is that there is not more than one parallel line that passes through a given point not on that line. Saccheri first assumed that there is not even one parallel line that goes through the point, and he succeeded in finding a contradiction between that assumption (of no parallel line) and results that can be obtained from the other axioms. He concluded that there was at least one line parallel with the original line that went through the point not on the original line. He then assumed that there were two or more parallel lines that passed through the point. By means of a construction based on those two parallel lines he found such planes, however, with very strange properties, so much so that it was clear that they did not exist in the physical space we see around us. That was enough to convince Saccheri, and he declared that he had found the contradiction he was looking for and that the parallel-lines axiom is derived from the other axioms (because if we assume it is incorrect and we find a contradiction, then it is correct).

Saccheri, however, did not take the trouble to examine whether the fact that the strange properties do not exist in our space can be derived from Euclid's other axioms. Only if that is done can it be concluded that there is a contradiction. Otherwise, the only conclusion that can be drawn is that Euclid's axioms also allow the existence of strange spaces. That flaw derived, of course, from the belief that the axioms described the space around us, and, to establish a contradiction, it was sufficient to find a property that our physical space does not have in order to conclude that the property does not exist in the mathematical space. A short while later it was found that there was no *mathematical* contradiction between the properties discovered by Saccheri and the other axioms, and the question of whether the parallel-lines axiom derives from the other axioms was declared still open.

The German mathematician Georg Klügel (1739–1812) made a conceptual contribution to solving the problem. His doctoral thesis at the University of Göttingen, Germany, was devoted to a detailed review of the parallel-lines axiom and the hitherto failed attempts to reconcile it with the other axioms. Klügel completed his thesis with the hypothesis that Euclid's fifth axiom was based on the experience of our senses and hence may not be correct. In other words, there may be geometries that satisfy the other axioms but not the parallel-lines axiom. That declaration itself made researchers try to construct geometries in which that axiom did not hold, and this they did quickly. One was by Abraham Kästner (1719–1800), who constructed a geometry with properties similar to those discovered by Saccheri about a hundred years earlier, but this time the conclusion was the opposite, that is, the parallel-lines axiom does not depend on the other geometrical axioms, meaning there are mathematical spaces in which that axiom does not hold while the other axioms do. The conclusion drawn was that Euclid's axioms do not describe the physical space completely, a possibility that, if Saccheri had thought of it, might have led to the solution of the problem some hundred years sooner.

The geometries developed by Kästner and others at that time had such strange characteristics that it was clear they were not relevant in the context of a description of nature. The obvious conclusion was that new axioms had to be added to Euclid's original ones that would characterize the space we experience every day. Such an attempt was made by a student of Kästner's, Carl Friedrich Gauss (1777–1855), one of the greatest mathematicians of all time. Gauss, who worked for most of his life in Göttingen in the then Kingdom of Hanover, Germany, was born to a poor family, but the exceptional mathematical abilities he exhibited even at an early age brought him to the attention of the Duke of Braunschweig, who used his influence with the University of Göttingen to get them to accept Gauss as a student. Gauss made an enormous contribution to mathematics, which we cannot describe here. Most of his work related to the theory of numbers, but he contributed greatly in other areas too and was also one of the great philosophers of the natural sciences. Gauss learned from his tutor, Kästner, that the parallel-lines axiom did not derive from the other axioms, but at least at the outset he still thought that the axiom described the world around us. For years Gauss tried to suggest "correct" axioms, meaning obvious ones, from which the parallel-lines axiom could be proven. After years of unsuccessful attempts, his faith was shaken and he started to look for other axioms to replace the parallel-lines one. He saw that the parallel-lines axiom, that is, the property that parallel lines do not meet, applies in our daily experiences, experiences based on the measurement of small distances. Thus geometry that describes the world around us complies with a property similar to the parallel-lines axiom when dealing with small distances. For example, the parallel-lines axiom leads to the result that the sum of the angles of a triangle is 180 degrees. In other geometries that were found to fulfill the other axioms, not the parallel-lines one, the sum of the angles of a triangle was greater than 180 degrees. Gauss added the requirement that as the triangles get smaller and smaller, the sum of the angles has to come closer and closer to 180 degrees, or, put differently, for small distances, the geometry must be similar to the geometry we experience day to day. None of the geometries discovered previously that fulfilled the axioms of the plane apart from the parallel-lines axiom satisfied that requirement.

The question still remains unanswered whether the parallel-lines axiom can be proved from Euclid's axioms together with Gauss's new requirement, or whether perhaps even with that new requirement the parallel-lines axiom does not hold. If the latter is the case, the question of what is the true geometry of the physical space arises even more forcefully. In this connection, the following story is told about Gauss. Gauss, as a mathematician, developed a method for measuring the state's land. He thus served as a guide to the state's official surveyors and actually carried out measurements himself. According to the story, which has no historical authentication, Gauss tried to measure the angles of a triangle formed by three mountains in Germany that were situated a long distance from each other. If he would have found that the distances between them were large enough for the sum of the angles of the triangle to be more than 180 degrees, he would have proved that Euclid's mathematical geometry does not correctly describe the physical reality. The measurements did not reveal such a triangle. Be that as it may, the story is consistent with the fact that Gauss himself solved the open question, showing that the parallel-lines axiom cannot be proven from the plane axioms even if his new condition was applied, but that revelation he kept to himself. Gauss did not divulge whether he thought the parallel-lines axiom applies in the physical world or not.

An example that shows that a geometry can exist that fulfills all the plane axioms, including the new condition that Gauss proposed but excluding the parallel-lines axiom, was discovered by two young mathematicians independently of each other. One was a Russian from Kazan University, Nikolai Lobachevsky (1793–1856), and the other was a Hungarian, Johann Bolyai (1802–1860), an officer in the Hungarian army. Bolyai was the son of a well-known mathematician who had corresponded with Gauss and who had received letters in which Gauss wrote of his doubts about the geometry of the world. Lobachevsky and Bolyai both constructed geometries with the desired properties, namely, small triangles with the sum of their angles close to 180 degrees, but this did not apply to large triangles. When the young Bolyai revealed his discovery to Gauss, the latter showed him that he, Gauss, had already arrived at such a geometry, but he was generous enough not to claim the right to be acknowledged as its discoverer.

The mathematical problem was solved: the parallel-lines axiom is not derived from Euclid's axioms even if the requirement is added that in small distances space must behave as we experience it in our daily lives. The physical problem remains, however: Which of the various possible geometries according to the axioms is the one applicable to our world? This is not a trivial question. We must remember that Newton's theory, including his original equations and all the equations and other developments since Newton, was based on space as defined by Euclid's axioms, including the axiom about parallel lines. Is it possible that everything derived from this mathematics is not relevant in physical space?

At this point Bernhard Riemann comes into the picture. Despite his short life (he was born in 1826 and died at the young age of forty), he made crucial contributions to mathematics and physics. Georg Friedrich Bernhard Riemann was a student of Gauss, but even as a student he worked independently. Born to a poor family, he was a very sick child and youth. He started studying theology with the intention of becoming a priest. At the same time he showed great mathematical ability and tried to integrate the study of the Bible with mathematics, even attempting to examine the book of Genesis from a mathematical standpoint. His father, who recognized the young Bernhard's talent for mathematics, urged him to apply to the University of Göttingen, where he chose to work for his doctorate under the guidance of Gauss. The method of study and research required the candidate to submit three research topics, which the tutor and the thesis committee were supposed to approve. The tutor and committee were then to set for the student a predetermined time to write a thesis on them. The third topic proposed by Riemann brought about, many years after his demise, a change in the perception of the geometry of the world.

Riemann's approach was also to formulate axioms, but instead of looking for axioms that would describe what we feel and see, he developed a system of axioms that a physical space "ought to" satisfy, with "ought to" meaning that the concepts of "closest" and "shortest" make sense in the system. The technique is related to the angular structure of lines and planes and the curvature of different planes. This mathematical subject is called differential geometry.

It is not necessary to specialize in the subject to understand the concept underlying it, and that is the concept of a geodesic, the shortest line between two points. In a Euclidean space, a straight line is the shortest path joining two points. In general geometries, that is not necessarily the case. It may be that in the some geometries there may not be straight lines in the Euclidean sense, but there will be a shortest route that joins the two points. To illustrate, consider the following: in the geometry on the face of the Earth there are no straight lines, but there are geodesics. Airplanes flying between two towns along a similar latitude, for example between Madrid and New York, choose a route that takes them far north because that is the shortest. Geodesics in general space constitute the building blocks of Riemannian geometry. It is unclear where Riemann got his inspiration to define such a structure, but what is clear is that he was aware of the difficulties in describing and determining the geometry of nature and that he was familiar with the work of his teacher, Gauss. At the same time, he was aware of the least action principle and of Fermat's principle that preceded it. He therefore apparently wanted to construct general, not necessarily Euclidean, spaces in which the principles of shortest distance and least action could be applied. Riemann died before he could clarify his intentions. The mathematical tools he bequeathed, in particular the geometries based on shortest distances, served Albert Einstein in the construction of the new geometry of nature.

28. AND THEN CAME EINSTEIN

Let us recall where matters stood toward the end of the nineteenth century with regard to the mathematical description of nature. On the one hand, Newton's mechanics based on arithmetic and Euclidean geometry had attained enormous success in both celestial mechanics and earthly engineering problems. The success of part of Newton's mechanics may be attributed to the existence of the mysterious matter, the ether. On the other hand, Maxwell had presented equations that predicted the existence of electromagnetic waves, which had indeed been found. The ether could not constitute the medium in which those waves moved, and no other medium was known. In addition, if Maxwell's equations were applied in Newtonian geometry, they lacked one very important element of Newton's theory—that the measurements could be taken in every inertial system. At the same time, mathematicians and physicists of the time started questioning whether Euclidean geometry was the right one for describing the world we live in. The feeling was, and Gauss stated it explicitly, that arithmetic could be relied upon as a tool for describing nature, but as it was not clear what the geometry of nature was, geometry could not be relied upon as such a tool.

This was the situation with which Einstein was familiar. Several additional discoveries and hypotheses were made that could have influenced him, but it is not clear to what extent he was exposed to and aware of them. One such discovery was the famous experiment performed by Albert Michelson and Edward Morley, two American physicists. At that time the scientific community still believed in the existence of the ether as the medium through which forces are exerted (Maxwell's theory had not as yet been accepted; the Michelson-Morley experiment took place before electromagnetic radiation was discovered in a laboratory). One of the questions that arose was: In what direction does the ether move relative to the Earth? The idea of the experiments was to utilize the fact that the ether was the medium through which light waves propagate.

The principle was simple. Assume that a beam of light travels from a source, *A*, on Earth toward a mirror at point *B* and back, with the mirror arranged such that the direction from *A* to *B* is the direction of the Earth's orbit. At the same time, another beam of light was sent from the same source *A* to a mirror at point *C* and back, with the direction from *A* to *C* at right angles to the direction of the Earth's orbit. A simple calculation shows that the first beam of light would get back to *A* after the second beam. A calculation is needed because the claim is not at all intuitive. In order to understand that it is correct, imagine that the speed of light is just twice the speed at which the Earth is moving. In that case, the distance that the first beam has to travel to get from *A* to *B* is twice the actual distance between them. In the time that takes, the second beam will have already gotten back to *A*; that is, it will get back to the source sooner. In reality, the speed of light is much faster than twice the speed of the Earth, and the time difference between each of the beam's returns to the source would be minimal.

To carry out the measurements required a series of sophisticated experiments, which Michelson started in 1881; Morley joined him in 1886, and their joint efforts showed that there was no difference between the times the beam arrived back at the source. This finding raised doubts regarding the hypothesis that the ether was the medium through which light propagated.

The Dutch physicist Hendrik Lorentz proposed a mathematical formula describing the dynamics that fitted the results of the Michelson-Morley experiment. The formula corrected the assumption that until then had been considered self-evident, as it was based on what our senses perceive. The assumption was that if an object F is moving at a velocity v_1 relative to object G, which is moving at a velocity v_2 relative to object H, then F is moving at a velocity of $v_1 + v_2$ relative to object H. Lorentz proposed replacing this formula with another (whose detailed formulation is not important for the current discussion) that showed that, at small speeds, the speed of *F* relative to *H* is very close to what our senses perceive, that is, $v_1 + v_2$, but when *F* is moving at very high speed relative to *G*, say close to the speed of light, and *G* is moving at a speed close to the speed of light relative to *H*, then the speed of *F* relative to *H* is also close to the speed of light (and not twice that speed, as Newton's theory would infer). Lorentz also observed that if his formula is used in Maxwell's equations, the same equations are obtained in all inertial systems.

The French mathematician Henri Poincaré (1854–1912), who was then one of the most famous mathematicians in the world, stated that Lorentz's formulae gave the best explanation for the difference between Newton's laws and Maxwell's equations. He himself developed the mechanics deriving from Lorentz's transformation, and already in 1900 and later in several additional papers, he published a complete version of what later became known as the special theory of relativity, including a version of the connection between mass and energy, that is, $E = mc^2$. These developments of formulae presented by Lorentz were extremely close to Einstein's special theory of relativity, which we will describe below. Yet neither Lorentz, though a famous physicist who actually was awarded the Nobel Prize in Physics in 1902, nor Poincaré, nor others who tried with the help of those formulae to reconcile Newton's and Maxwell's theories drew the right conclusions from the formulae. It was Albert Einstein who did so.

Einstein (the extent of whose knowledge of these developments is unknown, as we stated above) adopted Lorentz's formulae and distilled from them a physical property that completely contradicts intuition based on our daily experiences. The property is that the speed of light is constant in every inertial system of coordinates. That is to say that if light travels at a speed of *c* relative to an object *G*, and *G* is traveling much faster than an object *H*, even at half the speed of light, the light will still travel at speed *c* relative to object *H*. This leads to the conclusion that it is not possible to get an object to move faster than the speed of light. In a certain sense, Einstein claimed that the geometry of the world, including formulae for combining speeds, is described better by Lorentz's formulae than by Newton's. As stated above, at relatively low speeds Lorentz's formulae are very close to Newton's, which is why our intuition, which is based on our daily experiences of only relatively low speeds, led us to accept Newton's formulae as describing the world.

The mathematics that describes the new relations between position and speed brought Einstein to the realization of the possibility of something never previously envisaged, and in particular something that was opposed to our intuition (in the wake of Maxwell's basic contribution, the realization that mathematics can lead to revelations of totally new phenomena had become accepted). The equation leads to the conclusion that mass is converted to energy. This was a mathematical statement that could have remained in the realm of a technical mathematical conclusion without any physical importance. Einstein, however, interpreted the equation as a physical truth and drew the conclusion that the substitution between mass and energy was possible (although at that point in time he did not see a way to control and exploit this possibility). He even derived the famous formula $E = mc^2$ (after several more-complex versions) from the equation for the mass-energy equivalence. Einstein published these discoveries in two papers that appeared in 1905. Three years later, in 1908, Hermann Minkowski (1864–1909), who was one of Einstein's tutors when the latter was studying at the Institute of Technology in Zurich, Switzerland, presented a new geometry in which the coordinate of time did not have special status but was to be treated like the other, spatial, coordinates. To formulate the rules of this new geometry, Minkowski developed what is called tensorial calculus, which extended Newton's system of derivatives to more complex relations. Thus the special theory of relativity became established from both the physical and the basic mathematical aspects.

An interesting question is why Einstein received all the credit for the special theory of relativity when, as noted above, its essence, including the formula $E = mc^2$, had been published by Poincaré before Einstein. If we ignore the conspiracy speculations that are raised from time to time, the answer has two parts. The first difference between Einstein's theory and Poincaré's is a conceptual one. Poincaré developed mathematics and did not notice, or at least did not declare and emphasize, that mathematics presents physics with new principles. Einstein focused on the physical principles from which the new mechanics are derived, of course with the help of mathematics. Therefore, to attribute the new physics to Einstein is completely justified. (Einstein worked on his theory while in the Bern patent office, before he learned of all of Poincaré's findings.) The second part of the answer is that Poincaré's papers are densely written and hard to read, whereas Einstein immediately concentrated on the essence and the new, and he presented his theory in an almost intuitive way. For example, Poincaré gave the mass/energy ratio as $m = \frac{E}{c^2}$, an expression that is harder to grasp than the more familiar version. Simplicity has a clear advantage.

The special theory of relativity unified Newton's mechanics with the electromagnetic mechanics of Maxwell; in other words, it gave a joint mathematical system to both physical phenomena. The mathematical analysis of speeds significantly below the speed of light in effect coincides with Newton's theory. The effects related to the theory of relativity emerge only at speeds close to the speed of light. In classic engineering situations, the use of Newton's formulae is accurate enough, and for many years the part of the theory that related to relativity was in the domain only of scientists. In the current era, when for example communication at the speed of light is relevant to everyone, the equations that describe relativity are in widespread engineering use.

One Newtonian law of nature, gravity, remains outside the mathematical framework of special relativity. Moreover, Newton's second law and the law of gravity both relate to mass, the same mass. If these are different laws there is no reason for the same physical quantity to serve both. Einstein suggested that the two laws were two aspects of the same effect. The example he gave to make the point was a free fall in an elevator, that is, an elevator whose cable has broken and is falling freely, inside which, although a passenger is accelerating due to the force of gravity, he is not aware of any force being exerted. That is, the exertion of the force is also relative. Unlike the case of special relativity, where the mathematics had already been developed and Einstein's main contribution was to present the physical interpretation, in the analysis of gravity Einstein first posed the physical hypothesis. Without mathematics, however, the hypothesis had no scientific value. Einstein devoted several years of research to the attempt to find a mathematical theory that would unify gravity and the other forces, published a number of articles with partial results, and finally, in 1916, published the definitive paper presenting the general theory of relativity. Once again the solution required a new presentation of the geometry of the world. The mathematical framework Einstein used was that proposed by Riemann some sixty years earlier (see the previous section), and the mathematical tool that describes the mechanics in this geometric world was the tensor calculus that Minkowski developed to show the geometry of the world of special relativity.

Einstein adopted Galileo and Newton's idea of inertia, but he adopted the version that stated that a body that has no force exerted on it will continue to move along a geodesic, that is, the shortest line between two points in a space. Then Einstein claimed that in the geometry of the physical space, the shortest line between two points in this space is not a Newtonian straight line but a line that is seen in Newtonian space as a curved line. Moreover, the factor causing the curvature of the line is the existence of mass. According to this description, the gravitational attraction is only the result of a geometrical feature. For example, we think that the Sun attracts the Earth, and that is why the Earth does not continue in a straight line but revolves around the Sun; in effect what happens is that the Sun distorts the space in such a way that the elliptical orbit of the Earth is actually a "straight line" in the sense that it is the shortest route in the geometry of the space. Is this merely playing with words, or are we dealing with a physical feature that our senses did not perceive previously? The test will be whether the theory explains things that cannot be explained any other way, and it will be even more convincing if it predicts new effects.

One aspect that Einstein explained using the new geometry was the very small changes discovered in the path of the planet Mercury. Astronomers had proposed other explanations for those variations, such as the effect of another as yet undiscovered planet. Einstein also proposed a new prediction. If it is geometry that is the determining factor, then a physical body on which ordinary gravity ought not to act would follow the curved line that Einstein forecasts and would not follow a Newtonian straight line.

Light itself is such a physical object. If physical space curves around the Sun, then the light of a star that reaches us and that passes around the Sun will reach us on a curved route and will seem to us to be in a different location. We do not generally see the light of stars coming from the direction of the Sun. The best time to identify and measure the direction from which the light is arriving is during a full eclipse of the Sun. A number of scientific expeditions tried to check Einstein's prediction over several years but failed either because of bad atmospheric conditions when the sky was hidden by clouds precisely at the time of the eclipse or because of political events, such as when an eclipse of the Sun occurred in the middle of a war between Germany and Russia and the expedition's astronomical equipment was confiscated by the Russian authorities who were concerned about espionage.

The confirmation came on May 29, 1919, when there was one of the longest total eclipses of the Sun, lasting almost seven minutes. The eclipse started in Brazil and moved to South Africa. Two expeditions were organized by the British Royal Society; one went to Brazil, and the other to a small island off the coast of South Africa. Both of them managed to measure and corroborate Einstein's general theory of relativity. Later, doubts were expressed whether the equipment they had used was accurate enough to enable the conclusions they had reached to be drawn. In any event, since then the general theory of relativity has been confirmed many times. Einstein's equations correctly describe the geometry of the world. As was the case with the special theory of relativity, the general theory was also for many years relevant only to scientists. Today, with the wide use of outer space, for instance for GPSs (global positioning systems), effects that are part of general relativity are also relevant to engineering.

But even Einstein, the master of intuitive interpretations of mathematics, was not immune from the trickery of intuition. The common perception was that the gravitational force will, eventually, cause the universe the collapse. Einstein's intuition told him that the universe is static, and he corrected the equation by adding a constant, the cosmological constant, that stabilizes the equations. Later, Edwin Hubble discovered that the universe is indeed expanding, and with a constant rate. Einstein removed the cosmological constant from the equations, referring to the addition of the constant as the biggest mistake in his scientific life. This can be interpreted as saying that, had he believed in the original equations, he could have predicted the expansion of the universe before it was discovered by analysis of experimental data. Recently it was discovered that the expansion itself is accelerated. This can be explained by putting back the cosmological constant into the equations, this time to account for the acceleration of the expansion. Thus, it could turn out that the removal of the constant from the equation was another mistake by Einstein.

Albert Einstein was by all accounts the most famous scientist of the modern era and, hence, the most written about. Here we will present just a few of the central facts about his life and work relevant to our account. Einstein was born in 1879 in the town of Ulm, then in the Kingdom of Württemberg, part of the German Empire, and he moved with his parents to Munich when he was one year old. As a child and youth, he did not stand out as a student, but nor did he lag behind (as rumored later on). When he was fifteen years old, his family moved to Italy for economic reasons. Albert joined them but did not acclimatize well to the new environment, and he was sent to complete his secondary schooling in Aarau, in northern Switzerland. In 1896 he was accepted into the Swiss Federal Polytechnic (known today as the Swiss Federal Institute of Technology) in Zurich, and he graduated in 1900. In his university studies he did not shine either, mainly because he concentrated on subjects that interested him, physics, mathematics, and philosophy, and in these too he did not persevere with the studies themselves but invested his time and energy in independent reading. On completing his studies he tried for some years to obtain a teaching post, unsuccessfully, and eventually, in 1903, was given the position of examiner in the Swiss patent office in Bern. There he had to evaluate many patent applications for electromagnetic devices, whose uses in engineering were constantly increasing.

Einstein had come across Maxwell's theory in his studies, and he continued to be interested in it and to involve himself in the scientific side of the theory, as he did in other scientific and philosophical subjects, but not in any formal academic framework. At the same time, he studied and carried out research at the University of Zurich, being awarded his doctorate in 1905. In that same year, while still working as a patents examiner, he published four groundbreaking papers that have left a deep imprint on science. The first paper gave an explanation of the photoelectric effect, and we will return to this in the next section. Two of the papers dealt with what is now referred to as the special theory of relativity: the first dealt with the laws of mechanics that were derived from the new geometry that we described above, and the second with the equivalence of energy and matter. The fourth paper in the series, which resulted in the year 1905 becoming known as Einstein's annus mirabilis (miracle year), described the mathematical basis of the motion of particles similar to Brownian motion. This is the random motion of microscopic particles reported in various situation, named after the eighteenth-century Scottish botanist Robert Brown. This paper of Einstein's served as the springboard for the mathematical subject called random motion, a ramified area of mathematics still of active interest today.

These outstanding contributions brought Einstein broad academic acclaim, which led to his being offered an associate professorship at the University of Zurich. His reputation in the academic world did not filter through to the general public fast enough. When Einstein resigned from the patent office in 1909, four years after his annus mirabilis, explaining that it was because of the offer of a teaching post in the University of Zurich, his superior in the office reacted by saying, "Einstein, stop fooling around. Tell me the real reason for your resignation." At that time he had already started to work on the theory of gravity, to which he devoted about ten years that ended with the publication in 1916 of his paper presenting the general theory of relativity. Meanwhile he served for short periods as a professor at the University of Prague and at the Swiss Polytechnic in Zurich.

In 1913 he received a personal invitation from two of the best-known scientists in the world at that time, the physicist Max Planck and the chemist Walther Nernst, who came to Zurich to persuade Einstein to accept the position of head of the Kaiser Wilhelm Institute of Physics in Berlin, and Einstein moved there in 1914. As stated in the previous section, in 1919 the general theory of relativity was confirmed, an event that spread Einstein's fame worldwide.

He was awarded the 1921 Nobel Prize in Physics for his contribution to the understanding of the photoelectric effect, not for the theory of relativity. The Royal Swedish Academy of Sciences of course does not publish reasons why it does *not* award the prize for a particular achievement, but unofficially it was explained that according to the will of Alfred Nobel, the prize is supposed to be awarded for achievements of practical value to the welfare of humanity, and the theory of relativity was not considered to have such value. Obviously even such a narrow guideline was incorrect with regard to the theory of relativity. According to other rumors, some of the members of the prize committee of the Royal Swedish Academy of Sciences, although recognizing the greatness of the achievement, were still not convinced that the theory of relativity was correct.

Einstein spent his time in Berlin mainly in research in attempts to find a single theory that would explain both the mechanics of gravity and the quantum theory that had been developed in the meantime. He continued with these attempts, without much success, up to his final years in the United States, where he had moved due to the Nazis' rise to power in 1933. Luckily he was not in Germany (he was on a visit to the United States) at the time of the regime change; under the Nazi government his property was confiscated, he lost his German citizenship, and his theory was declared an incorrect Jewish theory. He spent some time at the California Institute of Technology (also known as Caltech) in Pasadena, not far from Los Angeles, and later joined the Institute for Advanced Study in Princeton, New Jersey. He became an American citizen in 1940.

Experimental confirmation of mass-energy equivalence came only in the 1930s. The outbreak of World War II led to the accelerated development of the technique of converting mass to energy, a development whose peak was the manufacture of the atom bomb. Dropping the atom bombs on Hiroshima and Nagasaki led to the end of the war.

In general Einstein distanced himself from politics, but he did not hesitate to express his liberal pacifist views. Nevertheless, in the Second World War he signed a letter in favor of the development of a nuclear bomb in order to achieve nuclear power before the Germans. Einstein, who was a secular Jew all his life, identified with his Jewishness and the Jewish people, supported the creation of the State of Israel, and was even invited to become its president after the death of its first president, Chaim Weizmann. He turned the offer down politely, as he did not consider himself suitable to the position, stating that he lacked "both the natural aptitude and the experience to deal properly with people and to exercise official functions." He died in Princeton in 1955.

29. THE DISCOVERY OF THE QUANTUM STATE OF NATURE

Aristotle believed that matter was continuous. Other Greek scientists, with Leucippus and Democritus at their head, claimed that matter consists of atoms that cannot be split. The approach of the Greeks who supported the atomic structure was based on philosophical considerations with no experimental substantiation. The approach of the opponents of the atomic theory was consistent with what our senses teach us, and therefore their view held sway until the sixteenth century. At the end of that century and at the beginning of the seventeenth, experimental results led to the recognition that matter was not continuous. The best-known contributors to this revelation were the British chemist and philosopher Robert Boyle (1627–1691), who identified the atoms and introduced the concept of the molecule that consists of atoms, and the chemist and physicist John Dalton (1766–1844), who also was British and who developed the theory that all matter is composed of atoms, and the type of atom determines its properties. Dalton also introduced the concept of molecular weight based on the relative weight of the atoms that make up the molecules, which enables us to identify, and sometimes refine, different materials.

Another significant breakthrough was made by the Russian chemist Dmitri Mendeleev (1834–1907), who constructed the first version of the periodic table of the elements. In Mendeleev's time, sixty elements were known, and based on their properties he created a partial table and managed to predict the existence of other chemical elements, which were discovered soon after. Mendeleev's story is indeed wonderful, but from our point of view it is important to state that his discovery was based on the assumption of aestheticism, symmetry, and simplicity. He did not suggest a physical explanation for this periodicity. The electrons, whose paths currently explain the periodic table, were as yet unknown, and it was believed that atoms could not be divided.

The picture changed with the discovery of particles with negative electric charge, that is, electrons. It was then understood that an electric current consists of the movement of electrons whose source is in the atoms, so the atom has different parts. Moreover, different atoms have different numbers of electrons, but in general their charge is balanced by an equal number of particles with positive electric charge, called protons. The number of protons did not explain the ratios of the molecular weights of different atoms. The British physicist Ernest Rutherford (1871–1927), who was awarded the Nobel Prize in Chemistry in 1908, proposed in 1910 both the existence of particles without an electric charge (neutrons) and a model of the atom, a model still used today: a nucleus with protons and neutrons in it and electrons moving around it (the existence of neutrons was not confirmed by experiments until 1930).

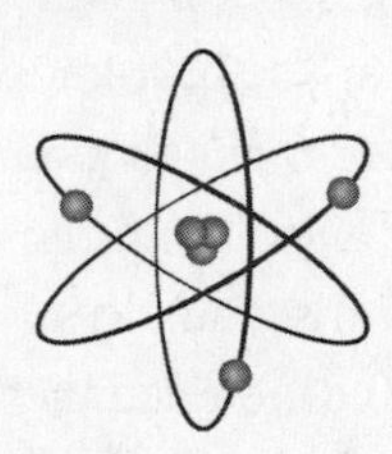

The number of protons and electrons determines the electrical properties of the atoms, while the number of neutrons explains the difference between the atomic weight and the number of protons. Rutherford presented the appropriate mathematical calculations together with his model, but it was not a mathematical model that could serve to explain the situation, but rather a metaphorical model based on intuition derived from the solar system. (It is interesting to contemplate what Rutherford would have suggested if Ptolemy's model had still prevailed.) The need for such a model is clear. The human brain needs to arrange relevant information in patterns with rules, and the framework for those patterns is generally taken from patterns known previously.

Rutherford's model, however, suffered from important shortcomings. The main one was that if the electron is a particle with a normal electric charge, the act of its constant revolving around the nucleus would create radiation and loss of energy until eventually it would collapse into the nucleus, something that was not seen to occur in reality. A far-reaching hypothesis was proposed by the German physicist Max Planck (1858–1947), who in his research into electromagnetic radiation came across a surprising fact. He found that the energy is transmitted only in quantities that are multiples of one basic quantity, still today called the Planck constant. His discovery of energy quanta earned him the Nobel Prize in 1918.

The idea was so innovative that it was hard to adopt it, until Einstein used that hypothesis to explain the photoelectric effect. The effect was that when a beam of light illuminates a metal board, electrons are ejected from the board not continuously but according to jumps in the energy of the light. Einstein's explanation, for which he was awarded the 1921 Nobel Prize as mentioned previously, was that the electrons around the nucleus of the atom can exist only at pre-given levels of energy, which are multiples of the Planck constant, and the light itself consists of discrete photons with the same energy level.

Einstein's explanation, together with the results of other experiments, led the Danish physicist Niels Bohr (1885–1962) to put forward an improved model of the atom, for which he was awarded the Nobel Prize in 1922. In Bohr's model the electrons can be found around the nucleus of the atom only at certain energy levels or on certain paths, and on those paths they do not lose energy. The transition from one level to another depends on the loss or gain of energy from outside at a quantity that is a multiple of Planck's constant. Bohr went on to calculate the number of electrons at every level and the different levels themselves. The calculations matched the data obtained until then and were also used to predict the results of experiments, which naturally increased faith in the model. Bohr's model related to photons and electrons as particles and ignored the wave motion of light. The French scientist Louis-Victor de Broglie (1892–1987) tried to bridge this gap, claiming that all types of matter have properties both of waves and of particles, and the properties of matter as particles predominate the greater the size of the matter. He even gave a formula for the amplitude of the wave as dependent on the size of the element and showed that indeed for large bodies the amplitude of the wave is so small that it is impossible to perceive it. This was an important finding that explains the fact that although we are all to some extent waves, we do not feel that we are. De Broglie was awarded the Nobel Prize in Physics in 1929.

All these findings and insights gave a detailed description of the known facts about the structure of the atom and the particles, including numerical calculations that matched the observations. Yet the model did not provide a mathematical explanation, and without mathematics, as we keep stressing, there is no understanding.

30. THE WONDER EQUATION

It was Erwin Schrödinger (1887–1961), an Austrian physicist and mathematician who studied and lived in Germany, who offered a mathematical explanation of the quantum effect. He presented what became known as Schrödinger's equation in a paper he wrote in 1926, when he was a professor at the University of Zurich, an equation that earned him the Nobel Prize in 1933 jointly with the British physicist Paul Dirac (1902–1984). Later he moved to Berlin but left Germany in 1933 as an anti-Nazi gesture and moved to England. From there he went on to the United States, but after a short stay in Princeton he moved to England, then to Scotland and finally, in 1936, decided to return to Austria, to the University of Graz. There he had to issue a retraction of the censure of the Nazis he had published three years earlier. Yet, his position in Graz was canceled and he moved in 1940 to the Republic of Ireland, which was neutral during World War II, where he founded the Institute of Advance Studies in Dublin. He later apologized to Einstein for his apparent support of Nazism, for Einstein was and remained his good friend.

The explanation, if it can be called that, suggested by Schrödinger for the various effects was a differential equation of the wave, or string-type, equation (see section 22) and as such was similar to Maxwell's equation. Schrödinger's equation was more complex and contained new mathematical elements in its description of nature (the details of which are not relevant here). Just as the wave equation has solutions with "clean," natural, characteristic frequencies, Schrödinger's equation also has such solutions. The characteristic solutions of the original string equation relate to effects that we feel, for example the "clean" vibrations of the strings of a musical instrument. Erwin Schrödinger proposed that the characteristic solutions to his equation describe the electrons of the atom, with the characteristic frequency describing the energy, while the length of the wave is related to the momentum of the electron, that is, the multiple of the speed and the mass. Just as the different elements in the string equation apply to different strings, here too different elements in Schrödinger's equation apply to different particles. A mathematical analysis of the equation that Schrödinger proposed for the known atoms showed a perfect match with all the results and experimental measurements obtained until then.

Again the information is in the form, the foundation of which was set by Maxwell: all the available information is explained using an equation whose relevance is not clear, but whose solutions enable new effects to be predicted. If the electron is the solution to Schrödinger's equation, then it is a wave. Yet no one had seen the wave of an electron. What does "the electron is a wave" mean? What is the medium in which the wave propagates? The answer is that the electron is a wave only insofar as it solves the Schrödinger equation, and even that solution we cannot perceive with our senses but can only measure certain quantities related to the frequency of the wave and its momentum or to the fact that a wave undulates. However, even in Schrödinger's days, science had already become accustomed to the fact that physics was mathematics that describes phenomena that cannot be perceived directly by the senses, and it can be understood only via its indirect effects, and therefore the equation was accepted as an equation that describes physics, meaning, it was accepted as physics.

The human brain, however, cannot absorb and deal with abstract quantities without an intuitive picture, and such intuition can be founded only upon known concepts. This limitation brought about attempts to clarify the nature of the solutions to Schrödinger's equation. Schrödinger himself interpreted the fact that the electron is described by a wavelike function to mean that the charge of the electron is in effect spread around the nucleus, and the wave describes how the charge is spread. Another interpretation, innovative and original, was given by the German scientist Max Born (1882–1970), earning him the Nobel Prize in Physics in 1954. He suggested that the wave describes the probability of finding the electron in a particular spot. The probability itself is given by the square of the height of the wave at every location. Thus Born introduced a completely new element into the description of nature: randomness, lack of determinism. It is not lack of knowledge that brings about the effect that appears random, nor a statistical approximation, such as mechanical statistics uses statistics because of the inability to analyze every single particle, but randomness that is inherent in nature itself. Moreover, according to Born, when a force is exerted on an electron, such as when measuring its location, the randomness disappears. The electron behaves like a particle, and the wave focuses, or in the language of physicists, collapses, in the exact location of that particle.

One of the conclusions drawn was that an electron can pass simultaneously through two holes, but when its location is measured it loses that property and will "decide" through which hole it has passed. It is difficult to absorb these properties because they are formulated in terms of day-to-day objects, and in our daily lives we do not encounter such properties. The reason is that electrons are not objects about which evolution has taught us how to develop intuition.

Born's interpretation was not accepted unquestioningly. Einstein, for instance, opposed it vigorously, stating that "God does not play with dice." After a while the idea spread in popular literature that Einstein was opposed to the whole quantum theory. That is incorrect. On the contrary, as we saw in the previous section, he was one of the founding fathers of the theory and certainly fully accepted Schrödinger's equation as describing physics as it is. Einstein was just opposed to the interpretation proposed by Born, which assumed a law of nature that permitted randomness. Nonetheless, Born's interpretation proved to be a reliable instrument for analyzing physics and as a source of proven hypotheses. Today Born's interpretation is accepted as a correct description of nature, of course at the subatomic level.

A mathematical analysis carried out by a student of Bohr and Born, the German scientist Werner Heisenberg (1901–1976), which earned him the Nobel Prize in Physics in 1932, yielded the *uncertainty principle*. This principle stated that it was impossible to determine accurately either the location or the momentum of an electron. Mountains of interpretations and implications, even including analysis of our daily lives, have been derived from this principle, most of them with no logical basis. The principle is a mathematical one. To formulate it mathematically requires some concepts to be developed that are beyond the scope of this book (for the mathematicians among the readers, the principle drives from the fact that the location operator and the momentum operator do not commute), but its implications for physics can be described.

The electron has both wave and particle properties. As a particle it is located at a defined spot, but to calculate its momentum we need to consider the wave as a whole. As soon as the location of the electron is measured, it stops being a wave and its momentum cannot be determined exactly. When calculating the momentum, the wave aspect of the electron is used, and its exact location is not known. This applies to electrons and to other particles. It is not relevant to situations and objects that are not simultaneously waves and particles. The principle is relevant to other situations in which the effect has two complementary aspects, such as signal processing. The signal can be described both by its frequency and by its progress in time. Both are parallel to the location and the momentum of the particle, and the uncertainty principle states that it is impossible to simultaneously describe both exactly (for the mathematicians, the product of the second moment of the function and its Fourier transform is bounded from below). These are mathematical principles that can be interpreted in mathematical uses, but caution must be exercised in trying to apply them in other situations.

It is worth repeating and emphasizing the new system of relations that appeared and developed at that time between nature, mathematics, and the description of nature. The basis consists of nature itself. We describe it with the help of mathematical equations. The equations can describe quantities whose properties are not directly accessible to us and of which we have no perception or feeling. The justification for the correctness of the equations is purely and simply that there is a match between the effects they predict and experimental results, and they constitute a source for additional discoveries. In order to analyze the behavior of the solutions to the equations, our brains need an interpretation in terms that they can imagine. The interpretation is "correct" only to the extent that with its help we can perform an analysis of the solutions to the equations. We have no alternative but must learn about nature in terms of the interpretation that we have developed. We must always bear in mind, however, that it is merely an interpretation of the mathematics that describes nature; nature itself can produce surprises.

31. GROUPS OF PARTICLES

At the beginning of the 1930s the subatomic situation was relatively simple: the atom was made up of a nucleus, inside which were neutrons and protons, and around which revolved things that were waves and also particles, that is, electrons. Light particles were also known, that is, photons. But it soon became apparent that subatomic reality was far more complex. First, following precise experiments that analyzed the frequency of the radiation it was found that there was not just one type of electron. In effect there are two types. The difference between the two types of radiation that they created was explained by the suggestion that the electron, while revolving around the nucleus, also revolves on its own axis, and the direction of this movement, to the left or to the right, is what gives the different types of radiation. The physicists called this turning on its axis "spin." Again, there is no guarantee that the electron particle, which is also a wave, does actually spin on its own axis. This property, however, would provide a good explanation for the difference between the two types of electrons.

Then the positron was discovered, a particle similar to an electron but with a positive charge. The positron was discovered as a result of a mathematical analysis of Dirac's equation, which is a version of Schrödinger's equation adapted for the electron, yet with an additional solution with a positive charge. This solution indicated the possibility that an "antimatter" particle existed. If this particle meets the "matter" particle, they will both disappear and become energy. Within a short time it was indeed found that the mathematical solution, somewhat exotic we might add, was realized by a real particle in nature, the positron. Within a few years other "matter" (and "antimatter") particles were discovered, which were then given the collective name of elementary particles. Initially these particles were studied by examining cosmic radiation and the results of the cosmic particles striking the earthly atoms. Cosmic radiation has great energy, but a large part of it is absorbed in the atmosphere. Research into the elementary particles was therefore then carried out by raising photographic plates to a great height and documenting the impact between the cosmic radiation and the earthly particles. Later, other means were developed, such as bubble chambers and later still particle accelerators, that recorded the collision between the accelerated particles and other particles and the changes that the collisions caused, including the creation of new particles. These were characterized by their energy, the frequency of their wave, their mass, and their spin, the level of which increased and now not only reflected direction but was also given values of a half, a third, and so on. The episode of creating a picture of the subatomic world is fascinating but falls outside the scope of this book. Suffice it to say here that a list was formed of the elementary particles, but to understand the order underlying it required mathematics. Schrödinger's equations, other equations that developed from them, Born's interpretation, and what was derived from them were sufficient to describe properties of the particles but did not explain their allocation according to the various properties. For that, a mathematical element that had not been used before in this area was incorporated, namely, groups.

The classification of the various particles by their properties led to their being arranged in tables according to their type. A particular type of particle is called a hadron. Two scientists, Murray Gell-Mann of Caltech in California and the Israeli Yuval Ne'eman (1925–2006), who was then (in

1961) at Imperial College, London, and later at Tel Aviv University, both noticed in that year, independently, that hadrons can be placed in several tables according to the characteristics of their spins so that each table consisted of exactly eight particles. Moreover, the relation between the particles in every table matched what the mathematicians had for a long time called the SU(3) group.

It is not necessary to get involved deeply in group theory to understand its role in describing the physical situation. A group is a collection of mathematical elements and the relations between them, for example, revolving the plane through either 90, 180, 270, or 360 degrees, with the last of those alternatives bringing us back to the starting position. This is a group whose elements are the turns, and the action between every two elements is the turn obtained after two consecutive revolutions. That is to say the relationships are: a turn of 180 degrees followed by a turn of 270 degrees is equal to a turn of 90 degrees, and so on. This is a group with four elements. To describe such turns, there is clearly no need to call them a group and to use sophisticated mathematics. What is special about mathematical terminology is that it enables us to describe more-complex systems. For instance, turning a die through 90 degrees in one direction around one of the three axes will define a more complex group. Mathematics studies even more complex groups and groups in which the relations between the elements are more complicated.

In the nineteenth century a Norwegian mathematician, Sophus Lie (1842–1899), found groups that described symmetries of differential equations. These are today referred to as Lie groups, and one of them is the SU(3) group, which Gell-Mann and Ne'eman suggested constituted the basis of the arrangements of the hadrons.

Yet not all the elements in the group were represented by particles that were known when Gell-Mann and Ne'eman put forward their hypothesis. In particular, in a lecture delivered by Gell-Mann at the European Organization for Nuclear Research (CERN) in Geneva, which currently has the largest particle accelerator, he presented the SU(3) group and its properties and noted that a particle was missing; he called it omega minus. That particle would have completed the group model. Moreover, its existence would have absolutely contradicted another model for classifying hadrons proposed by Japanese researchers. Gell-Mann, however, was apparently not familiar with the literature on experiments then being performed. One of the people who attended Gell-Mann's talk was Luis Alvarez, of the

University of California, Berkeley. He was a famous scientist who later received the Nobel Prize in Physics in 1968, and among his many scientific achievements he put forward the hypothesis that dinosaurs became extinct due to a meteorite impacting on the Earth. Alvarez stated that the missing particle, omega minus, had been identified seven years previously by Yehuda Eisenberg, a physicist at the Weizmann Institute of Science in Rehovot, Israel. Eisenberg had made his discovery using the technique of placing photographic plates at the height of the atmosphere, but that was an isolated occurrence not backed by a mathematical explanation, so the experimental discovery of the particle entered the catalogue of particles without attracting any great attention. Following Gell-Mann's lecture, several groups of experimenters started again to look for that same omega minus using the bubble-cell technique until in 1964 a team of physicists led by Nicholas Samios at the Brookhaven Laboratory in New York eventually again succeeded in identifying the missing particle several times and thus corroborated Eisenberg's finding.

Therefore, it can be claimed that in a certain sense, ex post facto, the existence of the particle was predicted by group theory. Enlisting group theory to characterize and classify elementary particles scored its first success. The order in which the elementary particles are recorded moved up from the level of simply a table to a mathematical theory, which enables sophisticated predictions to be suggested and examined.

It should also be noted, however, that there is no fundamental or logical explanation for the fact that group theory and the structure of the elementary particles match each other. From a conceptual viewpoint, that agreement might remind us of the match that Plato found between the world and the four elements of nature that it comprises, and the five perfect solids, or the matching, with the detailed calculations, that Kepler found between the paths of the six celestial bodies and the perfect solids (see sections 10 and 17). Will we in the future assess the role of group theory in describing the elementary particles in the same way as we assess Kepler's model of the perfect solids?

The combination of the existing mathematics, physical principles, and technology of experiments in physics resulted in the discovery, mapping, and understanding of the structure of a larger number of elementary particles and the interaction between them. In addition to the two forces already known, that is, gravity and electromagnetism, two new forces were discovered: a strong nuclear force and a weak nuclear force. Gell-Mann used mathematical principles and put forward the hypothesis that particles exist that are parts of protons and that have a fractional charge. These are the quarks, combinations of which make up the various protons. Although quarks cannot be isolated, their existence was proven beyond all doubt, and this earned Gell-Mann the Nobel Prize in Physics in 1969.

Since then the picture has broadened, and other experimental findings have been added, as well as many mathematical items. Yet the picture is still not complete or final. Right now an experiment is underway, at enormous expense, at the Geneva particle accelerator, intended to reveal the particle known as the Higgs particle (or Higgs boson), named after the British physicist Peter Higgs, who predicted the existence of the particle already in 1964. Initial reports indicate that a new particle has been found in the range of mass and energy in which the Higgs particle is predicted to be. Higgs and his colleague François Englert, who independently made the same prediction, won the Nobel Prize in Physics for 2013. If the initial findings are confirmed, this will corroborate the model known as the standard model. If it transpires that the particle discovered is not the Higgs particle, the physicists will have to rethink the picture of the subatomic world and perhaps will have to adopt a new mathematics to do so.

32. THE STRINGS RETURN

Physicists who deal with elementary particles are engaged in completing the picture of the subatomic world of these particles, but the model on which they are working is not consistent with the order that prevails between the collection of these particles and the theory of gravity. Moreover, to describe different particles, different versions of Schrödinger's equation must be used. In light of past successes in finding mathematical models that brought together different theories, physicists today feel obliged to find one theory, one equation, that will enable them to explain the whole of the subatomic world. The attempt to incorporate all subatomic phenomena within one equation is the development of a mathematical system known as string theory.

From the aspect of the relation between mathematics and nature, string theory presents another stage. Maxwell's revolution presented mathematics that described physics whose components could not be perceived directly but whose effect on other physical quantities could be measured and whose predictions, such as the existence of electromagnetic waves, could be confirmed or disproved. String theory is mathematics that describes physics whose components cannot be perceived and whose effect on other physical elements also cannot be measured at this stage. Moreover, at this time, the theory does not provide predictions that can be confirmed or denied, and it does not seem that it will be able to provide such predictions in the foreseeable future. Is this the picture of the world? Is this physics?

Some of my physicist colleagues deny that this is physics and state that those involved in the theory are just mathematicians. Others are prepared to include that community under the umbrella of physics (also because their work is theoretical and does not compete for expensive research resources). There are also physicists who believe that despite the fact that currently it is difficult to even imagine such a situation, the day may come when the way will be found to examine string theory with experimental methods and to obtain benefit from it.

So what is string theory? The mathematical system of string theory is essentially similar to the systems that define the world of elementary particles, combined with geometric elements. The solutions to these equations are the basic elements that the theory presents. Since the brain cannot analyze mathematics without recognizable metaphors, string theory is described and examined via interpretations of those solutions. We will also relate only to the interpretation of the theory.

The strings are firstly particles of miniature size. They are hundreds of thousands times smaller than the elementary particles (this explains why they cannot be perceived, as the means for perceiving the tiniest particles are based on the elementary particles). These strings are wave solutions, but unlike the electron, for example, which is described as a point particle that revolves around the nucleus of the atom, a string is described as a body that has a length, all of which vibrates and moves like a wave. Hence its name "string." Once a particle that has a length is permitted, the question immediately arises as to if its ends are joined, like a ring, or are they attached to a plane, or are they free? It turns out that all of these are solutions to the equation, solutions that give strings of various types. The different strings create structures from which, so it is hoped, the subatomic structure can be derived. However, it transpires that for those structures to exist, the physical space must have certain surprising properties. For example, the space must have more than the four dimensions in which we can perceive, that is, three spatial dimensions and time. This means actual physical directions, but the distance along each direction is too small for us to perceive or measure in any way. The number of additional dimensions depends on the specific theory. The latest versions refer to ten or eleven dimensions. Moreover, those equations that describe the strings also have solutions of another type, sorts of membranes that are likely to be huge. Do these solutions have a physical interpretation or perhaps even implementation? If so, those membranes could contain worlds in addition to our own, worlds that we cannot perceive or communicate with, although the distance between us and them might be the smallest. Furthermore, collisions between those worlds are possible and could perhaps cause huge explosions, like the big bang that led to the transformation of energy into mass and, according to the accepted theory, created the world we know today.

To the reader whose reaction at this stage is "I don't understand," I would say you are not alone. The writer of these lines does not understand much more, if to understand means to be able to translate the metaphors into mathematical language with real implications. This "understanding" means attempting to build a bridge between intuition about the world around us, intuition based on evolutionary development over millions of years and therefore limited by our senses, and the mathematical product that comes to describe situations that are so alien to what our senses teach us. Will this mathematics rise to the challenge of describing a real world? Time, apparently a very long time, will tell.

33. ANOTHER LOOK AT PLATONISM

We return now to the discussion of the connection between mathematics and nature and first remind the reader of the main difference between Plato's and Aristotle's (and their successors') approaches to the essence of mathematics as a description of nature. Plato claimed that mathematics has an independent existence in the world of ideas in which mathematical truth is absolute. Humans can reveal this truth via logic. The starting point of research that will reveal the truth is axioms, which must be derived from nature, but it must be borne in mind that nature itself only mimics the ideal mathematical truth. Aristotle, however, claimed that mathematics itself has no independent significance or even existence. If axioms are found that reflect the truth of nature, the conclusions that can be drawn by means of mathematical formalism will apply to the description of nature. Specifically, the closer the axioms are to the truth of nature, the better the mathematical conclusions to describe nature will be. Plato and Aristotle agreed that there would be differences between mathematical conclusions and actual observations of nature. For Plato these differences were disturbances. For Aristotle the differences derived from inaccurate axioms, or in more modern terms, the inaccurate construction of the mathematical model. Neither of them discusses the essence of the difference between mathematics and nature.

The Aristotelian approach to the applications of mathematics can be summarized by the statement that *mathematics is a very good approximation of nature*. Furthermore, when a mathematical model does not describe nature as it really is, the model must be corrected. As stated, this is Aristotle's approach according to which mathematics does not have an independent existence. To reach the correct description of nature, we start with an approximate model and correct it and make it consistent with reality by comparing the results derived from the model with empirical data.

New research into mathematics and its applications suggest another way of looking at the relation between mathematics and nature, one that is closer to Platonism and may be called Applied Platonism. This approach can be summarized by the statement that *nature is a very good approxima-*

tion of mathematics. Moreover, in some cases Platonic mathematics in the world of ideas inherently contains contradictions to basic laws of nature. Nevertheless, nature tries to copy it and has no choice but to approximate mathematics. The following is an example of this.

We have referred to the least action principle as the purpose underlying motion in nature. In the same way the minimization-of-energy principle also serves a purpose. Objects in nature strive to reach a situation of minimum energy, at least a local minimum; in other words, they will stay in the local minimum state unless they are subject to an external force. John Ball of the University of Oxford, England, and Richard James of the University of Minnesota in the United States examined the structure of an elastic object under stress. Their approach was a mathematical one. They wrote the expression for the energy of a body under stress and looked for the structure that would minimize the energy. They succeeded in solving the mathematical problem, and the result was that the mathematical solution is not applicable in nature. Mathematics required that the molecules in the elastic body simultaneously arrange themselves in two different ways. Clearly that is impossible in nature. Does it mean that the minimization-of-energy principle is not correct in this case? Laboratory experiments provided the surprising answer: The structure of the object in nature is an *approximation* of the mathematical result. The volume that the object occupies divides into microscopic parts so that in each of the tiny parts the molecules arrange themselves in one of the forms that constitutes the mathematical solution. Specifically, in each relatively large microscopic volume, both arrangements that together constitute the minimum energy appear and in the right proportions so that the average over the macroscopic surface is very close to the mathematical minimum. Nature tries to converge to the ideal solution that mathematics found, but it is not achievable.

Image courtesy of Hanuš Seiner.

This effect, that is, nature *approximating as closely as possible* to a mathematical solution that is impossible from a physical aspect, has since then been discovered in many situations and has provided a mathematical interpretation for previously observed effects. The approach also succeeded in predicting new effects that were then confirmed in the laboratory. Such a mathematical achievement is illustrated in the picture above, from the laboratory of Hanuš Seiner, of Prague. The lower part of the picture shows alternate microscopic layers of two states of metal, layers constituting a microscopic approximation to the required mathematical solution in which both layers appear simultaneously (the length of the picture represents an actual length of 2 millimeters). The upper left part of the picture shows the classic state of the metal. The possibility that there would be anastomosis between the mathematical approximation and the classic state of the metal (i.e., that they would have a common edge) is not self-evident. The transition from one state to another was predicted by mathematics, by John Ball and his colleagues, and was identified in Seiner's laboratory.

34. THE SCIENTIFIC METHOD: IS THERE AN ALTERNATIVE?

As we have seen, the scientific method that developed over thousands of years to describe the physical world is based on mathematics. This dependence on mathematics is common to all parts of the scientific world. Professor Herbert Jäckle of the Max Planck Institute in Germany, in a critique of the quality of research in a particular department of biology, wrote in 2010, "The aim is to go to the ultimate step, just as physics has done before: to describe development in mathematical terms and to model the system. Otherwise we have not understood the process and it remained a phenomenon, a miracle." That is a pictorial way, a variation of a claim by Aristo, of paraphrasing the view that without mathematics it is impossible to understand the effects seen in nature.

The question arises nevertheless: Are there other ways to describe nature, ways that will not depend solely on a mathematical model? And there is a similar question: In order to understand those effects that current mathematics does not describe well enough, or to which it is even not applicable, do existing mathematical methods have to be developed further, or must a more appropriate mathematics be found? Or, in order to understand those effects, should we perhaps adopt nonmathematical methods?

It is difficult to give definite answers to these questions. It is clear that mathematical methods have not been exhausted, and it is also clear that there is room for new mathematical methods beyond the discovery of new equations or new criteria such as the least action principle.

A novel approach being tried by various researchers is to change the equations that describe the state of nature for algorithms that bring about that state of nature. Darwin's theory of evolution is an example of an algorithmic law of nature. To predict what the state of nature will be at a particular time, it will then be necessary to perform a simulation, no doubt with the aid of computers. This approach is still in its infancy.

Another recent approach is to employ the so-called fractals. These geometrical objects have the property that a portion, even a small portion, of the body looks like a reduced-scale version of the entire body. Such bodies are called self-similar. Of course, a segment of a line is self-similar, but already in the beginning of the twentieth century it was discovered that such geometric figures may have very complicated structures, and their mathematical properties were investigated. It was Benoit Mandelbrot (1924–2010), a Polish-born who got his education in Paris and in the

United States and worked most of his scientific life for IBM, who discovered the resemblance of these figures to objects in the real world and coined the term *fractals*. Mandelbrot pointed out that, for instance, a cloud, a forest, a coastline, all have the property that a part of the object looks like the entire object. Of course, a self-similar body in nature lacks the property that any small part of it is similar to the entire body, as the mathematical analysis requires. A too-small portion of a cloud will contain only few molecules and will not resemble a cloud. Yet, the mathematical theory of fractals helps our understanding of the real-life objects. We may consider it as another example of Applied Platonism: nature is trying to mimic the fractal structure to the extent possible. Beyond a mere description of an object in nature, the approach has some concrete successes. For instance, it is difficult to even define the length of a coastline, since the self-similarity of the coastline of a rocky structure implies that the naive definition of a length would result in the coastline being infinitely long. The theory suggests an alternative measure, the fractal dimension of the line (that is too technical to review here). In the future the theory of fractals may provide more useful tools, helping to examine complicated structures in nature.

With regard to the possibility of complementing or even substituting mathematics with alternative approaches, we would observe first that the status that mathematics achieved in describing nature was not achieved easily or without competition. We could even claim that mathematics reached its current state almost as part of the evolutionary struggle. The competition in ancient times was against the various oracles, for example. Later, the competitor was astrology. The practice of using astrology displays some signs of a scientific activity. There are rules according to which predictions are reached. There are predictions even backed by claims of success in statistical tests regarding their reliability. The "missing" part is a mathematical model, a model that would explain how the planetary system affects us. Despite that lack, some people today still believe in and use astrology to explain and predict various effects, including effects in nature. As an argument supporting the validity of astrology, its proponents point out that the best scientists in the early and even later modern era believed in astrology. That is correct. But the argument ignores the fact that astrology was abandoned by science because of its lack of success.

Over the years several attempts were made to formulate other methods to complement the method of mathematical models. One example is that of Immanuel Velikovsky (1895–1979). Velikovsky was born in Russia, studied physics in the Hebrew University of Jerusalem, and studied psychiatry under the guidance of Sigmund Freud. He did not deny the scientific method but proposed adding to it; the essence of the addition was reliance on ancient literary and religious sources, including the Bible, and archeological finds. The most famous of his far-reaching hypotheses was his claim that the planet Venus was different than the other planets in the solar system as it was not created together with the rest of the system but was previously a wandering body in the galaxy that was caught by the gravity of the Sun. The biblical narrative of the "Sun standing still upon Gibeon and the Moon in the valley of Ayalon" was understood by Velikovsky as a physical occurrence that supported his claims. In addition, Velikovsky made several physical predictions, some of which were substantiated by measurements. For example, of all the hypotheses about the temperature on Venus, before that was measured by the spacecraft sent there, Velikovsky's was the most accurate. Nonetheless, his approach to scientific research was not accepted (to use an understatement; in fact, it was rejected vigorously by the scientific establishment). His opponents said that there was nothing wrong with using ancient sources or any other means to obtain inspiration, but those texts cannot be allowed to serve as a substantiation of a theory.

An area in which mathematical models have not yet proved themselves is that of life sciences. Much effort has been invested in attempts to apply the mathematical approach to biology and its scientific subgroups, some of which have met with much, but still only very partial, success. Experimental findings show that biology is far more complex than physics. Is this a temporary situation that will prevail until the correct mathematics will be discovered to describe biology? We should remember that until Kepler and Newton, the physics of the celestial bodies seemed like an enormous collection of findings independent of a simple mathematical framework. It may be, however, that the complexity in biology cannot be solved using the simple mathematics that we know, and new principles in biology need to be deciphered before mathematics can be successfully applied.

Mathematics also shows only partial success in the social sciences and humanities (we will present some approaches in the next chapters of this book). Here too there are competing theories, some respected, like the diagnosis of a person's mental state using psychological analysis, and some less respected, such as graphology to determine someone's character and palm reading for foretelling the future. Some of these also have some "scientific" aspects and apparent statistical support. Such support has no basis in science because until a mechanism, preferably mathematical, has been suggested as a basis for the method, it has no scientific validity. The existence of such methods, in particular the respected ones, raises the following question: Is the mathematical approach the right one for analyzing the social sciences and humanities?

CHAPTER V
THE MATHEMATICS OF RANDOMNESS

Can birds calculate probabilities? • How can you calculate the chances of winning an unfinished game? • Is it worthwhile to believe in God? • Why did the municipality of Amsterdam almost go bankrupt? • Who murdered Mrs. Simpson? • Why didn't dinosaurs develop gills against dust? • What are the chances of the Ayalon Highway being flooded? • Is there a "hot hand" in basketball?

35. EVOLUTION AND RANDOMNESS IN THE ANIMAL WORLD

The title of this chapter does not refer to the random part of the process of evolution but to the question that has been with us throughout this book, and that is, to what extent did evolution prepare us to analyze intuitively and understand situations in which randomness plays a part? The question is reasonable. Indeed, uncertainty and randomness appear frequently in nature and were part of the environment of species in the evolutionary struggle. We may assume, therefore, that the evolutionary competition gave rise to the development of intuition with respect to uncertainty.

First, however, we will examine more closely the difference between randomness and uncertainty. A situation of uncertainty is one in which we do not know what the result of an occurrence will be and we do not know the circumstances prevailing at the time of the occurrence. Randomness occurs when we do not know for sure what will occur, but we do know that what happens is controlled by a process with given probabilities. For example, when we throw a die with six faces, we do not know in advance which face will be uppermost, but we do know that each of the six faces has the same probability of being the one. The same applies to tossing a coin, with each of the sides, "heads" or "tails," having an equal chance of falling uppermost. The probabilities do not have to be equal, but the process will still be random. Thus, if four sides of a cube are blue and the other two sides are red, in a random throw of the cube the chances of a blue side appearing uppermost will be two-thirds, and of a red side, one-third. The probabilities are not the same, but the process is a random one. In contrast, in a situation of general uncertainty the result may not be determined by a probability process. For example, a committee is about to make a decision, and we have no idea of the procedure through which it reaches its decisions. That is a situation of uncertainty where, usually, we cannot assign probabilities to the outcome of the decision-making process.

One way of examining whether the relation to uncertainty in general and randomness in particular are rooted in evolution is to study reactions to such situations in the animal world. We used the same approach with the relation to arithmetic in section 2. The result with regard to randomness is very clear: animals succeed in discerning random states. Moreover, they sometimes use randomness to improve their situation. They also manage to recognize situations of uncertainty unrelated to randomness. Furthermore, they are sometimes aware of the fact that there is something they do not know, and then their behavior is similar to that of humans in similar situations. The conclusion to be drawn is that the relation to randomness and uncertainty derives from evolution. We will describe briefly a number of experiments from among the many that confirm this conclusion.

In experiments performed to show that animals react correctly to random situations, food was provided over a long period in one of two rooms, say *A* and *B*. The room was chosen randomly but with different probabilities. For example, room *A* would be chosen with a 40 percent probability, and room *B* with a 60 percent probability. A bell was sounded to indicate to the animals that food had arrived, without indicating in which room. If the animal went to the wrong room, it missed out on that meal.

Many subjects of the experiments, such as rats and pigeons, quickly discovered that the process was random and chose room *B* significantly more than room *A*.

One aspect of randomness is the adoption of strategies of random search for food, or random sampling of places to look for food. If the location of the food is random with known probabilities, a mathematical calculation can help to formulate optimal strategies of searching. Such a calculation shows that in many instances a random search is optimal. Extensive studies of food-search strategies among animals have shown that they do indeed adopt optimal random searches. That is not surprising. Even without learning and using the mathematics of optimal search, evolution nurtured those animals that adopted optimal-search policies. An animal that adopted the optimal-search strategy had an advantage in the evolutionary struggle, and natural selection gave preference to those animals that developed the ability to deal with randomness correctly.

Certain species of birds showed more sophisticated behavior related to randomness, such as, for example, the black-capped chickadee, which is the national bird of the states of Massachusetts and Maine in the United States. The chickadee, like other species, spends most of its time in thick shrubbery for safety, but it has to peck for its food in open spaces, and therefore from time to time it has to leave the shelter provided by the bushes. When searching for food it is thus exposed to the dangers of various predators, mainly larger birds of prey that cannot get into the bushes. If the chickadee leaves the bushes to search for food according to a set routine, the birds of prey would soon learn the pattern, and its chances of surviving would be low. Observations have revealed that it exits its shelter randomly, making it difficult for predators to predict when it will be exposed. The birds of prey also adopt a strategy of randomness, otherwise the chickadee would learn the hunters' patterns and would leave its shelter only when there were none around. The chickadee's strategy of leaving the bushes, that is, the average frequency, the time it spends "outside," and so on, takes into account the hunting strategy of the birds of prey. For example, if the average appearance of the bird of prey is short, the chickadee can allow itself more time away from its shelter. The ecologist Steven Lima, of Indiana State University, carried out an interesting series of experiments (the results were published in 1985). Lima confronted the black-capped chickadee with a situation in which the parameters of the predator birds' search strategy changed from time to time, for instance, the chance that it would be at a certain location. The chickadee quickly recognized the changes in the predators' hunt and adjusted its own random parameters for leaving its shelter. In other words, in the evolutionary process, not only did a species of bird develop that knows how to behave in an environment of given random parameters, but it can also identify changes in the parameters that define the randomness, and it changes its conduct accordingly. (More details on this and related research can be found in the monograph by Mangel and Clarke.)

People behave differently in situations of randomness than they do in states of uncertainty deriving from lack of clarity. For example, many people are prepared to buy lottery tickets, although the chances of winning are low. Yet they will hesitate before buying a ticket if they do not know how the winner is chosen. A study (published in 2010) on chimpanzees and bonobos (pygmy chimpanzees) by Alexandra Rosati and Brian Hare of Duke University showed that the apes exhibited patterns of behavior similar to that of humans. They could distinguish between situations of lack of knowledge due to randomness with laws of probability and uncertainty not necessarily related to randomness. Moreover, their reactions to these two situations were similar to those of humans, that is, lack of clarity led to a reaction that was more hesitant than the reaction to randomness.

A series of experiments on macaque monkeys and dolphins carried out by David Smith of the State University of New York at Buffalo and David Washborn of Georgia State University (with results published between 1995 and 2003) showed that these two species were aware of the fact that there were things they did not know. The study of the dolphins consisted of training them to react by pressing one of two pedals when they heard a note that was higher or lower in pitch than a certain note, but they also had the opportunity to press a third pedal if the height of the note was unclear to them. Not only did the dolphins press the "not sure" pedal at the right times, their behavior and body language also showed signs of hesitation and lack of decisiveness.

These and many other studies on similar issues show that evolution prepared many species of animals, and certainly humans, to understand and to react intuitively to situations in which they find themselves facing uncertainty and randomness. As we shall see in the next chapters, however, there are aspects of randomness for which they and we are less well adapted.

36. PROBABILITY AND GAMBLING IN ANCIENT TIMES

The clearest expression of randomness in human behavior over thousands of years of human history is connected to gambling and games of chance. There is a wealth of evidence about games of chance in ancient times. The small anklebones of sheep, called astragals, were used by the Egyptians and the Assyrians for games of chance. The bones were thrown randomly, and bets were made on which side of the bone would appear on top (the bone had four sides), comparable to throwing dice in our days. The bone could fall on any of the four sides, but the chances of its falling on each side were not the same and changed from one bone to another. Archaeological finds from about six thousand years ago show evidence of these games of chance in ancient civilizations, including bones that had been polished and shaped for purposes of the game. The ancient Greeks adopted the game, and the shape of the bone appears in statues of men and women playing a game of rolling the bone. The Romans also played a game with such bones, which were called tali.

The point of interest from a mathematical aspect is that the ancients knew how to draw conclusions regarding what we would today call the chances of the bone falling on each of the sides. The evidence is indirect and is derived from tables of numbers that correspond with prizes for the correct guess as to the uppermost side. These calculations were made with no awareness of the concept of chance or probability of any particular result. These concepts appeared only in the seventeenth century. We do not know how the ancients calculated the numbers. It is reasonable to assume that they were calculated by observation and using intuition acquired by watching very many instances of throws of the bone, but there is no evidence regarding an ordered method used to calculate the prizes.

Later, at the time of the Greeks and more commonly among the Romans, the six-sided dice that we know today appeared, as well as dice with other geometrical shapes, such as four-sided pyramids, each side of which was an equilateral triangle. Various materials were used for these dice, including animal bones, stones, ivory, and lead. Much effort was invested in polishing and working the dice to obtain maximum symmetry, interpreted of course as meaning that there were equal chances for a die to fall on any of its sides. Then the faces of the six-sided dice were given the numbers from one to six, and the object of the most common game, still common today, was to obtain the highest score in rolling the die. These games of chance captivated both the ordinary populace as well as the rulers in Greece and Rome. The games are mentioned in the mythology as well as in reports of rulers who were addicted and who would hire aids to carry the accessories for games of dice wherever they went and to calculate the ruler's winnings or losses. Gambling and the use of stones for throwing to create randomness were so common that Judaism found it necessary to prohibit those activities explicitly, as is stated in Deuteronomy 18:10–11: "Let no one be found among you who sacrifices his son or daughter in the fire, who practices divination or sorcery, interprets omens, engages in witchcraft, or casts lots, or who is a medium or spiritualist or who consults the dead," where "casting lots" means throwing cubes or dice and guessing the outcome. The commandment not to gamble shows that gambling was a social problem even then!

In contrast, there are instances in the Bible of the use of randomness as a positive mechanism. For example, Proverbs 18:18 says, "The lot causes contentions to cease, and parts between the mighty." This means that the casting of lots can resolve disputes between litigants. The Torah also recounts several stories in which randomness served as an instrument with which to reach fair decisions. When the Land of Israel was divided among the tribes it says (Numbers 33:54): "And you shall divide the land by lot for an inheritance among your families: and to the more you shall give the more inheritance, and to the fewer you shall give the less inheritance: every man's inheritance shall be in the place where his lot falls; according to the tribes of your fathers you shall inherit." "The lot" means casting lots, that is, the allocation of the land as inheritance between the tribes was carried out by lots. The Torah does not specify the method used to cast the lots, but the Talmud (the oral law eventually written down, the Mishna and the Gemara, interpreting and elaborating on the Bible) describes at length how the lots were cast. Note that the Mishna and the Gemara were written almost two thousand years ago. The procedure of casting the lots indicates (see, e.g., Bava Batra 122, Yerushalmi, Yoma 4.1) that two pitchers were shown to the people. In one were the names of the tribes, and in the other the names of the parcels of land. One name was taken at random from each pitcher, and that combination determined the allocation of the land. One might ask why wasn't one pitcher enough, holding the names of the tribes, and then for each parcel of land one name could have been drawn. From the mathematical aspect of the law of chance of today, there is no difference between the two methods. That was apparently understood even at that time, and the interpretations of the Talmud explain that the intention was to reinforce the fairness of the method (or in less politically correct parlance, in that way it was harder to cheat). In any event we see that the Bible and the commentaries saw drawing lots as a fair system.

This understanding also appeared in other cultures, such as in Greece. The Agora museum in Athens has an exhibit of a carved stone with a network of holes. The stone was used for selecting jurors for court cases held in the city. In the first stage, the men of the city would each insert a wood chip into a hole. Then the representative of the city, who had not been present at the first stage of the procedure, would come and randomly break off a number of the chips corresponding to the required number of jurors. Whoever's chip was broken had to serve as a juror for that day. In the next chapter, on the mathematics of human behavior, we will expand further on the use of a random process as a mechanism for achieving fairness.

In the two examples above, and in many other references to randomness both to achieve fairness as well as in relation to games of chance, the reference was based on intuitive understanding with no logical mathematical foundation, despite the fact that the logical approach and the use of axioms for mathematical analysis were quite developed. For some reason the contemporary scientists did not consider probability to be worthy of mathematical analysis. The notion of "probable" was used as early as in the days of Aristotle, but no proper attempt was made to develop or even to formulate the relevant mathematics. The absence of mathematical analysis of probability persisted until the beginning of the modern era. A number of factors resulted in the growing interest in the nature of the effects of probability, interest that eventually brought about the beginning of mathematical probability theory.

The popularity of games of chance did not wane for many years, and gambling houses spread all over Europe in the fifteenth and sixteenth centuries. The gamblers included some well-known mathematicians who apparently wanted to exploit their arithmetic abilities to become wealthy. The concept of chance or probability did not exist yet, but preliminary questions about the idea of probability had been asked; for example, out of a given number of throws, how often would two dice show the desired pair of sixes? Galileo was asked why, in a game consisting of throwing three dice, gamblers prefer to bet that the sum of the upper faces would be eleven rather than twelve, when both eleven and twelve can result from the same number of combinations of smaller numbers. In the same way they would prefer to bet on a total of ten rather than nine. Galileo's answer, which was correct, was that the number of ways in which the desired total can be described as a sum of three numbers between one and six is not important; what matters is the relative frequency of times the given total would appear when the three dice are thrown. The two computations are not the same. That was a mathematical explanation of the punters' behavior, behavior that developed through experience.

The Italian mathematician Gerolamo Cardano (1501–1576), of Pavia, near Milan, was foremost in the development of formulae dealing with questions like the above. Cardano studied at the University of Padua, and he was a physician, an astrologer, a mathematician, and also an inveterate gambler. In the field of mathematics he was well known in particular for his development of methods of solving cubic and quartic equations. He fell out with Niccolò Tartaglia (1499–1557), among others, who showed Cardano his method of solving those equations. Cardano did not hide the fact that he had learned the method from Tartaglia, but the latter claimed that he had made Cardano swear not to publicize his method, as doing so would harm him. The ability to solve equations was a way of making money by winning public equation-solving competitions.

Cardano also suffered from shortage of money, and he tried to get it via games of chance and gambling, and for that purpose he developed mathematical methods of calculating what is today referred to as the relative frequency of numbers coming up in throws of dice, for example, as we mentioned above, calculating the number of times a pair of sixes would be thrown or how often the sum of the two upper faces would exceed ten. Cardano gathered these methods and other studies related to gambling into notes for a book, but it was not published until after his death. Cardano's and his contemporaries' methods were limited to methods of calculation; in other words, they translated intuition into arithmetic without any logical mathematical basis. This changed very soon.

37. PASCAL AND FERMAT

The laying of the mathematical foundation of the theory of probability is generally attributed to correspondence in 1654 between two of the most famous mathematicians of their time, Blaise Pascal (1623–1662) and Pierre de Fermat (1601–1665). That claim contains an element of exaggeration, but, as we will see, that correspondence over several months, followed by the publication of an argument eventually known as Pascal's wager, did indeed provide the basic conceptual cornerstone of the theory of probability.

It was Pascal who initiated the correspondence when he asked Fermat for his reaction to his proposed mathematical solution to a question related to gambling. Pascal was shown the problem by his friend Chevalier de Méré (Antoine Gombaud), an amateur mathematician and a lover of gambling. The question had been publicized many years earlier, among others by the Italian mathematician Luca Pacioli (1446–1517) in a book titled

Everything about Arithmetic, Algebra and Proportions. The book was published in 1494, more than 150 years before the correspondence between Pascal and Fermat. The mathematical question was given the name the Unfinished Game. We will present here a version of the original question related to tossing a coin. The versions appearing in Pacioli's book and in the Pascal-Fermat correspondence are not significantly more complex. The difficulty in solving the problem was not a computational one but a conceptual one, as we will see. The problem is as follows.

Two players are betting on a hundred-dollar stake. They decide to toss a coin five times. If heads is uppermost in most of the five throws, the first gambler will take the whole amount, and if tails appears more often, then the second one will take the whole amount. They start tossing the coin, but after three times, in which heads has appeared twice and tails once, they cannot continue. The question is, in light of this partial result, how are they to divide the money? To be able to answer the question, we must first define what is meant by "how are they to." Clearly the issue is not a moral or legal one, such as claiming that as the procedure was stopped, the bet is canceled. From the way the question was formulated by Pacioli and also Pascal and Fermat, it is clear that they are speaking of a fair division based on the outcome until the procedure was stopped.

Given the current understanding of the theory of chance, the answer to the question is simple. The chances of winning must be updated in light of the outcome of the first three tosses of the coin. At the outset, the chances of winning were equal for both players, and if the hundred dollars had been divided before the first toss of the coin each would have received fifty dollars. That would also be the position after two throws, if heads had appeared once and tails once. Calculating the chances in more general cases is not much more complicated. For example, in the situation described in the question, if the coin would have been tossed a fourth time, the chances of heads appearing would be 50 percent, and of tails, 50 percent. If heads appeared, the first gambler would take the whole amount, and if tails came up, then a fifth throw would be needed to decide the outcome. In a fifth throw, again the chances of heads and tails are equal. Thus, in the fourth throw, only if the outcome was tails (with a 50 percent chance) would a fifth throw be required. So at the fourth throw, with a 2:1 lead, heads has a 50 percent chance of winning the whole game if heads comes up again, and another 25 percent chance if the fourth result is tails and the fifth is heads (i.e., a total of 75 percent), while tails has only a 25 percent chance of winning the game, that is, if the result of both the fourth and fifth throws are tails. So the fair division after the first three throws would give the first player (heads) seventy-five dollars and the second player (tails) twenty-five dollars.

The above analysis uses the concept of chance, a concept that did not exist at the time of Pascal and Fermat. A number of eminent mathematicians addressed this problem before the Pascal-Fermat correspondence, including Cardano, who did not come up with a complete solution, and his rival Tartaglia, who put forward an incorrect solution. The latter also claimed that this was not a mathematical but a legal problem. No agreed solution based on mathematical arguments was reached.

Then, in the first (no longer extant) letter, Pascal described the problem to Fermat and proposed an incorrect solution. The suggestion he put forward was that if the game had continued for the full five throws, there were three possible outcomes, shown by the black circles in the diagram below, which illustrates the fourth and fifth tosses of the coin. Thus, if at the fourth throw heads comes up, then heads has won the game. If tails comes up at the fourth throw and then again at the fifth throw, tails wins the game. Lastly, if tails comes up at the fourth throw, and if heads comes up at the fifth throw, then heads wins. Thus, according to Pascal, because heads has twice as high possibilities of winning the game as tails, the player with heads should get two-thirds of the pool, and tails, one-third. Note that this proposal does not take into account the chances of the different outcomes.

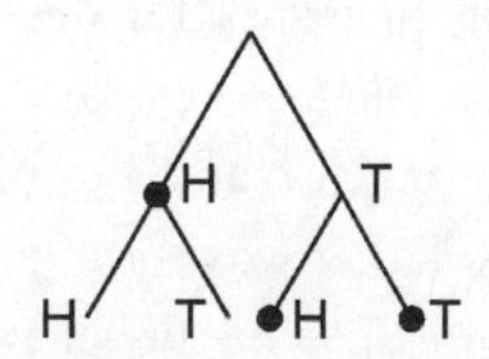

Fermat wrote a reasoned reply, saying that all the possible outcomes of tosses of the coin had to be taken into account, even if the game was decided at the fourth throw, or in today's terminology, the case in which the game was decided by the fourth throw should be given greater weight, of 50 percent. Pascal immediately realized his mistake, and in the brief correspondence that extended over a few months, the fundamental concepts of the theory were developed. Even then the concept of probability did not appear explicitly in their letters, but the essence of the relevant mathematics was presented.

At this stage a few words about Pascal and Fermat are in order. Pierre de Fermat was a lawyer who described himself as an amateur mathematician. He did not depend on mathematics for his livelihood and could therefore choose subjects for research at his leisure. He also did not feel a great need to publish. Only recently was the problem known as Fermat's last theorem solved. The proof was completed in 1995 by Andrew Wiles. It is said that in the margin of one his books Fermat wrote that he has the proof of his theorem but that there was no room to write it in the margin. Even if the story about writing in the margin is true, it is hard to believe that Fermat actually had the proof of his theorem. Number theory was his main field of interest, and he contributed greatly to the subject, but the proof of the theorem requires techniques not available at those times. He also addressed other mathematical questions, for instance, problems related to topics that were eventually developed by Newton and Leibniz into differential and integral calculus, as described in chapter III. We also saw in section 21 Fermat's principle that explained the mathematics underlying Snell's law of refraction. All this resulted in Fermat being one of the best-known scientists in Europe at that time.

Pascal, who at the time of the correspondence with Fermat was less known, approached Fermat in respectful deference, and his admiration only grew with his understanding of Fermat's solution to the unfinished-game problem. Pascal very much wanted to meet Fermat, but this never came about. Blaise Pascal's father was a public official, a judge, and a tax collector. The young Pascal helped his father with tax collection and even invented a mechanical calculator to help with the calculations needed for the job. He made a number of such machines and tried his hand at selling them, a sort of start-up for those days, but did not meet with much success. As a mathematician he showed ability in a variety of subjects but also occupied himself with philosophy and, at a later stage in his short life, became more and more religious, apparently due to an illness he suffered, and stopped dealing with mathematical research. Yet even his theological works had a mathematical feel to them. In one of them he touched on the question of the existence of God, as follows.

Pascal put forward a claim that became known as *Pascal's wager*: We accept, he argued, that there is no ultimate proof of whether God exists or not, but we have to decide whether we believe in his existence or not. We will examine the results of our decision. If God does not exist, the difference between our decision to believe or not believe in his existence would be expressed in only slightly different lifestyles, fulfilling commandments, praying, and so on; that is, a small change only. If on the other hand God does exist, the difference between our decision to believe in his existence or not to believe in it would be enormous. It would mean the difference between eternal life in heaven or perpetual torment in hell. Therefore, just as decisions on how to bet in games of chance, if there is even the smallest chance that God exists, and as stated there is no proof that he does not, the obvious conclusion is that it is worthwhile to believe in the existence of God.

Pascal was interested mainly in the theological aspects of the argument—that indeed were quite novel at the time—yet the analogy with games of chance was in the original text (still without the explicit mention of the concept of probability). This was perhaps the first time that an analysis of a one-time decision, whether to believe in God or not, was made with the help of mathematics and aided by concepts that we call probabilistic concepts. Pascal's analysis became known before it was published in written form in 1670, some seven years after his death.

38. RAPID DEVELOPMENT

The analysis of the unfinished-game problem presented by Fermat and Pascal came just at the time when the need for mathematics that would describe randomness was growing. Possible uses of the concept of randomness led to a growing number of mathematicians involving themselves in the subject, and the theoretical aspects proposed by Fermat and Pascal played a central role in the rapid development of the mathematics of randomness.

There were three main reasons for the increased interest in this area. One, mentioned above, was the spread of gambling and games of chance, with quite a number of scientists and mathematicians among the participants.

The second reason was related to the bankruptcies of many European cities as a result of liabilities for payments of pensions and allowances they had taken upon themselves. Governments and municipalities in Europe, mainly in Holland and England, had for a long time followed the practice of financing their expenditures by taking loans from their citizens, in exchange for which they undertook to pay the lenders fixed amounts as long as they lived. The problem was that the mathematics required to calculate the proper amount of the repayments, and in particular to avoid incurring liabilities that the borrowers would not be able to meet, did not exist at that time. The amount of the regular repayments was determined intuitively or by using calculations that were biased by the borrowers' desire to raise more and more money. The government of England, for example, issued an ordinance in 1540 that the amount of the annual repayments on a loan had to be such a sum equivalent to the entire amount of the loan that would be repaid in seven years, whereas the annual payments continued until the death of the lender. The age of the lender and the life expectancy of potential lenders was not taken into account. Newton included in his lectures in 1675, and later published in a short paper, a description of how to use the binomial formula to calculate the current value of an interest-bearing account, what is today known as capitalization. He did not even hint at the need for anyone guaranteeing the payments to relate to the average length of life of the recipients of the payments. The result of this practice was that more and more municipalities went bankrupt. The relevant branch of mathematics, unknown at that time, is of course what the actuarial profession deals with today, namely, calculating the life expectancy of the population of lenders, and hence the expected liability of future payments, and comparing those liabilities with the expected tax revenue. The data required for these calculations, such as mortality tables for different towns, already existed. These had been compiled for years out of curiosity and interest, and especially to understand the effects on mortality rates in different places of the various plagues that struck Europe. The basic concepts of probability and expectation, however, were not known. When municipality after municipality encountered financial difficulties and even bankruptcy, it created pressure to find a solution to the problem, and the development of the mathematics of probability and statistics was given a boost.

The third reason for the development of probability concepts was the development of jurisprudence. The awareness of the validity of legal arguments was increasing in Europe in those days, as was the recognition that legal proof that leads to conviction or acquittal beyond all reasonable doubt was almost impossible to obtain. Probability considerations in legal arguments had already appeared in ancient times, but with social progress the requirement arose that litigants base their case on quantitative assessments of the probability that their case was correct. In any event there was a growing need to develop terminology and methods of analysis that would help in such quantification.

The questions that arise in an analysis of chances and risks in ensuring payments are totally different from those that arise in, say, legal claims. Claims about betting, insurance, public-opinion surveys, and so on, relate to a random occurrence among many possibilities or random sampling from a large population, namely, statistics. In contrast, legal claims generally relate to a single, non-repeated event or to the degree of belief in the correctness of a certain argument. There is no apparent reason to expect the same theory and the same mathematical methods to apply to both types of situation, but nevertheless the same terminology of probability theory is used in the analysis of repeated and one-off incidents. This is the duality inherent in probability theory and statistics, and the pioneers of the theory were aware of it. We will discuss this in more detail later on.

A few years after the Pascal-Fermat correspondence, in 1657, Christiaan Huygens (1629–1695) published a book that summarized his own work on the theory of probability and the knowledge on the subject that had been accumulated until then. He was encouraged to publish it by Artus Gouffier de Boissy, Duke of Roannez, who was also a patron of Pascal. Despite his efforts, Huygens never met Pascal face-to-face, as Pascal was at that time more involved in theology than mathematics, but he was well aware of Pascal's work. Huygens's book was the first published on the subject of probability and was entirely devoted to an analysis of the randomness related to games of chance, repayments of loans, and so on, in other words, its statistical aspects. Among other things, he wrote about the methods of calculation developed over many years by mathematicians related to random draws, counting methods, and the like. Huygens, who was Dutch, was one of the most respected and esteemed mathematicians, physicists, and astronomers of the period. Apart from his contribution in various other fields, he was famous for his explanation of the propagation of waves before the wave equation had been formulated. He traveled widely on the Continent and to Britain, and in 1663 he was elected to the Royal Society and later to the French Academy.

Huygens's book was also the first to discuss the idea of an average and of expected value. He coined the term in Latin *expactatio* ("expectation"). Today, engulfed as we are by statistical data of all sorts on all subjects and in all locations, it is difficult to imagine that until the middle of the seventeenth century the concept of the average was not in widespread use as a statistical quantity. At that time averages were calculated by physicists to obtain a good estimate of inaccurate measurements, for instance the paths of the celestial bodies, but not for statistical analyses. The mathematical analysis of games of chance and expected future payments led naturally to the development and use of the concept of the average. The first well-ordered presentation of that concept appeared in Huygens's book.

The discussion was incisive. To illustrate: Christiaan's brother Ludwig used the existing tables of mortalities and found that the average lifetime of someone born in London was eighteen years. "Does that number have any significance?" Ludwig asked his brother. It is known, he claimed, that infant mortality is very high, and many children die before they are six years old, whereas those who do survive live to the age of fifty or longer. Indeed, a good question, and in our days the answer would be that different indices are used for different purposes. Christiaan Huygens did not deal with these situations but focused on the subject of gambling, and he stuck to his opinion that expectation was the right measure for estimating the value of bets, with the value being the expected payments weighted according to the chances of winning.

Here we will repeat the precise mathematical definition of expectation. If the monetary payments A_1, $A_2, \ldots, A_n$ in a draw are won with probabilities $p_1, p_2, \ldots, p_n$, respectively, the *expectation* of the draw is $p_1A_1 + p_2A_2 + \ldots + p_nA_n$. Just as with the concept of the average, questions about the justification and interpretation of the idea of expectation were asked, forcefully, because the idea of probability had not been sufficiently clarified. It was recognized only in the context of relative frequency, in other words, p_1 in the definition is the approximate proportion in which the win of A_1 will occur if the draw is repeated many times.

One basic question that was asked was where do these probabilities come from? Also, how are they calculated? In throws of a die, for example, as there is no reason that one face should be on top more than any other, the probability of each face coming on top can be calculated; but how, Huygens had already asked himself, can the probability of catching a particular disease or being injured in an accident be calculated?

Despite the difficulty in establishing the concepts, the relevance to what they could be used for was clear. The first use was statistical analyses. The word *statistics* is derived from the word *state*, in the sense of country, and statistics did in fact deal mainly with subjects related to the management of a country. The most rapid progress was made in Holland, in relation to the pricing of the repayment of loans. This was because both the politician responsible for that subject, Johan de Witt, who was a leading figure in Dutch political life, and the *burgomaster* (mayor) of Amsterdam, Johannes Hudde, a town that was hard hit because of promises of unrealistic payments, were good mathematicians. They both engaged in and contributed to Descartes's Cartesian geometry. They consulted Huygens himself, who was then in Amsterdam, and in 1671 de Witt published a book in which he set out the theory as well as the practice of calculating loan repayments under various conditions with detailed exemplary calculations. The book came with a confirmation by Hudde that the calculations were correct. The method soon spread across all of Europe. Of all countries it was England that was slow in adopting this use of the theory, and even a hundred years later the government authorities there were still selling pensions at prices that were not based on proper mathematical costings.

Alongside the development of the practical statistical use of the mathematics of randomness, progress was also made in theory regarding the connection between probability and legal evidence. Leibniz was the leader in this field, and in 1665 he published a paper on probability and law, with a more detailed version published in 1672. Leibniz's interpretation of probability was similar to the sense in which Aristotle viewed it, that is, the likelihood of an event in light of partial information. Leibniz, who came from a family of lawyers, tried to present quantitative measurements for the correctness of legal claims. After a visit to Paris and after becoming familiar with the Fermat-Pascal correspondence as well as Pascal's wager, Leibniz realized that a similar analysis could also be used in cases in which one had to assess the probability that a claim or certain evidence, even if it is one-off, that is, non-repeated, is correct or not. He analyzed the logic underlying the information brought before a judge and proposed measuring the likelihood of a conclusion, giving it a value of between zero (in the event that the conclusion is clearly wrong) and one (in the case when it is without doubt correct). Thus Leibniz laid the foundation of the analogy between the likelihood of an occurrence and the mathematics of situations that are repeated randomly. Herein apparently lies the secret of the great impact of the Fermat-Pascal correspondence and Pascal's wager. They used the same mathematical tools in discussing events that can be repeated, such as a game of chance stopped in the middle, as well as non-repeated occurrences, such as the question of the existence of God. Yet neither Leibniz nor others who dealt with the concepts of likelihood and probability reached an understanding or consensus of what these probabilities were derived from or where they were formed.

39. THE MATHEMATICS OF PREDICTIONS AND ERRORS

A crucial step forward in the establishment of the link between the concept of expectation and the practical use of statistics was made by Jacob Bernoulli (1654–1705), one of the most prominent members of a family that had a great influence on mathematics. He first analyzed repeated tosses of a coin, assuming that the chances of its falling on either side were equal. By sophisticated use of Newton's binomial formulae, Bernoulli analyzed the following: he examined whether in repeated tosses of a coin the chance that it would fall with a particular side up, say heads, would be close to 50 percent of the total number of flips. He found that the chance grew closer and closer to certainty the greater the number of flips of the coin. Clearly in a given series of tosses of the coin, the proportion of the number of heads to the total number of throws could have any value between zero and one. Nevertheless, as Bernoulli showed, as the number of throws increases, it is almost certain that the number of times heads appears will be close to 50 percent of the total number of throws. These trials are still today called *Bernoulli trials*, and its mathematical law is called the *weak law of large numbers* (the formulation of the strong law of large numbers would later require twentieth-century developments).

Bernoulli himself and others who contributed to this innovative research path extended the results to more general cases than repeated flips of a coin, even to the case of repeated sampling from a large population, and to random errors in non-exact measurements. As we have noted previously, in order to assess a physical quantity, the measurement of which entails measurement error, physicists used an average of many measurements. The mathematical result confirmed that the greater the number of repeats or measurements, and on the condition that the repeats are carried out totally independent of each other, the average of all the measurements will be close to the true value with a likelihood that increases and converges to certainty. At the same time, Bernoulli discussed the question of what creates the different probabilities and how confident can we be in their numerical values. He was apparently the first to distinguish between *a priori* probability, which we can calculate and derive from the conditions of the experiment, and *a posteriori* probability, which we see after performing a series of experiments. He turned the development of methods to calculate a posteriori probabilities into an objective, and it played an important role in future progress.

Bernoulli's weak law of large numbers is one of the limit rules that refer to statistical aspects of large samples or many repetitions of trials. Already then discrepancies were found between claims regarding large numbers and human intuition. Further on we will discuss in greater detail the discrepancy between intuition and the mathematics of randomness, but here we will give just two examples.

The first discrepancy is generally referred to as *the gambler's fallacy*. Many gamblers continue betting even if they are losing, believing that the laws of large numbers ensure that they will get their money back eventually. Their mistake is that the law says only that on average their wins will be close to expectation, but it does not relate to the amount of the win or loss. Even if the average of the wins and losses in a series of games is one dollar, the loss itself, in the case of many repeated bets, may be ten thousand or a million dollars. The difference between the average and the actual value gives an enormous advantage to the gambler who is wealthy and who can finance the loss until the probability turnaround arrives. That difference led to the bankruptcy of many gamblers who did not have deep enough pockets. Evolution did not give us the intuitive understanding of the difference between the average and the value itself when dealing with large numbers; the reason, apparently, is that in the course of evolution humans did not encounter examples of so many repetitions of events.

The second discrepancy between intuition and probability concepts is known as the St. Petersburg paradox, named after the city in which Daniel Bernoulli, nephew of Jacob Bernoulli, presented the problem to the Imperial Academy of the Arts. Remember that Hyugens referred to the expectation of a lottery as a fair measure of the cost of participating in it. That approach proved itself as an accounting basis for calculating loan repayments or the price of participating in a lottery. Now consider a lottery in which a coin is tossed many times, say a million. In this, if the first time the coin falls tails up is on the nth throw, that is, until then it fell $n - 1$ times with heads up, the participant receives 2^n dollars. A simple calculation shows that expected winnings are one million dollars! Would you agree to pay one hundred thousand dollars, or even ten thousand dollars, to participate in this lottery? I do not know anyone that would agree to do so, and this difference between the theory and the practice is the paradox. Daniel Bernoulli had a social explanation for this, and we will discuss it in the next chapter. A different explanation, which we will also expand on in due course, is the gulf between mathematics and intuition. The latter tells us that the coin will not fall on the same side a large number of times and ignores the chance, albeit a small one but with high winnings, that such an event will occur.

The search for methods of calculating probabilities in situations with incomplete information at the outset, a subject raised by Jacob Bernoulli, led to the development of a technique based on a mathematical theorem known as the *central limit theorem*. The first move in that direction was made by the French mathematician Abraham de Moivre (1667–1754). He spent many years in England, much of the time with Newton, after being exiled there as a result of the persecution of the Huguenots in France. Whereas Jacob Bernoulli examined the extent of the deviations of the average from expectation and showed that most of the deviations are concentrated around zero, de Moivre decided to study the distribution of those deviations, in other words, how they divide between relatively large deviations and medium and small ones. He focused on Bernoulli's coin tossing experiments in which, let us say, heads gives a prize of one, and tails yields nothing. He found, using a mathematical calculation, that if the size of the deviation from the average is divided by the square root of the number of throws (i.e., instead of dividing the total winnings by n after n flips of the coin to find the deviation from expectation, he divided by $\sqrt{n}$), the distribution becomes closer and closer to a bell shape. If the coin is not equally balanced, with the chance of heads say a, the shape of the bell will depend on the value of a, but if the result is divided by $\sqrt{a(1-a)}$, which in due course came to be known as the standard deviation, the distribution obtained is bell shaped independent of the value of a. De Moivre actually calculated the shape of the bell obtained, shown in the diagram together with the formula (which is unimportant for our current purposes).

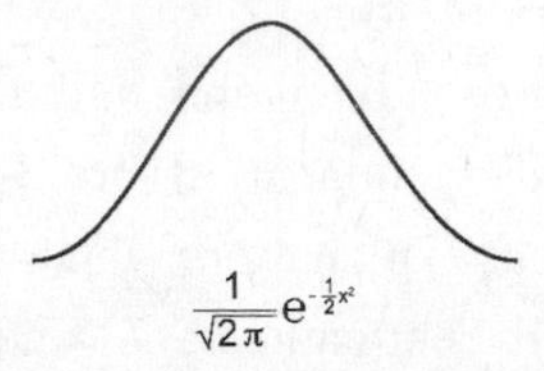

$$\frac{1}{\sqrt{2\pi}}e^{-\frac{1}{2}x^2}$$

It is not clear whether de Moivre realized the implications of his discovery for the theory of statistics and its practice, but several famous mathematicians generalized de Moivre's limit law and found that it applies far more widely. The research reached its pinnacle in the work by Pierre-Simon Laplace (1749–1827), who as well as proving the wider applicability of the central limit theorem also laid the foundations for its uses in statistical analyses. Laplace was born in Normandy, France, and his family intended that he become a priest, but his mathematical inclinations prevailed, and he was accepted to carry out research under D'Alembert. He completed a research project in mechanics, a study that earned him a position in the military academy as a mathematics teacher and an artillery officer. There he befriended Napoleon Bonaparte, a friendship that certainly did him no harm in the politically stormy atmosphere in France at that time. He survived the French Revolution by maintaining a low profile and even became head of the French Academy. His book on the analytical theory of probability, published in 1812, he dedicated to Napoleon.

At about the same time, in a book published in 1810, Gauss himself presented the same limit law. Gauss, who clearly was familiar with de Moivre's work, focused his attention on a different aspect than did Laplace.

Gauss was very interested in the results of measurements and in the question of how to find the closest value to the correct one in measurements that include random measurement errors. With that in mind, Gauss developed a system that still today is called the least-squares method, and, based on the assumption that the errors in the calculations are random and independent, showed that the average was the value that predicted the correct value with the greatest degree of accuracy. He extended the system to more complex calculations and, in that framework, even proved the central limit theorem. The bell-shaped distribution is today called the normal distribution and also the Gaussian distribution, in recognition of his contribution.

Laplace and Gauss, as well as many of their research colleagues, realized that the statistical methods could be used to answer questions beyond the field of gambling and flipping coins. If a particular outcome is the result of many occurrences that include randomness and those random occurrences are independent, the distribution of the outcomes will be similar to the normal distribution around the average value of the outcome. Laplace, who was also interested in and contributed much to astronomy, used this technique to analyze deviations of the planes of orbit of the planets from one, middle, plane. The planes of orbit of the planets around the Sun almost coincide, and their deviations from one middle plane are very small. Are those deviations random, or do they have another cause? By using the statistical technique he had developed, Laplace showed that the deviations are a close approximation to the expected distribution based on the central limit law, and therefore it was highly probable that they were random deviations from one orbit plane. The expression "highly probable" is an indication that this is a statistical result and not a mathematical certainty. At the same time Laplace showed that the planes of orbit of the various comets do not comply with the expectation based on the central limit law, and he therefore concluded that those deviations were not caused by random deviations from one plane. Gauss also used the least-squares method for calculations in astronomy. At that time the asteroid Ceres, which moved on an orbit that disappeared behind the Sun, was identified, and the question was when would it reappear on the other side of the Sun. Little data had been calculated regarding its path, and what there was included many measurement errors. Gauss applied his method and predicted with surprising accuracy Ceres's continued route, a prediction that justifiably brought him worldwide renown.

The work of Gauss and Laplace brought the mathematics of randomness and its uses in statistics to a center-stage position in science. Since then, the central limit law has been found to be correct, with minor changes, in more general cases than those analyzed by Laplace and Gauss. Especially worthy of mention are the Russian mathematician Pafnuty Chebyshev and his students Andrei Markov and Aleksandr Lyapunov. They were active in the second half of the nineteenth century and firmly established the central limit theorem without the assumption that the random events that lead to deviations from the average have equal distributions, that is, they have the same characteristics of randomness. Such a general rule was important to justify its uses for situations that appear in nature. The lack of interdependence in nature between the random events can be justified, but it is more difficult to justify the assumption that the random characteristics are equal for all events. The work of the Russian mathematicians reduced the gap between the mathematical theorem and its possible applications, and thus the central limit theorem, together with other limit theorems, became the norm for the various uses of statistics.

The main use of limit theorems is to estimate statistical values such as the average, the dispersion, and so on, of data that have random inaccuracy. It is generally difficult to assess whether the errors are random or not. Even if the errors are random, however, absorbing and understanding the technique required for the use of the mathematics that developed involves difficulties that can themselves cause errors. Again, the difficulties derive from the way our intuition relates to data. We will mention two such difficulties.

We are exposed to huge numbers of statistical surveys in our day-to-day lives. When the results of a survey are published, it is usually done in the following form: The survey found that (say) 47 percent of the voters intend to vote for a particular candidate, with a survey error of plus or minus 2 percent. Yet only part of the conditions and reservations about the results are presented in the survey report. In effect, the correct conclusion from the survey would be that there is a 95 percent chance that the proportion of voters who intend to support that particular candidate is between 47 percent plus 2 percent and 47 percent minus 2 percent. The 95 percent bound is quite normal in practice in statistics. The survey can be devised such that the chance that the assessment derived from the survey is correct is 99 percent (it will then cost more to carry out the survey), or any other number less than a hundred. The condition of 95 or 99 percent is not published. Why not? The fact that the declared limits of the results, that is, that the resulting interval of 45 to 49 percent, applies with only a 95 percent probability could have great importance. The reason would seem to be the difficulty in absorbing quantifications.

There is another aspect of statistical samples that is difficult to understand. Let us say that we are told that a survey of five hundred people selected at random in Israel is sufficient to ensure a result with a plus or minus 2 percent survey error (with 95 percent accuracy). The population of Israel is about eight million. What size sample would be required to achieve that level of confidence in the United States with 320 million inhabitants? Would it need to be forty times the sample size in Israel? Most people asked that question would answer intuitively that a much larger sample is needed in the United States than in Israel. The right answer is that the same size sample is required in both cases. The size of the sampled population affects only the difficulty of selecting the sample randomly. Once we have sampled correctly (and most survey failures derive from the inability to sample correctly), the size of the survey error is determined only by the size of the sample. This is another example of the discrepancy between human intuition and mathematical results. Evolution indeed did not prepare us for large samples.

40. THE MATHEMATICS OF LEARNING FROM EXPERIENCE

Let us go back to a development that started in the eighteenth century that had an aura of logic about it and is therefore "responsible" for serious errors in the application of the theory. The statistical methods described in the previous section help us when the particular probabilities are known, and we need only to calculate the chance of a specific event occurring or, when it is possible, to estimate the statistical parameters. The technique does not teach us how to improve the assessments when new information is supplied to us. It was de Moivre, in his book on methods of dealing with randomness, who asked the question of how to act when new information is added. It was Thomas Bayes who answered the question, and the system whose fundamentals he put forward is known as the Bayesian method.

Thomas Bayes (1702–1761) was born in England but studied mathematics and theology at the University of Edinburgh, Scotland. He was more interested in theology and, following in the footsteps of his father, who was a Presbyterian minister in London, served in a similar capacity in the Mount Zion Chapel in Tunbridge Wells, Kent, England. In his lifetime he published only two works. One was on religious matters. The other was an attempt to defend Newton's approach to infinitesimal calculus against the severe attacks that claimed that fluxion had no logical basis. The criticism was published by the famous Irish philosopher Bishop George Berkeley (after whom the University of California, Berkeley, is named). Bayes did not see it fit to publish his formula in his lifetime, and it was published only after his death by his friend Richard Price, who was bequeathed Bayes's writings, and who realized the importance of that work.

Bayes's formula is very easy to understand technically but is also very difficult to absorb and apply intuitively. We will discuss the reasons for this and its sometimes-serious results in the coming sections. Here we will just present and explain the Bayesian formula itself (the calculations can be skipped without disturbing the overall picture).

We will start with an example based on a question that was asked in the school matriculation examination in probability in Israel in 2010. Of three boxes, the first contains two silver coins, the second has one silver coin and one gold coin, and the third holds two gold coins. One of the boxes is chosen at random, and one of the coins in it is chosen at random. The simple question is, what are the chances that the coin left in the box is a silver one? For reasons of symmetry one could conclude that the chance is 50 percent because the question does not differentiate between the roles of the two types of coin. That was the answer that would have been given before the revolution of Fermat and Pascal (without using the concept of "chance," which did not exist then). One can also carry out the following calculation: Each box has a probability of one-third of being selected. If the first is chosen, the probability of the coin that is left in the box being silver is one (i.e., a certainty). If the second box is chosen and one coin is chosen randomly, the chances that the one left is silver is one-half. If the third box is chosen, the remaining coin after the choice of one is not silver. Now calculate $\frac{1}{3} + \frac{1}{3} \cdot \frac{1}{2} = \frac{1}{2}$, and we obtain that the chance is 50 percent.

 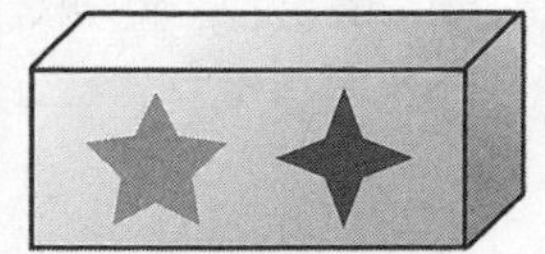 

Now we ask a more complex question. A coin is removed from a box that was chosen randomly, and it turns out to be a gold coin. What are the chances that the coin left in the box is silver? It is simple to formulate the question, but try giving an intuitive answer (without having to use formulae that you may have learned in lessons on probability). A simple analysis shows that from the information that a gold coin was taken from the box we can conclude that the box chosen was not the first (which held two silver coins). The other two boxes have an equal chance of being selected, that is, 50 percent. If the second box is chosen, the remaining coin will be the silver one (as the gold coin was taken out). If the third box is selected, the remaining coin will be the second gold one in that box. Thus, in the situation as described with a gold coin removed from a randomly selected box, the probability of the remaining coin being silver is one-half. Although this analysis is simple, it is incorrect (there was a reason that de Moivre did not arrive at a satisfactory answer to the question of how to solve such problems and left it in his book as an open question). The error is similar to the one committed by Pascal in his first letter to Fermat, as described in a previous section. In other words, the "solution" ignores the probabilities of a gold coin being selected in the different scenarios and therefore is in error in the precise implication it draws from the information. The correct analysis is: the gold coin that was drawn from the selected box came either from the second box (with a gold and a silver coin) with a probability of $\frac{1}{3} \cdot \frac{1}{2} = \frac{1}{6}$, or from the third box (which held two gold coins) with a probability of $\frac{1}{3}$. Only in the first of these two possibilities will the remaining coin be silver. The weight of that happening in the number of occurrences in which a gold coin is selected in the first draw is $\frac{1}{6}$ divided by $\frac{1}{2}$, that is, one-third.

The principle underlying the above calculation is simple. If you wish to draw a conclusion based on new information, you have to take into account all the factors that are likely to result in that information reaching you, and to weight all those factors according to their probabilities. Specifically, relating to our above example, assume that you want to find the probability that event *B* will occur, given that you are told that *A* has occurred. First, find the chance that you will be told that *A* has occurred if *B* occurs. Then find the chance that you will be told that *A* occurred if *B* does not occur. Then calculate the weight of being told that *A* occurred if *B* occurred relative to the total chance of being told that *A* occurred. This scheme can be written in the form of a formula, which we will set out in the next section. The principle underlying the weighting is the essence of the Bayes's scheme. We will present several other examples that will make the situation even clearer.

The principle presented by Bayes enables the probabilities to be updated whenever new information is received. Theoretically the probabilities could be updated continuously until exact assessments are obtained. That refinement of Bayes's scheme was developed by Laplace. Laplace apparently arrived independently at a formula similar to Bayes's and then developed the complete formulae of updates that become more and more exact, but when he heard of Bayes's previous work, he gave Bayes's name to the method. That name, the Bayesian inference or Bayesian statistics, has prevailed still today.

But the approach has a fundamental drawback. In order to apply Bayes's formula we need to know the probabilities that the events we are referring to will take place. The problem is that generally, in our daily lives, the information regarding these probabilities is not known. How then can we learn from experience? Bayes had a controversial answer: if you have no idea of the probability that *A* will occur or will not occur, assume that the chances are equal. Once you assume the initial probabilities, also called the a priori probabilities, you can calculate the new probability, called the a posteriori probability, with a high degree of accuracy. The question arises: Can we allow the use of an arbitrary assumption about the values of an a priori probability?

The dispute between the supporters and the opponents of the system was not limited by space or time. The statistics of frequencies and samples had a firm theoretical basis, but to use it required very many repetitions of the same random occurrence. This type of statistics was not applicable to statistical assessments of non-repeated events. Bayesian statistics is a tool for analyzing isolated events, but without reliable information on a priori probabilities, the results depend on subjective assessments, and these do not constitute a reliable basis for scientific findings, its opponents claimed. It is better to rely on subjective assessments than to ignore the advantages offered by the method, replied its supporters. Moreover, they added, the more information that is added, the greater the reduction in the effect of the arbitrary assumptions, until it reaches a minimum, a fact that gives scientific validity to the Bayesian approach. This dispute spilled over onto a personal level, and for many years the two methods developed side by side. Even today statisticians are divided into Bayesians and non-Bayesians, but it now seems that the borders and limitations of the two methods have been drawn more clearly, and each occupies its proper position.

41. THE FORMALISM OF PROBABILITY

The mathematical developments and increasing uses of the concepts of probability theory and statistical methods resulted in the accumulation of great expertise in the practice of the mathematical theory at the beginning of the twentieth century. This development, however, was accompanied by much unease, the roots of which were mentioned previously. First, there was the duality in the subject matter. The same terms and considerations were used both in the analysis of repeated events, in which the probability can be interpreted as the proportion of total occurrences in which the event takes place, as well as in cases of assessing the probability of a nonrepeated event. Second, no understanding or agreement had been reached regarding the source of the probabilities. Even in coin-flipping experiments, the only reason for thinking that both sides of the coin had equal chances of falling uppermost was that there was no reason for thinking that the chances were not equal. Is that argument strong enough to convince us that the calculations represent nature? In addition, there was no general logical mathematical framework for dealing with the mathematics of randomness. For example, no one had proposed a precise general definition of the concept of independence. The reader will no doubt have noticed that we have used the term *independent* several times, and the intuitive feeling is that even without a formal definition, we know when events are independent. That feeling, however, is not enough for mathematical analysis, and a definition that met the strict criteria of mathematics did not exist.

George Boole (1815–1864), a British mathematician and philosopher, tried to present a general mathematical framework. He claimed that mathematical logic, and in particular the union and intersection of sets to present information, is appropriate for the analysis of events involving probabilities. To this end Boole constructed the basis of logic through the use of sets and defined what is today known as Boolean algebra. These efforts did not result in much success, however, among other reasons because Boole's works contained discrepancies that resulted from a lack of consistency in the model he used. For example, Boole related in different and conflicting ways to the concept of independence. In one case independence meant the inability to imply a conclusion from one event to another, and in another case it meant that events do not overlap. Thus, at the beginning of the twentieth century, the mathematics of randomness did not provide a satisfactory answer regarding how to analyze events involving probabilities and the sources from which those probabilities originated.

It was Andrey Kolmogorov (1903–1987) who proposed the complete framework of logic. Kolmogorov was a preeminent mathematician of the twentieth century. In addition to his contribution to mathematical research, he was interested in the teaching of mathematics in schools and held various administrative positions in universities and Russian academia. Kolmogorov made important contributions in a wide range of mathematical subjects: Fourier series, set theory, logic, fluid mechanics, turbulence, analysis of complexity, and probability theory, which we will turn to shortly. He was granted many awards and honors, including the Stalin Prize, the Lenin Prize, and, in 1980, the prestigious Wolf Prize, the award ceremony of which he did not attend. This led to a change in the rules of the prize, so that in order to be awarded the prize, the recipient must attend the ceremony.

Kolmogorov adopted the Greeks' approach. He drew up a list of axioms with which the concepts that had previously been used only intuitively could be explained. We will discuss the connection between the axioms and nature after we have presented them. Kolmogorov's general approach adopted George Boole's proposal from several decades earlier, that is, the use of logic operators on sets to describe probabilities. The axioms that Kolmogorov wrote in his book in 1933 are quite simple and are set out below (they can be followed even without previous mathematical knowledge, but even if they are skipped, the text that follows can still be understood).

---

1. We choose a *sample space*, which we will call Ω. This is an arbitrary set whose members are called *trials* or *samples*.
2. We select a collection of sets, all of which are partial sets of the sample space Ω. We will denote this collection of sets Σ, and the sets within it we will call *events*. The family of sets Σ has several properties: the set Ω is within it (i.e., Ω is an event). If a sequence of sets (i.e., events) is within it, then the union of these events is also within it. If an event is in the collection, also its complement, that is, Ω minus the event, is an event.
3. For the collection of events we will define a *probability function*, which we will denote *P*. This assigns to each event a number between 0 and 1 (called the probability of the event). This function has the property that the probability of the union of a sequence of events that are pairwise disjoint is the sum of the probability of the individual events. Also, the probability of event Ω is 1.

For those unfamiliar with the jargon or terminology of mathematics, we will state that two events (two sets) are disjoint if there is no trial (member in Ω) that is in both events. The union of two sets is the set that includes the members of both sets. Thus, the second axiom says, among other things, that the set that contains the trials in both events is itself an event.

There is a reason for the statement that the collection of sets Σ of the events does not necessarily contain all the partial sets of the sample space Ω. The reason is essentially technical, and there is no need to understand it in order to follow the rest of the explanation. (The reason is that when Σ consists of all the subsets, it may be impossible, when the sample set is infinite, to find a probability function that fulfills the requirement of the third axiom.)

One of the innovative features of the axioms is that they ignore the question of how the probabilities are created. The axioms assume that the probabilities exist and merely requires that they have certain properties that common sense indicates. Following the Greek method, when you try to analyze a certain situation, you must identify the sample space that satisfies the axioms and describes the situation. If your identification is accurate, you can continue, and with the help of mathematics you can arrive at correct conclusions. Kolmogorov went further than the Greeks, however. They claimed that the "right" axioms were determined according to the state of nature. Kolmogorov allows completely different spaces to be constructed for the same probability scenario. An example follows.

The framework defined by the system of axioms enables a proper mathematical analysis to be performed. For instance, we wish to calculate the probability of an event B in a sample space that includes only partial events of event A, which has a probability of $P(A)$. This new probability of B will be equal to the probability of that part of B that is in common with A (we are concerned only with that part of B) divided by the probability that A occurs. This can be written as the following formula. Denote the part that is common to A and B by $B \cap A$, called A intersect B. Then the probability of the partial event of B that is in A is $\frac{P(B \cap A)}{P(A)}$. This is called the *conditional probability*. The two events are *independent* if it is impossible to draw any conclusions regarding the existence of the second event from the existence of one of them, even a probabilistic conclusion. The mathematical formulation of independence states that the updated probability of B equals its original probability, or $P(B \cap A) = P(A)P(B)$. We have obtained a mathematical definition of independence. The same can be done with respect to other concepts used in probability theory.

This is an appropriate place for a warning: Many texts refer to the expression $\frac{P(B \cap A)}{P(A)}$ for conditional probability as the probability of B *given* A. This in turn leads to the interpretation of the conditional probability as the updated probability of B when one is *informed* that A has occurred. Such an interpretation may lead, as we shall see later, to errors when applying the formulae. While in plain language the two expressions, given and informed, are not that different, in applications, when we are informed of an event, the circumstances in which the information is revealed should be taken into account. When we are informed that A has occurred, we can by no means automatically conclude that the conditional probability of B given A depicts the updated probability of B.

And now we present, as promised, the formula for Bayes's theorem (this can be skipped without rendering the text that follows it less understandable). Assume that we know that event A has occurred, and we wish to learn from that the chances that event B will occur. For the sake of the example we assume the conditional probability, which we denote as $P(B \mid A)$, describes the desired probability of B when we know that A has occurred. Bayes's formula as we described it verbally in the previous section is

$$P(B|A) = \frac{P(A|B)\,P(B)}{P(A)}$$

Moreover, as we explained above, $P(A \mid B)$ is $P(B \cap A)$ divided by $P(B)$. (If we wish to conform with the wording of the principle as displayed in the previous section, we should write the denominator as $P(A \mid B)P(B) + P(A \mid \sim B)P(\sim B)$ where $\sim B$ indicates the event B does not occur. (This is the way most texts write it.) Does that sound complicated? Perhaps so, but the framework provides a proper mathematical basis for the analysis of randomness.

Notice the assumption we made: The circumstances are such that $P(B \mid A)$ is the correct updated probability. Otherwise we should resort to the original Bayes's scheme as described in the previous section, namely, we should calculate the ratio of the probability that A has occurred when we are informed that B has occurred to the entire probability that we are informed that A has occurred. In many applications the assumption does not hold, that is, the probability that we are informed that A has occurred is not $P(A)$.

The above framework provides an outline for the construction of probabilities, but the events that appear in the axioms do not necessarily have significance in reality, significance that we can identify or calculate. Take as an example one toss of a coin. The sample space may be made up of two symbols, say a and b, with equal probabilities. If we declare that if a occurs this means (note that this is *our* explanation!) that the coin falls with heads showing, and if b comes up in the sample it means that the coin fell with tails uppermost, we have a model for one flip of the coin. We cannot analyze two consecutive tosses of the coin in the framework of this sample space because there are four possible outcomes of two flips of the coin. For that case we have to construct another sample space. To arrive at a model that permits multiple flips of the coin, the sample space has to increase. To enable any number of tosses of the coin, we will require an infinite sample space. The technical details will interest those who deal with the mathematics (and students), and we will not present them here. We will just say that in a sample space in which an infinite series of flips of the coin can take place, events occur the probability of which is zero.

This is certainly intuitive. A ball spinning on a continuous circle stops at a point. The chances of its stopping on a predetermined point is zero, but the chance of its stopping on a collection of points, for example, on a complete segment, is not zero. With this end in view, Kolmogorov used mathematics that had been developed for other purposes and that explained how it was possible for a section to have a length while it consists of points the length of each of which is zero; the explanation had not been available to the Greeks when they encountered a similar problem. Furthermore, Kolmogorov's model can be used to explain and prove Bernoulli's weak law of large numbers (see the previous section), and even to formulate and prove a stronger law, as follows. We will perform a series of flips of a coin. These can create many series of results. We will examine those series of outcomes in which the proportion of the number of heads to the total number of throws does not approach 50 percent as the number of throws increases. This set of series, says *the strong law of large numbers*, has a probability of zero. (The careful reader whose mathematical education included Kolmogorov's theory will have noticed that although that event has zero probability, it can nevertheless occur. Indeed, there could be samples in which the proportion does not approach a half, but these are negligible.)

Another aspect of Kolmogorov's axioms is that it gives a seal of approval to the use of the same mathematics for both types of probability, that is, probability in the sense of the frequency of the outcomes of many repeats, and the probability in the sense of assessing the likelihood of a non-repeated event. Both aspects of this duality are described by the same axioms. Indeed, another look at the three axioms above will show that common sense will accept both interpretations of probability. As the mathematics is based solely on the axioms, the same mathematics serves for both cases.

What then does assessing the probability of a non-repeated event mean? The mathematical answer is given in the axioms and their derivatives. The day-to-day implications are a matter of interpretation, which is likely to be subjective. It is interesting that late in life Kolmogorov himself expressed doubts about the interpretation of the probability theory related to non-repeated events, but he did not manage to propose another mathematical theory to be used to analyze this aspect of probability.

Kolmogorov's book changed the way mathematics dealt with randomness. Concepts that had been considered just intuition became subject to clear mathematical definition and analysis, and theorems whose proofs had also relied on intuition were now proved rigorously. Within a short time Kolmogorov's model became the accepted model for the whole mathematical community. Nevertheless, as the reader who has not previously come across this mathematics can guess, the method Kolmogorov suggested was not easy to use. Moreover, the formalism failed to overcome the many difficulties and errors in the intuitive approach to randomness, because it is a logical formalism that the human brain is not set up to accept intuitively.

42. INTUITION VERSUS THE MATHEMATICS OF RANDOMNESS

When we react to a situation involving random events we use intuition that the human race developed over millions of years of evolution. As we claimed in the first chapter of this book, evolution did not provide us with the tools to think intuitively about situations involving logic. It is not just the difficulty in using an intuitive assessment of a situation in which logical consideration is the correct device to use; the naive use of intuition can lead to errors and even mental illusions similar to the visual illusions we examined in section 8. In this and the next section, we will analyze some of the common errors and illusions related to randomness.

We will start with a real-life example. Every blood donation is of course tested to ensure that the donor is not suffering from the HIV pathogen that causes AIDS. The error in the test is minimal, but it exists, and is about 0.25 percent. That means that there is a 99.75 percent chance that an HIV carrier will be correctly identified as such, but there is a quarter of one percent chance that the test will yield an incorrect result and the carrier will in error be declared healthy. There is an equal chance of a quarter of one percent that a perfectly healthy person will be incorrectly diagnosed as being an HIV carrier. A potential donor in a blood donation unit was tested, and the result showed him to be an HIV carrier. What is the chance that he really is a carrier?

The great majority of respondents answering that question (and I have asked it in different forums and of various audiences) assess that the chance that the person tested is a carrier is 99.75 percent, that is, in accordance with the possible error in the performance of the test. Few assess the chance as slightly smaller, generally without explaining why their assessment is lower than the figure we have set as the error. They presumably think that the correct answer is not 99.75 percent, otherwise why would they be asked such an apparently simple question? Very few give the right answer (and in general they have come across this or a similar question previously). What is the right answer? To arrive at the correct answer we must first examine what situations could have yielded a positive result in the test. The subject may indeed be an HIV carrier, and the chance of the test giving a positive result is very high, 99.75 percent. The subject may, however, be completely healthy, and the chance of the test giving an incorrect positive result is very small, 0.25 percent. Yet the population of healthy people is large compared with the population of HIV carriers, and the number of those whom the test wrongly identifies as carriers could be very large, larger than the entire population of actual carriers. In order to be able to assess the chances that the subject really is a carrier, we need to know one additional fact, and that is the proportion of carriers in the whole population. The figures published by the World Health Organization (WHO) show that in the developed countries HIV carriers constitute about 0.2 percent of the population, that is, one carrier per five hundred people. If we accept that figure, we can use Bayes's formula as described in the previous section. The formula weights the chances that a carrier will be identified as such in the test, compared with the chance that anyone, healthy or a carrier, will be found to be a carrier in the test. The calculation is

$$\frac{0.9975 \times 0.002}{0.9975 \times 0.002 + 0.0025 \times 0.998}$$

giving a probability of about 0.44, that is, the chance is only 44 percent that a positive result correctly identifies a carrier. If carriers were only 0.1 percent of the population, the chances fall to about 28.5 percent.

The fact that a donor's blood sample is tested, in other words the subject, that is, the donor, is chosen almost randomly, is an important fact of the analysis (it pertains to the assumption we made when introducing the formula in the previous section). If the subject had been sent for the test because he was suspected of being an HIV carrier, for example because he showed certain symptoms, the probability that he was really a carrier would be different; to find it we should use the original Bayes's scheme.

Why do most people asked the question consider that the chances that someone who gets a positive result in the test (as a carrier) really is a carrier is 99.75 percent? The reason lies in the way the brain analyzes the situation, a way that is inconsistent with the mathematical logical understanding of it. The brain perceives certain data and decides intuitively which are important, without undertaking an orderly analysis of the information. It does not look for missing information. Evolution instilled in us, or more precisely instilled in our subconscious minds, the recognition that it is generally not worthwhile to devote the effort necessary for a rigorous analysis of the problem. Therefore the brain concentrates on one prominent piece of information: the chance of an error is only a quarter of one percent.

This type of error is not confined to medical tests. Courts tend to convict someone who confesses to murder even without corroborating evidence. The reason that judges give is that the probability that someone will confess to a murder he has not committed is negligible. That fact is correct, but the statistical conclusion is not. To illustrate: assume that only one person in one hundred thousand will confess to a murder he has not committed (and taking into account the conditions of police interrogation that suspects undergo, that assumption is certainly not an overestimate). Assume also that someone is arrested randomly from a population of four hundred thousand, and he confesses to a murder committed the previous day. The chance that he is the real murderer is only 20 percent! The chance that the real murderer was found is only one in five (the murderer himself, if he confesses, and the other four of the population who would confess even if they have not committed the murder). With a larger population, the chances are even lower. The mistake the judges make lies in the fact that they only examine the chances that someone has been arrested who has a tendency to admit to a crime he did not commit; but once the suspect confesses, that probability has no significance. Once a suspect has confessed, what matters is how to differentiate between those who would make a false confession and the real murderer. That can be done by means of additional evidence or suspicious circumstances. Judges often overlook this distinction. In an article published in the Israeli journal *Law Studies* (*Mechkarey Mishpat*) in 2010, Mordechai Halpert and Boaz Sangero analyze the case of Suliman Al-Abid, who confessed to the murder of a girl and was convicted despite the fact that there was hardly any independent circumstantial evidence implicating him. Eventually it was found that he had been convicted wrongly. The article discloses the error in probability made by the judges and brings other examples from the legal area.

As an introduction to the next example, here is an amusing anecdote. A man is caught at an airport trying to board a plane with a bomb in his suitcase. His argument was as follows: "I did not intend to blow up the plane, but I heard that the chances of two people who do not know each other both taking a bomb onto a plane is much smaller than the chances of one person trying to do so, so I took a bomb with me to reduce the chances that the plane will be blown up." Clearly the logic employed by this passenger is flawed, but is it easy to pinpoint the mistake? Such errors occur, however, not as a joke, but in many situations we encounter in our daily lives. Here is another example.

This occurred in the trial of O. J. Simpson, the famous American football player accused of the murder of his ex-wife and her boyfriend. To support its case, the prosecution cited evidence that Simpson had beaten his ex-wife previously and had threatened several times that he would kill her. These instances had been brought to the attention of the police, and they had had to intervene on a number of occasions and remove him from her house. The defense brought the following probability counterargument. Reliable statistics covering thousands of cases showed that of those instances registered by the police, less than one-tenth of those who beat their spouses and threaten to kill them actually attempt to do so. Even fewer, less than one in a hundred, actually kill their spouse. The conclusion to be drawn, according to the defense, is that the chances that O. J. Simpson actually killed his ex-wife are less than 1 percent, a probability that constitutes reasonable doubt. The jury apparently accepted that claim, and even the judge did not comment on the basic error made by the defense. The defendant was acquitted. A proper analysis shows that the probability calculation presented by the defense did not take into account a basic relevant fact: Mrs. Simpson *was* murdered. If that undeniable fact is taken into consideration and the question is then asked that if a woman has been threatened by her ex husband and is then murdered, what are the chances that she was murdered by that ex-husband, the chances are very different from those presented by the defense. It would be wrong to conclude that the jurors were uneducated or ignorant. Evolution did not prepare us for proper analysis of issues in which the claims are logical arguments.

Here is another example that reflects a real-life situation. Six equally attractive girls reached the final round in a beauty contest. Each has the same chance of winning. The winner has been selected, but it has not been announced who it is, and the girls are on their way to the stage for the declaration of the winner. The last girl in line on the way to the stage, number six, cannot suppress her curiosity, and asks the guard, who knows who the winner is, to tell her. He answers that he is forbidden to reveal the result, but he does tell her that number one is not the winner. Number six is happy; her chances, she thinks, have just risen from a sixth to a fifth. Is she right? Most people asked the question answer that she is right, or they say that her chances remain as they were, one-sixth. The argument put forward by the first group is that five contestants are left, and the fact that number one is not the winner does not add any information about the others, so the remaining contestants' chances are equal, that is, a fifth. The others argue that as the guard had to mention one contestant who had not won, the fact that he indicated one of those who had not won did not add any information, and the chances of the remaining contestants remained as they were, that is, one sixth. These are intuitive answers. Very few people notice that one vital piece of information is missing, without which no reliable answer can be given! The story does not reveal the method adopted by the guard, that is, the algorithm or formula according to which he acted and that determined the *a priori* probabilities that he would indicate number one as a non-winner. Try to complete what is needed in Bayes's scheme, and you will see that there is insufficient information available for it to be used. Without information that details the possibilities in which the guard would point out the number one contestant as a non-successful participant, it is impossible to answer the question. It is easy to make up a story describing how the guard would behave so that the chances remain one-sixth (for instance, the guard chooses the girl he points out at random among those who have not won, excluding number six—we leave out the computation). But it is also possible to make up another story about his behavior that would increase the chances to one-fifth (for instance, the guard picks, among those who have not won, the one who gets on the podium first). It is even possible to depict other scenarios that will lead to other consequences. The human brain is not constructed in such a way that it searches for missing information. Such a search is wasteful from an evolutionary standpoint. The brain fills in what is missing from the story of the event with reasonable data. Sometimes this turns out to be correct, and sometimes not. This efficiency of the brain is justified in most day-to-day problems, but it is far from consistent with mathematical analysis.

The errors deriving from the difference between logical mathematical analysis and the intuitive way in which the brain handles such problems of probability can sometimes carry a heavy penalty. In many articles and in a book in the bibliography at the end of this book, the German psychologist Gerd Gigerenzer cites shocking examples of failed medical procedures and patients' traumatized reactions resulting from incorrect information regarding the outcome of laboratory tests. The message of Gigerenzer and others is that decision makers, including physicians, economists, politicians, and the like, must be taught how to act in situations of uncertainty, in other words, to absorb and implement Bayes's thought process. Is that possible? Gigerenzer's opinion is that Bayesian logic can be absorbed if, instead of analyzing problems using concepts of probability events, we train ourselves to think in the framework of repeats of a situation. In other words, we should exchange events as in the Kolmogorov model for considerations of relative frequency. Thus, according to Gigerenzer, if in the example above of the blood donor and his blood tests, instead of considering the one individual, we check a series consisting of many subjects, we would note that many of them, more than a half, are healthy subjects whom the tests incorrectly showed as being carriers. Gigerenzer actually presented figures showing a great improvement in groups of physicians who had learned to analyze in this way situations of randomness.

I find it difficult to accept Gigerenzer's conclusions. I think that the errors are basic and stem from intuitive thinking. People erring in a certain situation are less likely to make the same mistake if they encounter exactly the same situation again. Their decisions will not improve, however, if the uncertainty appears in a slightly different guise. Someone noting that most of the subjects recorded by the tests as being carriers are actually healthy could to the same extent implement Bayes's original formula. The only solution that I can think of to the problem of errors is that in those cases in which the error can result in much damage—in medicine, economics, assessments of intelligence data, and so on—one must strictly analyze the situation using mathematical tools explicitly and avoid intuitive thinking. If it is not particularly important that the right answer is reached, that is, if the potential error is bearable, it may be that the way evolution teaches us to react is acceptable and even preferable and more efficient. That may lead to errors in a number of instances, but it may solve other situations correctly and may save time and effort.

43. INTUITION VERSUS THE STATISTICS OF RANDOMNESS

Although evolution did not prepare us to analyze intuitively with logical elements situations of uncertainty, we could assume that we would react correctly to statistical situations. Throughout the whole of the evolutionary process, humans have been exposed to random occurrences. Nevertheless, even in these cases errors related to statistical randomness are repeated again and again; we mentioned some of them in section 39 on the mathematics of predictions and errors. Some of the errors can be explained by evolution itself. We will give a few examples.

The Ayalon Highway in Israel that traverses Tel Aviv and its suburbs is intended to enable vehicles to cross the city quickly. Shortly after a central section of the highway was opened with due pomp and ceremony, the Ayalon River overflowed due to very heavy rain, and the highway was flooded. This led to severe traffic jams, and the CEO of the Ayalon Highway Company was invited to appear on television to explain the reasons for the flooding. His explanation was convincing: To build a highway that would be immune to any possible flooding would be prohibitively expensive. The engineers therefore took a calculated risk and constructed a road with a wide margin of error, so that flooding was expected only once in twenty-five years. It was bad luck, he explained, that this flooding occurred only a short time after the highway was inaugurated, but that was the nature of randomness. He went on to calm the viewers that now they could look forward to a long period of flood-free driving on the highway. Exactly three weeks passed, and the highway was flooded again. The CEO was again invited to appear on television and with a crestfallen face mumbled something about independent and dependent events, without managing to persuade the interviewer that the engineers' calculations were accurate under the circumstances. The reason for the mistake is clear: in his first broadcast the CEO gave insufficient weight to the most important piece of information, that the highway had just been flooded. If a flood occurs because of extremely heavy downpours, the ground is saturated with water and even light rain may then cause a flood, in other words, the next flooding is not an event independent of the first.

The attitude to the significance of numerical values that the law of probability attributes to various events is not uniform or consistent. Some years ago there was a danger that the Sea of Galilee would flood its shores. The executive responsible for Israel's water sector explained on television that the chance of such flooding was 60 percent and went on to say that a miracle was needed to avoid it. Is an occurrence that has a 40 percent chance of taking place considered a miracle? I doubt it. And indeed, that year a "miracle" occurred, and the Sea of Galilee did not flood. To many doctors, an 80 percent chance of a patient's recovery and a 97 percent chance of recovery may seem similar, but to the patient who understands the law of probability, the difference is huge. A 97 percent chance of recovery means that the treatment is successful in all but a few cases. A 20 percent chance of failure indicates that failure is a systemic possibility.

The attitude to events with very low probability is also inconsistent. On the one hand, people buy lottery tickets although the trouble it takes outweighs the probable winnings. The reason is apparently the positive personal feeling of looking forward to a possible win, even though they know it is unlikely to be realized. On the other hand, the intuitive tendency is to ignore events that have very little chance of being realized. Sometimes this tendency is crucial, particularly in financial, economic, political, and similar matters. This tendency to ignore unlikely events may also be traced to evolutionary sources. In the broad framework of the struggle for survival, the means devoted to facing up to occurrences with a small chance of happening are made at the expense of the major efforts needed in the struggle for survival. For example, if dinosaurs would have developed gills that enabled them to breathe dusty air, they would have survived the meteoric dust that according to the generally accepted explanation engulfed the Earth and resulted in their extinction. On the other hand, a species of dinosaur that would have devoted efforts to developing such gills at the expense of the struggle for day-to-day survival may not have survived and may have become extinct before the meteor collided with Earth. The evolutionary struggle is one of here and now, that is, it takes into account only current conditions and ignores possible future events or events with a low probability of occurring. This fact has filtered down into the way we react to dangers that have a low probability of being realized.

Another error that may be called a mental illusion is related to the interpretation of statistical data, and it too may be traced back to evolutionary origins. As we explained in section 4, identifying patterns is an innate ability. Moreover, it is preferable to err on the side of overidentification. Failure to identify an existing pattern may bear a heavy price, compared with the damage that may be suffered through identifying a nonexistent pattern. The psychologist and expert on decision making Amos Tversky (1937–1996), together with his colleagues Thomas Gilovitch and Robert Vallone, decided to examine the "hot hand" belief in basketball. Every basketball fan knows of this phenomenon. When a player scores a number of baskets in successive throws, he, his coach, the opposing team, the spectators, all feel that he has a "hot hand" and that it is reasonable that he should also try to score in the future. In terms of the law of probability, the hot hand rule says that a number of successful shots at the basket increases the chances that the next throw will also be successful, in contrast with the case in which the same player under the same conditions did not score in his previous attempts at the basket. The situation can be explained, and the usually accepted explanation is the combination of achieving self-confidence with the psychological effects following a run of successes.

Tversky and his colleagues decided to examine the hot hand concept, and over a whole NBA basketball season in the United States they watched one of the most successful teams at that time, the Philadelphia 76ers, and recorded each shot at the basket and monitored the runs of successful shots. They discovered, to the surprise of many, that the hot hand was an illusion, a fallacy. In a random series, a run of successes can also occur without the chances of success in the next attempt increasing. The runs (or "streaks") of successful shots in the 76ers' games did not differ from those of a random series. The parameters of the series, that is the percent of successful shots, are likely to change from player to player and from one game to another, but under the same conditions the chances of a successful shot at the basket does not increase after a run of successful shots.

This finding should have immediate implications because a player's hot hand, if it does not exist, has a direct effect on how the coach manages the team during the game. Tversky and his colleagues' findings met with a mixed reception. They had no effect on the spectators, the players, or the coaches, who continued to believe in the hot hand, and who continue to act accordingly. Opinions in the scientific community are divided. Some accept the findings as they are, and others think that the hot hand phenomenon does exist but is expressed differently. I do not know whether it exists or not, but there is a simple explanation for the illusion: the need to look for and find patterns is deeply embodied in our genes, so much so that events such as a run of successes, or successful shots at the basket, or a few consecutive years of hotter than usual weather, or successive stock-exchange profits, a run that is consistent with the statistics of random events we interpret as valid nonrandom occurrences.

CHAPTER VI
THE MATHEMATICS OF HUMAN BEHAVIOR

Does the Consumer Price Index cause sunspots? • Are there optimal marriages? • Game theory or conflict theory? • How much would you pay for a lottery ticket that is expected win a million dollars? • Is it irrational to throw money into the trash bin? • Is someone who believes everything "simple"? • Can one arrive at a decision without preconceptions? • What is evolutionary rationality?

44. MACRO-CONSIDERATIONS

Since the dawn of history man's behavior has been the subject of analysis and debate in various spheres: literature, art, law, and political and philosophical studies. Yet, the use of a mathematical approach to describe and analyze people's conduct and decisions began only toward the end of the eighteenth century. In this chapter we describe some of these developments.

Human conduct, particularly in economic matters, can be divided into individual behavior and group behavior. Clearly the two are connected, as individual behavior determines group behavior. Yet, in economic issues it is still difficult to find a mathematical model that can provide a quantitative prediction of how global economic parameters follow from the decisions of individuals. It was the Scottish philosopher and economist Adam Smith (1723–1790) who coined the phrase *the invisible hand*. He presented the concept in his book *An Inquiry into the Nature and Causes of the Wealth of*

Nations, published in 1776. In this book Smith laid the foundations of the theory of capitalism: every individual tries to maximize his own welfare, without regard to the needs of the public, and an invisible hand translates those individual actions such that they improve the situation of the society. The nature of the invisible hand remained unexplained. The first explanations did not appear until the 1950s, when economists began systematically to base the theory of capitalism on defined fundamentals; however, this approach met with only very limited success.

On the face of it, behavior in which every individual is concerned only with himself is highly consistent with Darwin's concepts of evolution, as the evolutionary struggle leads to competitive conduct. A closer look, however, reveals that competition in nature is not between individuals but between species. The victorious species are those that survive for generations, and they are not necessarily the species in which every individual fends for itself. A species may be able to survive because its members are prepared to sacrifice themselves for the common good. Such an evolutionary analysis showing the link between individual behavior and the success of the group does not yet exist with regard to economic conduct of communities. Furthermore, the performance of a large economy is largely the result of the decisions of many individual decision makers, each one of whom has little or negligible effect. In this sense there is a similarity with the mathematical description of nature. No quantitative mechanism for the invisible hand has been discovered that combines the elementary particles that have wave characteristics into an element that fulfills Newton's laws.

The mathematical tools used today to analyze macroeconomic activity are not essentially different from those developed for understanding how nature functions. These tools include equations of different types, for example, differential or other equations that deal with economic quantities such as consumption, savings, and interest rates. The equations are meant to describe how an economy works and not what type of economy we might be interested in. An analysis of the model sometimes may help us understand what steps the fiscal or monetary policy makers ought to take to achieve the desired objective. The model itself, however, describes the economy as it is. As in a social world, we cannot perform proper controlled experiments, so economists use data provided by bureaus of statistics. The technique used to analyze these parameters is known as *econometrics*, which is a development of the statistics that we described in the previous chapter. The methods developed for purposes of economic analysis are very advanced, but the approach is not essentially different than that which developed over the years for use in the natural sciences and technology. Success in describing macroeconomic conduct is lagging behind the level of the success of mathematics in describing physics and its uses in technology. Is this merely a question of time, and will the gap be closed with the improvement of the existing models, or is there possibly a need for a new mathematics to describe human behavior? There is no unequivocal answer.

We will not describe the macroeconomic models in detail. We will just present two examples of considerations specific to social sciences. Both are related to the recipients of the Nobel Prize in Economics in 2011. The official name of the prize is the Sveriges Riksbank Prize in Economic Sciences in Memory of Alfred Nobel (the Sveriges Riksbank is the central bank of Sweden). Alfred Nobel did not specify the social sciences as one of the fields in which a prize would be awarded. We cite these examples because of the special way in which mathematics is used to describe the complexity of human conduct, and they do not reflect the entire range of uses of mathematics in macroeconomics.

An inherent element of human decision making is the assessment of what is likely to happen in the future, when in many cases the individual considers that his influence on the future is negligible. For years there was an understanding that macroeconomic developments are affected by individuals' expectation regarding the future, but that understanding was not translated into a component in equations. Robert Lucas of the University of Chicago, who was awarded the Nobel Prize in Economics in 1995, and colleagues who included Thomas Sargent of New York University, winner of the Nobel Prize in Economics in 2011, developed the theory of *rational expectations*. They found a mathematical way to incorporate market expectations in an equation determining the process of development of the economic parameters. These expectations are part of the variables in the model; they influence the other variables and are influenced by them.

Market expectations that affect the development of the market are clearly specific to social sciences. The formulation that describes in mathematical terms the role of such expectations in future developments was adopted by economists and constitutes an integral element in many macroeconomic models.

The second example is merely anecdotal and should not be considered as representing econometric practice. We chose this example because it teaches us something about the link between mathematics and uses. The joint winner of the Nobel Prize in Economics in 2011 with Thomas Sargent was Christopher Sims of Princeton University. The prize citation refers to Sims's contribution to the analysis of time series, that is, statistical series that change with time. The analysis of statistical series in general, and those that change with time in particular, have long interested scientists, and the basis for the mathematical methods of analyzing the errors in these series dates back as early as in the time of Carl Friedrich Gauss. Among other advances, systems were developed for finding whether two series of data were correlated, and that gave quantitative indices for the degree of correlation. In their use in the natural sciences, however, the question did not arise as to which of the two series was dominant, that is, which caused changes in the other series. For example, there is a causal relationship between the Earth's revolving on its own axis and high tide and low tide. Yet no one looked in the two data series for the answer to the question of what is the cause and what is the effect, does the tidal flow cause the Earth to revolve or vice versa. The answer is derived from the laws of nature themselves. In general, in natural sciences we do not try to derive the cause and effect from the two data series themselves, but we try to derive them from their underlying model. Unfortunately, the mathematical models of social and economic occurrences are not reliable enough to enable a similar analysis to be performed. It is natural, therefore, that attempts will be made to derive the cause and determine which of the two series is dominant from the series themselves. Among Sims's contribution to understanding time series was his enhancement of a method proposed by the British economist Clive Granger (1934–2009), a 2003 Nobel laureate in economics. The method was supposed to determine which of two data series was the cause of the other. The test is known as the Granger-Sims causality test and serves to test and examine causality in many areas of social science and economics.

In 1982 two economists, Richard Sheehan and Robin Grieves, published results of the use of the Granger-Sims causality test to examine possible causality between the appearance of sunspots and the business cycle in the US economy, related to both gross national product (GNP) and the price index. The article was published in the *Southern Economic Journal* (volume 48, pages 775–78). The results were statistically significant and showed that the business cycles in the American economy is the cause of the sunspots. It is clear that this result is inconceivable. Yet to make it quite clear, it does not disqualify the statistical test. What should be learned from it is that we should not rely on a statistical test that is not supported by an independent model. As there is no model that incorporates the effect of GNP on sunspots, the statistical analysis that examines this is not applicable. Such use of a statistical test is limited. The right approach is first to propose a model representing a causal relation, and then to leave it to statistics to confirm or refute the model. A statistical test alone without a possible model of the effect itself can lead to fundamental errors.

45. STABLE MARRIAGES

We will now present an example of a mathematical analysis of the possible actions of a group of people. We use the term *possible* actions and not *desired* actions or *recommended* actions. We will explain the reason for that later on, and here we will just note that following the successful experience of using mathematics in the natural sciences and technology, many hoped that in the social sciences too, mathematical analysis would show how a society should act. The current mathematics of human behavior is very far from being able to fulfill such hopes.

The example relates to an issue that arises in various situations in our lives. Graduates of medical school look for hospitals in which they would like to work in their chosen specialty, and the hospitals are looking for new interns. The graduates have their individual preferences regarding the hospitals, while the hospitals have their preferences relating to whom they would like as interns. How can and ought the aspirations of the two sides be matched? In a similar vein, how should candidates for a teaching position in a university be selected, how should footballers be taken on to join the team squad, and even how should brides and grooms be matched by a matchmaker?

The natural question that arises in this context is what is the best way of filling available positions. To answer this we need to define the criteria for "best," and then to find a way to arrive at the solution. Two famous mathematicians, David Gale (1921–2008), of the University of California, Berkeley, and Lloyd Shapley, of the University of California, Los Angeles, tackled this problem using the following mathematical approach. It earned Shapley the Nobel Prize in Economics in 2012.

As this falls within the sphere of mathematics, we should first clearly define the framework of the discussion. Here we will focus on a particular instance, but it is not difficult to extend the solution, when we reach it, to other cases that are closer to reality. Assume that we have a group of N women and a group of N men, and our task is to find a match for each member of the one group with a member of the other. Each man has his priorities regarding the qualities he would like his potential partner to have, and likewise with the women. There is not necessarily any correlation between the preferences of the members of the groups, and if we were to match them randomly, many would be unhappy. In effect, in any system of matchmaking between them, there are likely to be men and women who would not end up with their ideal match and who may even be matched with someone very far from their initial preferences. In such a situation, what is the optimal match? In our context, "match" means the overall matching of all the men and all the women.

The first contribution made by Gale and Shapley was to change the question! Instead of trying to find a criterion for optimality, they formulated a condition, which in the days of the ancient Greeks was called an axiom, that the match had to satisfy. They called this the stability condition. It can be described simply, as follows. The match will be considered not stable if a man and a women can be found who prefer each other to the partner that the system proposed for them. The condition that Gale and Shapley set was that the match should be stable, that is, it should not be unstable. The reason for that condition is apparent: a match that is not stable will not last. The man and woman who can improve their situation by getting together will do so, and it will disturb the order that the match-making had determined. The next stage is to define a match as optimal if every man and woman is matched with the best partner among the stable matches. In other words, first of all the match must be stable, and among the stable matches the match has to be the best for every man and every woman. In a paper published in 1962, titled "College Admissions and the Stability of Marriage," Gale and Shapley presented an algorithm that led to a stable match. The algorithm can be stated without formulae or equations, as can the proof that it results in a stable match. We will show the algorithm in a quasi-visual form, although clearly, in the computer age it can be produced in an instant, even if it relates to very large populations.

In the first stage every man places himself next to the woman who heads his list of priorities. Some women may have more than one man standing next to them. Each woman selects the man she prefers from those standing next to her, and she sends the others back to their places. In the next stage every man rejected by his first choice, his top priority, stands by the woman who is next in his list of priorities. Again there may be some women with more than one man near them, and each one of these selects the one she prefers from among those men. It may be that the one she chooses was the one she chose in the first stage, but it could also be that one of the newcomers around her now tops her list. The others she sends back to their places. This continues until there is only one man next to each woman. The algorithm ends, and the result is a stable match. The proof that this is so is that as long as there are at least two men by any woman, the one who is rejected will at the next stage go to the woman who is lower in his order of priorities. The number of such "declines" in the priorities of every man is finite, so that the process is completed when there can be no further declines, that is, there is only one man by each woman. The stability derives from the fact that at each stage the women either stay with the man who chose them previously, or they select a man higher in their own scale of priorities. If a man prefers a woman with whom he is not matched at the end of the process, the fact is that he had already offered himself to her at an earlier stage, but she had preferred to remain with a man who was higher in her own priorities. That woman will therefore not prefer him to the man whom the procedure matched with her, and hence the stability.

What about the optimality conditional on the stability condition, as defined above? Gale and Shapley showed that there are instances in which it is impossible to find an optimal match. They also showed that the algorithm we have described has the characteristic of partial optimality, meaning that every man gets the woman who is highest in his order of priorities from among those he could be matched within a stable match. Herein too lies the difficulty with this particular proposal. If at the outset, instead of every man approaching the woman he preferred we had decided that every woman would approach the man who headed her list, and so on, that procedure would also result in a stable match, which might be different from the final outcome in the process we described. It will be partially optimal in as much as every woman gets the man highest on her priority list from among those with whom she could have been matched as part of a stable match. Which of the two possibilities is preferable? Mathematics does not answer that question; it just offers alternatives and describes their features.

Gale and Shapley's analysis and algorithm had a marked effect on theoretical research into staffing and choice, and also on the practical side. The basic algorithm and its results were extended to more complex frameworks much closer to reality, and they were also extended conceptually, with proposals for changes and improvements. For example, it can readily be seen that when this algorithm is used, there is no advantage to either side in not revealing his or her real priority. This is in contrast to various situations in social conduct in which it is worthwhile to someone participating in a procedure to lie regarding his or her private preferences.

Several institutions have adopted classification and selection procedures that are consistent with Gale and Shapley's proposal, including hospitals selecting new interns. It is interesting to note that in both of the above instances the institutions chose to adopt the algorithm that was optimal from the candidates', interns' or students' point of view. Their reason for this choice was apparently social. The institutions rated it more important that the candidates should feel that they had obtained the best possible position, consistent with the definition of stability, than that the institution itself should feel the same. Other institutions select candidates using different methods. American universities, for example, send candidates for faculty positions offers to which they must respond within a relatively short period of time; they do this to enroll their chosen faculty before they receive offers from other universities that they might prefer. The result is instability, as Gale and Shapley's basic assumption predicts.

Joint winner of the Nobel Prize in Economics in 2012 was Alvin Roth of Harvard University, who has moved since then to Stanford University. He developed an application of Gale and Shapley's method to match candidates for organ transplants with organs available for transplant, taking into consideration the quality of the match, the chances of success, personal details, and so on. Roth's research addressed also the issue of market design, namely, examination of the rules and procedures related to organ transplants, aiming at rules that take into account public beliefs and faith, yet increase the number of potential donors, a nontrivial goal that has deepened the use of the mathematical principles.

It is also interesting to compare the algorithm derived by mathematicians with the solutions to the problem of matching that nature arrived at via evolutionary development, at least in those cases where the couple must invest effort in establishing an economic base for themselves, such as a home for a human couple, or a nest for mating birds. That is why we can observe the stability characteristic in many species of animals that remain with their mate for many years, and even extreme cases in which their original choice of mate remains their lifelong partner. Nature, however, found ways of reaching stability that differ from the algorithm proposed by Gale and Shapley. In many types of animals, stability, in the form of loyalty to one partner, is embedded in the animals' genes. Those that maintained their relationship survived, and now that whole species has the characteristic of maintaining the stability of their relationships. Another method, which can be seen in an acute form in human societies, is the imposition of difficulties or fines for breaching a partnership. If the very act of separating from a partner causes unpleasantness for the disloyal individual or for the couple that decides to split up, the tendency to preserve stability will grow.

46. THE AGGREGATION OF PREFERENCES AND VOTING SYSTEMS

The roots of the second example we will present trace back to the beginning of the use of mathematics to understand human conduct. The marquis de Condorcet, Marie Jean Antoine Nicolas (1743–1794), of the de Caritat noble family, was a mathematician and a political philosopher of social science in France. Together with his interest in mathematics, in which he excelled from an early age, he involved himself in questions of society and economics. In contrast to many political leaders and thinkers of his time, he publicly expressed radical liberal views, including support for the equality of women and blacks. He reached the exalted position of secretary of the illustrious French Academy, and he was one of the leading supporters of the French Revolution. He then came into conflict on ideological grounds with some of the leaders of the revolution, and like many of his intellectual colleagues he had to flee and hide from the authorities, but eventually he was arrested. The marquis de Condorcet died in jail under mysterious circumstances, and it may reasonably be assumed that he was murdered.

The marquis's mathematical background is quite apparent in his social and economic writings, and with some degree of justification he is credited with initiating the mathematical approach to social and economic issues. In the spirit of the time, Condorcet tried to examine different systems of election in a democracy. In his writings he constructed and presented a very basic example of the difficulty of agreeing on a good electoral system. The example is sometimes referred to as *Condorcet's paradox*. The situation he examined was that in which a group of voters have to reach an agreement on one individual from a number of candidates. Let us examine the criterion, said the marquis, in which the majority decides. In other words, one candidate is preferable to another if most of the electors prefer him or her. Does this relation of preference define a victorious candidate?

In his answer, Condorcet described the following example. Three electors have to select one representative from among three candidates, whom we shall denote A, B, and C. Each of the electors has his own rating of the three candidates. The first elector rates A above B, and B above C. The second elector prefers B to C, and C to A. The third elector prefers C to A, and A to B. Now we will see the result arising from the criterion we set. It can easily be seen that according to our criterion A is preferable to B (the first and the third electors, i.e., the majority, preferred A to B), B is preferable to C (the preference of both the first and the second electors, i.e., again the majority), and C is preferable to A (according to the second and third electors, a majority). The result is that despite each elector having a clearly defined order of preference, the system of the majority deciding leads to a circular position with no winning candidate indicated.

Condorcet's paradox illustrates the difficulty inherent in adopting the criterion of decision by majority. The marquis himself defended the system and tried to promote it in those situations in which it was applicable. He even suggested and strongly urged the adoption of the algorithm that would lead to a situation that if there is a preferred candidate according to this criterion, that is, a candidate who is preferable to all other candidates, then he should be the elected candidate.

A contemporary opponent of the marquis, Jean-Charles, chevalier de Borda (1733–1799), proposed another system. He suggested that every elector should rank the candidates according to his preferences, and the sum of the points, or their weighted combination according to an agreed system, would determine the victorious candidate. This method is still in current use in many situations in which candidates have to be ranked in accordance with the electors' preferences. Clearly, Borda's method does not comply with Condorcet's criterion.

Attempts to find a better or fairer system continued for many years, until in 1951 Kenneth Arrow, a mathematical economist from Stanford University, presented a result that puts the issue of aggregate preferences in a surprising new light. Arrow was awarded the Nobel Prize in Economics in 1972 for this result, among others. Arrow's result also has implications for an individual in complex situations. We will present this interpretation toward the end of this section.

Arrow chose the axiomatic approach. Instead of proposing and analyzing actual systems of choice, he formulated a number of requirements that a system of aggregate preferences should satisfy. The framework is similar to that of the marquis de Condorcet, that is, every elector has his own grading of the candidates. The system has to yield a rating of the candidates that will eventually reflect the will of the group of electors. The requirements, listed below, are quite minimalistic.

1. *Independence of irrelevant alternatives*: If one of the candidates withdraws, and this does not affect the ratings of any of the electors with regard to the remaining candidates, the rating that the system gives to the remaining candidates also stays unchanged.
2. *Unanimity*: If all the members of the group prefer candidate A to candidate B, the group ranking that the system yields should also prefer A to B.
3. *No dictator*: No individual elector, let us call him a dictator, is in the position that the system always chooses according to the dictator's preferences regardless of the preferences of the other electors.

Reservations may be expressed about these requirements, but it should be borne in mind that they really are minimalistic. Requirement 2, for example, is much easier to satisfy than Condorcet's parallel requirement, according to which it is sufficient for a majority of the electors to prefer A to B for that to become the group preference. Arrow requires only that if *all* the electors prefer A to B, then that will be the group choice. If it is not the case that all members of the group share the same preference for A over B, then the second requirement places no restrictions on the final rating. Similarly, if the withdrawal of one of the candidates does change the rating of one of the electors, the first requirement does not impose any restrictions on the final rating. The no-director condition prevents the situation in which the preference of the dictator is the decisive factor even in cases where the other electors, however many they are, have preferences different than those of the dictator.

Arrow's surprising result, referred to as *Arrow's impossibility theorem*, states that there is no system of selection that satisfies the three requirements (when there are at least three electors and at least three candidates).

The result, whose mathematical proof is not at all difficult, had a great impact on social scientists. Mathematics drew the boundaries of what could be achieved. Research is continuing, of course, in different directions. For example, in the framework discussed above, all possible private orders of priority are taken into account. We could consider a framework in which only certain orders of priority are considered, and then it might be possible to fulfill Arrow's axioms. Attempts have also been made to use other axioms. On the other hand, generalizations have been worked out that show that the impossibility theorem is far wider ranging than in Arrow's restricted example of methods of choice. There were also attempts to present selection methods in which the axioms are fulfilled with regard to the preferences of most of the electors. This subject became a popular area of research in the mathematics of social sciences; it is called *social choice*. The study and research have not led to any clear conclusion as to the ideal system of choice, but they have presented tools that help in the examination of the appropriateness of the different systems to given situations.

Did the research into methods of choice have an effect on day-to-day life? Not much of one. It is true that in a few instances, particularly when policy makers took the trouble to consult with experts, it can be seen that the limitations of the methods of choice were taken into account.

In most cases, however, that is not so. The scientific council of the institute in which I work comprises about two hundred professors, each one a leader in her or his scientific field. On one occasion this body had to decide between two alternatives, and from the discussion it appeared that both had the same chance of being selected. One of the professors, wishing to make the motion he preferred more attractive, proposed a small change in the wording of the proposal. The chairperson, who also favored the same course of action, immediately accepted the idea and announced that the choice would be between the three possibilities, and whichever received most votes would be accepted. The chairperson apparently did not realize that by so doing he blocked the chances of his preferred choice, because the votes of those who supported that choice would be divided between the two similar proposals (one of the original two and its slightly amended version). In the discussion someone pointed out that it would not be right to vote on three options, but in the hubbub that ensued it was impossible to explain why. The argument that helped to prevent the proposed vote taking place was a comparison with a parliament. Parliaments, it was explained to the chairperson, always choose between two alternatives, they either accept or reject a proposal, and the scientific council acted like a parliament. However, the system whereby one always chooses between two alternatives also has drawbacks. For example, the order in which the alternatives are put to the vote has a marked effect on the outcome. Let us look again at the example of the marquis de Condorcet. Assume that the first vote is between A and B, and the winner then stands against C in a second vote. The result will be that C is successful. If the first vote is between B and C, and the winner stands against A, A will win. This gives considerable power to whoever determines the agenda.

Some interpretations of Arrow's result are indeed far-reaching. In some cases the blame may be put on a formulation of the result that, from the logical aspect, is equivalent to Arrow's result but is such that it is likely to divert the users in very different directions. Instead of claiming that there is no system of voting that satisfies all three of the requirements listed above, it can be claimed that if we want to fulfill requirements 1 and 2, we must agree that the rating will be determined by a single elector, a dictator. As axioms 1 and 2 are so elementary, the distinction is interpreted as meaning that the options are either allowing for a dictator or ignoring the mathematical analysis. I can illustrate this with a real instance that I witnessed.

The public mood in Israel in 1976 brought about the establishment of a new political movement, the Democratic Movement for Change. In the course of its establishment, those active in setting it up had to decide on the method for selecting the list of candidates that the movement would put up in the forthcoming election for the Knesset (Israel's parliament). They turned to mathematicians and physicists for help in choosing the best system, and these advisors came up with a sophisticated system. We will not describe the details of the proposed system, but we will observe that even a quick look showed that it would be easy for an organized group within the party to obtain far greater representation on the list of candidates than their relative size warranted. When this was brought to the attention of the system's architects, they dismissed it offhandedly, repeating the claim mentioned above that there is no need to relate to Arrow's result, because if it were adopted, it would result in the list of candidates being chosen by a dictator, in total opposition to the founders' stated aim of creating a democratic movement. They added that they had carried out simulations of possible outcomes, and those did not indicate any minority group taking control. When the actual internal elections took place, two groups took advantage of the system and obtained representation far in excess of their relative size within the movement. This eventually led to the disintegration of the movement, although in the general election to parliament it was quite successful in the number of its candidates that were elected to the Knesset. One of the movement's founders summarized the events in a book and wrote more or less that the organized groups did not vote as expected. In effect, he and his colleagues who performed the simulations did not show much understanding of ways of voting.

As mentioned, Arrow's mathematical theorem can be interpreted also in the framework in which an individual has to decide on his priorities in complex situations. One way of dealing with a list of possibilities that we must grade, say a list of places to visit that we are checking before the next holiday, is to draw up a list of criteria against which we rate the options. The criteria will probably include such features as the cost, the amount of pleasure it will give us, the physical effort required to get there, and so on. Each criterion will have the list of possibilities as we have graded them, and we must decide on a combined rating based on those for the separate criteria. Arrow's requirements will then look as follows:

1. *Independence of irrelevant alternatives*: If one of the choices is ruled out without affecting the ratings of any of the criteria of the remaining options, the grading that the system gives to the other options also remains unchanged.
2. *Unanimity*: If one of the options, say A, is rated higher than option B in all the criteria, then the system will rate A above B.
3. *No dominance*: No single criterion is dominant, that is, such that the final rating will always be the rating in that criterion, regardless of the ratings in the other criteria.

The interpretation of the three requirements is similar to that given in the case of choices. In the current instance the third requirement has no social aspect or implication. Indeed, if one of the criteria, say the price, is so dominant that it determines the overall rating, the rating problem becomes simple. Arrow's theorem says that it is impossible to rate the options in such a way that the three criteria are satisfied at the same time. How then can alternatives be rated? The discussion above, regarding choices, applies here too. For example, we can adopt Borda's approach of assigning points to each of the criteria and combining the points of the different rankings to obtain a group ranking. That is the method commonly used in most combined ratings we usually encounter, even though in that method not all of Arrow's conditions are met.

47. THE MATHEMATICS OF CONFRONTATION

In this section we will discuss the mathematical analysis of methods of decision making by humans as individuals. What is special in a social framework is that the decisions of any particular person may clash with the activities of others, and the person making the decisions must take into consideration what his colleagues or opponents will do. At this point we can distinguish between two different scenarios. In one, the impact of the actions of the individual decision maker on say the market situation will be negligible. In the other, the actions of each one of the individual decision makers could influence the final outcome.

In the first case the decision maker assesses the circumstances he faces, and in light of those circumstances he assesses the results of the different decisions he can make and chooses what seems to him the best. The function of mathematics in this process is to construct a mathematical model that can be analyzed quantitatively, and to propose methods of reaching the optimal decision. The mathematical subject is called optimization. We will not expand on it here, because the mathematical methods it uses are not essentially different from those we have encountered previously. We will nonetheless mention the outstanding example of such a mathematical development that had a significant impact on decisions in the capital market. We are referring to the Black-Scholes model and the Black-Scholes formula, which analyzes the risks and chances incurred in investing in the capital market.

The economists Fisher Black and Myron Scholes published a paper in 1973 in which they put forward a mathematical framework that enabled them to analyze the risks involved in investing in options in the capital market. Following their paper, Robert Merton published a paper in which he extended the system and developed the mathematics to the level that it became a tool in the hands of investors in the stock market. The mathematical tool employed in the model is differential equations, which we came across in our discussions of models of natural phenomena, with the difference that the variables are not location, energy, and speed, but prices and rates of interest, and so on. The model was put to routine use by investors in the stock market, and in 1997 Merton and Scholes were awarded the Nobel Prize in Economics. Black died in 1995.

In the second of the two scenarios outlined above, the decision maker takes into consideration the reactions of other individual decision makers to his own decisions. They too act in light of their assessments of the actions of others. In this situation the use of the word *optimal* may sometimes mislead. For example, take the situation in which several people make their individual decisions, and the outcome is determined by their decisions combined. What is optimization in this case? Optimal for one participant may be terrible for another. If one of the group members knows or assesses the decisions of the others, he will be faced with the optimization problem described above. But then each of the others can estimate the choice of that same individual and can change his own choice accordingly. In which case, the first can alter his decision in light of the new situation, and so on. It is then not clear what the optimal decision is. The mathematical field that analyzes such situations is *game theory*. The rather simplistic name is likely to be misleading because this branch of mathematics is dealing with a very serious subject, the analysis of confrontation. On the other hand, the name *game theory* has become so accepted by the public that it is commonly assumed that the term *game* in this context means confrontation.

The situation that we have described is relevant also to conflicts between decision makers in daily life as well as to social games such as chess. A mathematical analysis of the theoretical possibilities in a game of chess was carried out in 1913 by the German mathematician Ernst Zermelo (1871–1953), famous for his contribution to the foundations of mathematics, whose paper titled "On an Application of Set Theory to the Theory of the Game of Chess" introduced new concepts to that area of research and provided the source of the name *game theory*. Other well-known mathematicians continued to develop that field, including the French mathematician Émile Borel (1871–1956), who in 1921 introduced the mixed-strategy concept, and John von Neumann (1903–1957) who in 1928 proved the minimax theory. We will meet these two concepts again later in this section. Zermelo's paper dealing with social games led to the development of the subject, which analyzes conflicts between individual decision makers, be they private individuals, company directors, or military and political leaders. The insights reached by game theory have been used since then to analyze situations of people and companies and to understand conflicts of interest between animals. With animals we cannot identify conscious decision making, but the process itself takes place as if someone were making deliberate decisions. In particular, viewing the evolutionary struggle itself as a conflict between species helps in the analysis of the evolutionary process.

Just as in other areas of specialization that mathematics uses to describe and explain phenomena, before using them we must specify precisely what framework of mathematics we are working in. One of the models in game theory is called *games in strategic form* (we shall allude to another model, namely cooperative games, later in the section). This is a game between several players, each one of whom must choose one of a given number of possibilities, called strategies. The choice is made simultaneously, and when each participant makes his decision, he does not know what the other players are choosing. The game ends when the decisions of all players have been received. Next, each player receives a "payoff" that depends on the combination of all the decisions taken. All the players know the payoff and its dependence on the strategies in advance. The payoff can be monetary, but it can also be in another form, assuming that the player has a full rating of preferences for the rewards he is entitled to receive. Clearly the strategic-form model does not cover all conflict situations between players. We will limit the mathematical analysis by imposing the condition that the number of strategies facing each player is finite (for discussion purposes only; the professional literature also analyzes situations with infinite possibilities). Our purpose at this point is to arrive at an understanding of what mathematics can offer at the conceptual level, and that can be achieved with a simple model. If we wish to use the results to analyze day-to-day situations we will need to check the extent to which the circumstances match the mathematical model.

Even at this level of presenting the definition of a game, we can propose "obvious" properties of the players' possible decisions. For instance, assume that a player recognizes the following property of one of the strategies facing him, say strategy A: for every possible move by the other players, strategy A yields him the highest payoff. It is then proper for us to refer to the decision to adopt strategy A as the optimal decision. In game theory, such a strategy is called a *dominant strategy*. It stands to reason that a player who identifies a dominant strategy among the options confronting him will adopt that strategy. If each of the players has a dominant strategy, we have "solved" the game. However, players do not always have a dominant strategy. In those cases the best response to one's opponents' moves is achieved via different strategies.

Another possibility that a player can choose is to look for a maximum-minimum strategy. In other words, the player can calculate the lowest payoff that every strategy might yield him and choose the best of those low payoffs. This concept describes behavior that settles for minimizing possible losses or that reflects concern over the worst-case scenario. In many situations the use of these strategies does not yield reasonable outcomes.

A crucial step in the analysis of possibilities in a game in strategic form was made by John Nash of Princeton University, New Jersey. Nash is known by the general public mainly because of the book and subsequent film, *A Beautiful Mind*. The book, written by Sylvia Nasar, recounts Nash's life history, from his graduate studies at Princeton University in 1948, his illness, his disappearance from the world of research, and his being awarded the Nobel Prize in Economics in 1994. Nash, who also made significant contributions in many other areas of mathematics, proposed the following definition.

> *Assume that every player chooses one of the strategies open to him. This collection of strategies is in equilibrium if no player can by himself earn a higher payoff by changing his strategy while the other players keep to their original ones.*

The rationale underlying his definition was that if the players agreed on a choice of strategies that are in equilibrium, or if they discovered somehow that the other players would choose strategies in equilibrium, there is no incentive for any of them to switch from that strategy unilaterally.

Nash did not put forward his definition in a vacuum. Exactly the same concept had been proposed more than a hundred years earlier by the French mathematician and economist Antoine Augustin Cournot (1801–1877), who made valuable contributions to economic theory in several areas. Cournot defined the concept of equilibrium between two firms that constitute a duopoly, that is, the two firms control the market. As Cournot formulated the concept in a relatively complex model, his definition lacks the simplicity and clarity of Nash's definition, but nevertheless many show their respect for him by referring to the concept as Cournot-Nash equilibrium. Others, after Cournot, also used the same concept in various forms, but it was Nash's precise formulation that led to the recognition and widespread use of the concept. Nash went further and proved the existence of equilibrium in a framework we will describe a little later, but first we will give three examples of the concept of equilibrium that are standard in any textbook on game theory.

The first example is known as the *prisoner's dilemma*. Despite its eye-catching name, the game reflects situations we encounter frequently in trade, economics, our social lives, and so on. The dilemma is between cooperation that would to some extent benefit both players, on the one hand, and noncooperation that would be advantageous to one player at the expense of the other. That is indeed a common situation, but in our case each participant must make the decision without any communication with the other one. The mathematical version of the tale is of two suspects who committed a crime, but the police have insufficient evidence for a conviction without one of the suspects incriminating the other. The police therefore make an offer that if one of the suspects will testify against the other and the other continues to claim he is not guilty, the cooperative suspect will go free and the other one will be sentenced to four years in prison, the punishment prescribed by law. If neither agrees to testify against the other, it will be impossible to prove the more serious charge against them, but they can both be convicted of a lesser offense that carries a one-year prison sentence. If they both accept the offer and each agrees to testify against the other, they will both be convicted but their testimony will earn them a reduced sentence of three years. Each suspect is faced with two possibilities, or in the terminology of game theory, two strategies. One option is to testify, and the other is to refuse the offer, that is, to deny the charges. Each must make the decision without knowing what the other will do. It is customary to summarize the possibilities in the following table.

	Prisoner 2	
Prisoner 1	0,4	3,3
	1,1	4,0

The rows in the table represent the possibilities, that is, the strategies, facing the first suspect, and the columns represent those of the second suspect. The top row shows the position when the first suspect testifies against the second, and the second row is the situation when he refuses to do so. The left column shows the position when the second suspect refuses to testify and continues to plead not guilty, the right shows when he agrees to testify against the first. The two numbers in the cells of the table show the number of years each suspect will be sentenced to as a result of their decisions; the number on the left refers to the first suspect, and that on the right refers to the second.

The reader will readily see that the table reflects what was described above verbally. Each suspect wants to minimize the number of years he will have to spend in prison. When the concepts described above are applied, it is easy to see that testifying against the other suspect is the dominant strategy for each of them. In other words, for each suspect, testifying against the other is worthwhile, regardless of what the other decides to do. Specifically, the pair of strategies in which each suspect agrees to testify is in equilibrium. According to our description of dominant strategies above, we have solved the game. Each suspect will agree to testify against the other, and the outcome will be that each will serve a three-year sentence, whereas if neither had agreed to testify they would each have served only one year.

Have we then solved the game correctly? We will return to this question further on, and here we will note that we have solved the mathematical game, but the mathematical game ignores many aspects of conflicts that we come across in our daily lives.

The second game is known as the *battle of the sexes*. Presented in table form, it looks as follows:

Wife

Husband		
	0,0	2,1
	1,2	0,0

The situation is that a husband and wife have agreed to go out together that evening. He would prefer a football game as their joint destination, and she would prefer them to go to the opera (the rows represent the husband and the columns the wife). Both would prefer to spend the evening together and not go to different events. As in the previous table, the left number in each cell is the "return" to whoever is represented in the rows, in this case the husband, and the right number in each cell is the return to the wife (represented in the columns). Each one would like to obtain as high a return as possible. It can easily be seen that in this game there are two possible results that are in equilibrium: either to go together to the football game or to go together to the opera. The mathematics does not indicate which of these is the better choice.

The third game is a version of the common and familiar *heads or tails*. One player writes down his choice of heads or tails, without revealing his choice to the other player. The second player has to guess the choice of the first. If he guesses correctly the first player pays him, say, one dollar. If he guesses wrongly, he pays the other player one dollar. In this game the table looks like this

Player 2

Player 1		
	1,-1	-1,1
	-1,1	1,-1

It can be proven quite simply that in this game there are no equilibrium strategies. This game is also called a zero-sum or fixed-sum game, as the payoffs merely pass from one player to the other. As can be seen in the table, the sum of the payoffs in each cell is zero, namely, the gain of one player is the loss of the other.

We have shown three games in one of which there was one equilibrium, in the second there were two equilibrium outcomes, and in the third, none. The explanation of equilibrium given previously still holds to some extent. For example, in the battle of the sexes, if the husband and wife agreed to meet at the opera, it would make no sense for one of them to arrive at the second performance without coordinating that in advance with the other. Game theory has no answer to the question of whether the opera or a football game is preferable; it does not rate the two options, per se. Neither does game theory recommend the adoption of an equilibrium strategy, even if it is the only one. Here is a table of a game that reflects this.

Player 2

Player 1			
	0,0	0,0	1,1
	3,4	4,3	0,0
	4,3	3,4	0,0

The numbers in the cells indicate, say, monetary payoffs. The strategy in which the first player chooses the upper row and the second player chooses the right column are the only ones in equilibrium, although it is clear that if the players restrict themselves to choosing the other rows and columns, the situation of both of them will improve, and moreover there will be no incentive for them to revert to their initial strategies.

Game theory developed another course of action for players in games presented in strategic form, and that is to use a *mixed strategy*, something that children do naturally when playing heads or tails. To prevent his rival from revealing whether he has chosen heads or tails, he flips a coin so that his choice is a random one. The mathematical expression of this is deciding not on a particular strategy but on a draw in which the chosen strategy is drawn according to the distribution of the chances that the player determines. In such a case what is the payoff? The actual result will be determined only once all the draws have taken place and it is then known which strategies were drawn in the lottery. At the decision stage the player knows only what lottery he has chosen, and likewise for the other players. For the purpose of deciding which lottery to choose, we determine that the outcome of the game will be the expectation (in the probabilistic sense) of payoffs, the expectation according to the chances that the players have determined. Although ultimately the player will receive the payoff that is the result of the various lotteries, the assumption is that at the stage of deciding on the lottery between the various strategies, the player is interested in achieving the highest possible expectation. We will examine this assumption in the next section.

As we noted, the concept of mixed strategies was introduced by Émile Borel, who also showed how to calculate such strategies in equilibrium for many instances of zero-sum games, but he did not believe that it was possible to solve all games in this way. John von Neumann, however, proved that in such games mixed strategies that were in equilibrium could always be found. The expectation of the payoffs is then the same in all the equilibrium states, independent of which equilibrium strategy is chosen, and that payoff is known as the value of the game. This is known as the minimax (or minmax) theorem, because the strategy minimizes the possible loss by each of the two players separately, leading to an equilibrium outcome. In the heads or tails game, the value is zero and the strategies selected by children, that is, to choose heads or tails with equal probabilities, are in fact in equilibrium. John Nash went further and proved that in every strategic game (that is a game in which every player has a finite number of strategies) there are mixed strategies that are in equilibrium.

If we adopt the possibility of mixed strategies, not only is it certain that an equilibrium exists, but in certain situations possibilities of reaching a more reasonable equilibrium arise. For example, we explained why the equilibrium in the 3 × 3 table above is not reasonable as a solution to the game, although it is the only equilibrium. If we allow mixed strategies, and assuming that the players want to maximize their expected payoffs, a new equilibrium comes to light. This is obtained when each player chooses one of the other options, that is, the two lower rows in the case of the first player, and the two left columns in the case of the second, with equal probabilities.

Note that by using mixed strategies we have moved from a situation in which a player receives an actual payoff, whether monetary, or the number of years in prison, or what have you, to a situation where the payoff is a lottery between several possible actual payoffs. In the next three sections we will discuss people's attitudes to states of uncertainty, and in particular to the use of lotteries.

Sometimes the suggestion to look for an equilibrium strategy causes unease, for example in the case of the prisoner's dilemma. On the abstract level we agreed that if a player has a dominant strategy he will use it. In similar day-to-day situations, however, we find it worthwhile to cooperate rather than to opt for the dominant strategy. The reasons for such a choice in normal daily life are not reflected in the limited model we presented. For example, our model does not take into consideration what friends of the suspect sentenced to years in jail are likely to do to the suspect who betrayed his partner in crime. This is a clear illustration of the fact that the mathematical game lacks very important elements that decision makers must take into account. Researchers were not unaware of this aspect, and they put forward models that take into account responses that are outside the table itself. One of the models allows an unlimited or unknown number of repeats of the game, so that every suspect will take into consideration future decisions of the other suspect, decisions that are likely to include retaliation for the current lack of cooperation. The analysis of such repeated games was one of the judges' reasons in the citation awarding the Nobel Prize in Economics in 2005 to the mathematician Robert J. Aumann of the Hebrew University of Jerusalem.

The concepts of game theory have been included in other areas, such as economics, in which the distribution of resources takes place via markets that bring prices of the different products into equilibrium. Such a model was proposed by Kenneth Arrow, whom we mentioned earlier, and Gérard Debreu (1921–2004), a mathematical economist of the University of California, Berkeley, who was awarded the Nobel Prize in Economics in 1983.

As a result of these developments, the concept of equilibrium as well as other concepts developed in game theory, such as a zero-sum game, became part of general public discourse, although the participants in the discussion do not always draw the appropriate conclusions. For example, the incentives for people or companies to honor signed agreements are the punishments incurred if they breach the agreement, whether punishments are imposed by the courts, such as prison sentences or fines, or by future boycotts. In the absence of any retribution or expected punishment for infringement, an agreement is no more than a statement of intention, and it will be broken as soon as it is worthwhile for one of the parties to do so. Therefore, in cases where there is no system of punishment, as in many international agreements, it is up to the parties to the agreement to see to it that, as far as possible, the agreements themselves should have the property of Nash's equilibrium. In other words, no party to the agreement should benefit by breaching it unilaterally. Despite this elementary insight that originates from game theory, it does not appear that politicians who are signatories to agreements, sometimes fateful agreements, even try to take this into consideration.

The model we have analyzed is only one of a multitude of possibilities suggested by game theory to reflect and analyze circumstances in which decisions must be made either in confrontation with or in cooperation with other players whose interests might be opposed to ours. One fundamental model is of *games in cooperative form*. We shall not elaborate on this approach but will just say that here games are not determined by strategies that the players must take, but by the payoffs to coalitions that the players may form. Different formations of coalitions give rise to different allocations of, say, wealth. The theory studies, for instance, what would be stable outcomes, namely, outcomes that would not result in the disintegration of a given coalition; or how to measure the strength of a player based on the various coalitions he can join. The theory has been promoted to a great extent by John von Neumann and his colleague Oskar Morgenstern (1902–1977). They displayed much of the theory in the book *Theory of Games and Economic Behavior*, published in 1944. Many contributions to the theory of games in cooperative form have been added since then. It may be mentioned that von Neumann himself saw in this theory the right building block for a mathematical analysis of human behavior, parallel to infinitesimal calculus as the building block for the analysis of the physical world (in particular, he was not too fond of games in strategic form). It seems that in recent years games in strategic form were found to be more attractive; yet the cooperative game models, including market design relying on what cooperative game theory tells us (recall the aforementioned Nobel Prize to Alvin Roth) has not been abandoned and continues to be a fruitful field of study.

Mathematics plays a crucial role in analyzing the situations we addressed. Yet unlike the mathematics of the natural sciences, which can predict what is likely to happen, the mathematics of decision makers in a situation of conflict does not indicate what may occur, nor does it advise how to act. The current models are far from accurate depictions of the complexity of confrontations in real life. The product of this mathematics consists of proposed concepts or methods that will give the decision maker a better understanding of what confronts him. Sometimes mathematics may indicate a way to arrive at a criterion that expresses what is meant by "best for all." In that case, mathematics may lead to the discovery of the best action. Mathematics may sometimes indicate what properties in the model limit the decision maker and, if possible, may lead to a change in the rules of the game. The mathematical analysis shown above is based on the assumption that the participants are rational and that they are acting to achieve the best, the best according to their subjective preferences, of course. Whether decision makers, either individuals or those making cardinal decisions at the national level, do in fact act in such a rational manner is a serious question that we will address in the next sections.

48. EXPECTED UTILITY

This section is somewhat technical. Its purpose is to describe rational considerations that people accept in regard to lotteries, yet, as we shall see in the next section, they do not follow them when acting intuitively.

Game theory allows the decision maker to act according to his own subjective preferences. In the case of mixed strategies, that is, using lotteries as a means of decision making, the subjectivity is likely to reflect also the attitude of the decision maker to the lottery. Nevertheless, in the previous section we assumed that the expected payoffs determine the value of the lottery for each of the players. This assumption does not reflect reality. Some people are sure that luck always lets them down, and therefore they will not agree that the expectation of payoff should determine the value. Others love risk, and for them the lottery is worth more than the expected payoff.

John von Neumann and Oskar Morgenstern studied this question in the aforementioned book. They proposed the following solution. Try to replace the payoffs listed in the table of the game with other numerical values without altering the ranking of the payoffs such that the new values fulfill the expectation condition set previously. In other words, the value of the lottery, which von Neumann and Morgenstern called the utility of the lottery, will be the expectation of the new payoffs. There is no a priori reason that it will always be possible to find numbers whose expectation will reflect the preferences of the players with regard to the lotteries. Von Neumann and Morgenstern proved, however, that provided the players' actions are in accordance with some simple characteristics that every sensible person would accept as reasonable, it is possible to find such a utility. We would note at the outset that people do not act in accordance with the characteristics identified by von Neumann and Morgenstern, and we will discuss that in the next section. Yet if we examine those traits in an abstract rational manner, it is apparent that they describe how we should conduct ourselves. In the context of our discussion, the characteristics, which in the spirit of Greek mathematics can be called axioms, are as follows.

1. The player knows which of any two possible payoffs, including the ones identified by lottery, is preferable to him, or he can decide that they are equal. This relation is transitive, that is, if option A is preferable to B and B is preferable to C, then A is preferable to C.
2. If in a certain lottery a player is offered the possibility of changing a payoff for another that is preferable to him, including a payoff that is a lottery, he will accept the offer.
3. The way the lottery is carried out, that is, the way in which the probabilities are formed, does not affect the value of the lottery, as long as the probabilities do not change.
4. For all three payoffs in which A is preferable to B that is preferable to C, there is a positive probability, which we denote by p and which may be very small, that getting C with a probability of p and getting A with a probability of $(1 - p)$ is preferable to getting B.

The axioms are indeed convincing. For anyone not motivated by superstition there is no reason not to accept the second and third axioms. The first axiom is correct theoretically but perhaps not practical. Von Neumann and Morgenstern, however, suggested ways of calculating the new utility, if it exists. The fourth axiom is also reasonable. To anyone claiming that his reservation about getting payoff C is so strong that he is not prepared to take even the small risk p that he will receive C, we would point out that he does leave his house, travel by car or train, and even fly occasionally, despite the fact that those activities bear a risk, which may be small but is certainly not zero, that he will suffer severe or even fatal injury.

A utility that has the property we mentioned, that is, that the utility of the lottery is the expected utility, is named after those who developed the concept and is known as *von Neumann–Morgenstern utility*. As stated, if the axioms are fulfilled, the von Neumann–Morgenstern utility exists. The possibility of changing the actual payoffs for others such that the expectation of the new values reflects the value of the lottery to the participant was, in fact, proposed by Daniel Bernoulli in relation to the St. Petersburg paradox mentioned in section 39. Bernoulli's explanation of the paradox was that the value of very large monetary payoffs are not reflected in their face value but in another value, which Bernoulli already called utility. That utility is a function that increases very slowly when the amounts of money keep rising. The value of the lottery in the St. Petersburg paradox should be measured, according to Bernoulli, by the expected value of the function, which explains why people are not prepared to pay large sums to participate in that lottery.

49. DECISIONS IN A STATE OF UNCERTAINTY

In this section we focus on the question of how people make decisions in states of uncertainty and on the fallacies we discussed in sections 40, 42, and 43. Also in this section we discuss the general question of how decisions are made intuitively as opposed to using mathematical analyses. It is perhaps not surprising to discover that the way decisions are made is not always consistent with what mathematics and orderly logical analysis would recommend. We will try to understand some of the reasons for that. Some questions will still remain unanswered, such as: Is it possible to develop mathematics that will describe human conduct that is not always rational? Can people be taught to behave rationally? Is it worthwhile trying to do so? Will decision makers behave rationally when faced with really important questions?

It is worth clarifying what we mean when we declare that certain conduct is irrational. Different people have different objectives. It would be wrong to state that someone who decides to inflict pain on himself or to lose money is irrational. The desire to own assets is a subjective characteristic, and throwing money away is clearly rational for someone who detests money. Likewise, choosing to hurt himself is rational behavior if the person likes doing so. A common expression used to describe actual subjective preferences is *revealed preferences*, that is, preferences revealed by your actions. According to this approach, everything you do is rational from your point of view.

The irrationality we are trying to identify here is different. We find that sometimes someone's behavior deviates from basic assumptions or axioms that are not subjective and that the decision maker himself agrees are the guidelines for conduct. Nevertheless, he sometimes acts in a manner opposed to those axioms. Why? We claim that to a great extent the reason for this type of irrationality is evolutionary. The way we think and respond is molded by millions of years of evolution that brought us to ways of making decisions that in the terms we have just described are irrational. Yet underlying this behavior there is a logic, and I therefore suggest describing such conduct as reflecting *evolutionary rationality*. In many cases of irrational behavior, it is possible to discover the underlying evolutionary rationality.

Two of the major contributors to understanding human behavior in the context of uncertainty and decision making in general are Amos Tversky and his colleague Daniel Kahneman, who started their work at the Hebrew University of Jerusalem and continued at Stanford University and Princeton University. Kahneman was awarded the Nobel Prize in Economics in 2002, and his cooperation with Tversky was mentioned in the citation of the prize committee (Tversky died in 1996). We cannot summarize here the findings and explanations of what Tversky, Kahneman, Maya Bar-Hillel of the Hebrew University of Jerusalem, and others discovered, but we will cite some examples.

We will start with a result presented by the French economist Maurice Allais (1911–2010), Nobel Prize winner in Economics in 1988. Allais performed an experiment that can easily be reproduced, and I myself have used it in several of my lectures. The experiment reveals behavior that can readily be agreed deviates from basic principles of rationality. To understand the deviation from rationality we will first discuss a somewhat-abstract example.

A person is asked to choose between two options:

(i) To participate in a drawing for a gift A with a 75 percent probability or gift B with a 25 percent probability.
(ii) To participate in a drawing for a gift A with a 75 percent probability or a gift C with a 25 percent probability.

In addition we know that the person prefers gift C to gift B. Which of the two options will he choose?

A rational person would opt for the second alternative, and most if not all people do that. Note that we did not say, for example, that C represents a more valuable asset or more money, as the decision maker's preference might be to lose money. We only said that in the order of priorities of the decision maker, C ranks higher than B; in other words, in a choice between B and C he would choose C. This was von Neumann and Morgenstern's second axiom in the previous section. If we improve the situation of the decision maker (improve in his eyes, that is) in one of the components of the lottery without worsening the payoff in any of the other components, a rational person will choose the improved payoff. Note that we do not discuss considerations about the expectation of the lottery. The gifts mentioned may have no numerical measurement.

The conduct that Allais found deviated from the rational choice we just described. His example was as follows.

A number of people were asked to choose between the following two options:

1. To participate in a drawing for three thousand dollars with a 100 percent probability of winning or for zero dollars with a 0 percent probability.
2. To participate in a drawing for four thousand dollars with an 80 percent probability of winning or for zero dollars with a 20 percent probability.

The same people were then asked to choose between the following options:

3. To participate in a drawing for three thousand dollars with a 25 percent probability of winning or for zero dollars with a 75 percent probability.
4. To participate in a drawing for four thousand dollars with a 20 percent probability of winning or for zero dollars with an 80 percent probability.

(We put the probability of 100 percent in the first option to emphasize that the certainty of winning three thousand dollars plays no part in the example. Indeed, an event with zero probability can occur.) Most respondents chose the first option of the first two, and option 4 of the second two. In accordance with the revealed preferences principle, we do not determine which of the possibilities is better, and we certainly do not presume to state that anyone who chose differently than we would have chosen is not acting rationally.

This is where the surprise comes in. Those who chose options 1 and 4 were acting contrary to what we agreed above, which was that the better option, that is, option (ii) in the abstract example above prior to the presentation of this concrete example, is the rational choice. In the same way, those who chose options 2 and 3 in the above example are also deviating from the conclusion above, to which they agree, that option (ii) in the abstract example is the one to choose. We would emphasize that it is not irrational to choose either option 1 or option 2, but to choose option 1 and then option 4 reflects irrationality in the sense that we have indicated. The argument is that in a probabilistic sense option 3 consists of 25 percent of option 1 and another 75 percent chance of receiving zero, while option 4 consists of 25 percent of option 2 and 75 percent chance of receiving zero (it takes some calculation, which we leave out, to verify that). Thus, if you preferred option 4 to option 3 (and if in the abstract example you always opt for choice (ii)), you must prefer option 2 to option 1. One can try to excuse the deviation from rationality by claiming that the example is confusing, the calculations are complicated, and so it is difficult for the decision maker to explain his deviation from the principle to which he agrees theoretically. This argument does not explain why participants consistently choose options 1 and 4.

Nevertheless, there is a reason for this tendency, and in my opinion it is inherent in the way we relate to numbers representing probabilities. In section 43 we discussed the principle that evolution led us to ignore events with low probabilities. That is entirely rational from an evolutionary perspective, and it helps the human species, as well as others, to survive in the evolutionary struggle. The difference of 20 percent between option 1 and option 2 causes most respondents not to risk the almost-certain win of three thousand dollars by taking an 80 percent chance of winning four thousand dollars. The risk of "only" 5 percent between options 3 and 4 seems reasonable because intuitively the result with a 5 percent probability can be ignored. Intuition does not grasp the mathematical fact that, from the aspect of decision making, the 5 percent are equivalent to the 20 percent in the other alternative.

Evolutionary rationality expressed by underestimating the importance of events with low probability keeps appearing. It is difficult to include all aspects of this attitude in the category of irrational behavior, but the discrepancy between it and the mathematical calculations is revealed time and time again. For example, someone is offered the choice between a trial in which he will have a 60 percent chance of success or a trial repeated independently five times (i.e., performed six times in all) in which his chance of success is 90 percent each time. If this choice takes place in a mathematical context, say in a lesson on probability theory, most of those present will carry out the calculation and will find that 0.9 to the power of 6 is smaller than 0.6 and would choose the first option of the single trial. If the situation and the probabilities are within the description of an event in which the required mathematical exercise is not emphasized, and the decision is made by "gut feeling" and not by calculation, the decision will tend markedly toward the repeated trials with a higher chance of success in each one. Likewise, if offered the choice of overcoming an obstacle several times, with a low probability of failure each time, or overcoming a more serious obstacle once with a higher risk of failure, most people would tend to go for the first option, independent of any mathematical calculation.

There is no discrepancy between the fact that there is a tendency to ignore events that have a low probability and the fact that people continue to buy tickets in the national lottery or bet on the results of sporting events, knowing that their chances of winning are very small, even minimal. The explanation is in the attitude to lotteries in general, which is likely to reflect the tendency to seek risk or to be risk averse, and these can appear in different forms in different circumstances. A reasonable person may buy a lottery ticket because the expected loss is small relative to the high prize he could win, and also because the good feeling he gets from the possibility of winning, until the lucky numbers are announced. The same person would not risk all his property on such a lottery, even if the prize he could win was hundreds of thousands times larger. Being risk averse or risk prone does not contradict rational behavior.

The discrepancy between behavior and probabilistic assessments on the one hand and mathematical logic on the other comes to light also in other types of situations. In experiments carried out by Kahneman and Tversky in the 1980s they discovered the following. A number of people were asked, each one alone, to assess what is the probability that in 2018 Russia would break off diplomatic relations with the United States. At the same time a group of other people was asked, again separately, to assess the chances that in 2018 Russia would be in conflict with the Ukraine, the United States would intervene, and as a result Russia would break off diplomatic relations with the United States. Mathematical logic says that the probability of the first scenario is higher than that of the second. The empirical evidence was that people thought that the second scenario had a higher probability. The explanation lies in the way people arrive at their assessments. The second group was presented with a plausible, realistic scenario that would lead to Russia breaking off diplomatic relations with the United States, whereas the first did not present such a clear-cut scenario. The more realistic-sounding possibility overcame logic. Evolutionary rationality overcame rationality. Kahneman and Tversky called the mechanisms that result in these deviations the availability heuristic and the representativeness heuristic. The deviations that these cause can also be seen in other areas.

We will give few examples in which the relation to uncertainty is affected by preconceptions and not necessarily by logical and mathematical arguments. We saw earlier that in biblical times and in ancient Greece it was recommended to use, and indeed use was made of, random events to arrive at just or fair outcomes, just at least in the probabilistic sense. We have no way of knowing how the public viewed the fairness of the method. In modern times such methods are not always treated with the proper academic equanimity.

A lottery as a fair instrument of recruitment to the army has been used by several countries for a long time, including the United States. In 1970, in the call-up for the Vietnam War, there was a central lottery of birth dates, and it was decided that young people within a given age range whose birth dates were drawn in the lottery would be drafted into the army. The outcome led to a strong opposition to the whole idea of the draft. The system was that chits representing all the possible dates were placed into a container. The numbers from 1 to 365 were placed in a second container. Then, one by one, a chit with a date was drawn from the first container, and a chit with a number from the second. The number determined the order of the draft. For example, those born on the date drawn from the first container together with the number 8 from the second were in eighth place in the list for the draft, and so on, until the required number of recruits was reached. Those with a high number were apparently not drafted at all. The reason for not using just one container and say drawing only dates of birth and agreeing that the order in which they were drawn would determine the order of call-up was, it seems, the same reason mentioned in the Talmud and its interpreters (see section 36), that it would be easier to establish the fairness of the method. The outcome of the 1970 lottery provided the basis for critics of the system. We show below the numbers of the average position in the list of recruits of those born in each month, as came up in the draw (the data are taken from an article by Stephen Fienberg in *Science*, January 1971).

January	201.2
February	203.0
March	225.8
April	203.7
May	208.0
June	195.7
July	181.5
August	173.5
September	157.3
October	182.5
November	148.7
December	121.5

Thus, for instance, those born in January were, on average, 201.2 in line for the draft while the average place for those born in December was only 121.5. It paid to be born in the first half of the year. The average position over the whole year is about half the numbers of days in a year, that is, 183. The list clearly shows that the average position in the list for the draft of those born in the months August to December, about 157, is significantly lower than the position of those born in the first months of the year, about 203; that means that those born in August to December were more likely to be drafted.

Does this mean that the procedure is unfair? Not necessarily. The bias resulted from the details of the way the chits were put into the container, and we will not go into that now. However, before the drawing started, all months had exactly the same chance of being drawn. The mathematical claim of fairness and equality did not help. The results gave rise to criticism of the randomness of the system of drafting, and, although there was an attempt to change it in the 1971 draft, the strong opposition to the system itself was one of the factors that led to the canceling of compulsory drafting by means of a lottery and the formation of an army based on professional soldiers.

Another example is related to polling methods. It is very difficult to collect reliable statistics about, say, drug users, alcoholics, tax evaders, and so on. People do not trust that their answers will be kept in secret. Statisticians, among them, Tore Dalenius (1917–2002) of Brown University and the University of Stockholm, who was one of the pioneers of considering psychological aspects when designing public polls, suggested a way of overcoming this distrust. Assume that the question is "Are you a tax evader?" Before responding, the subject should secretly perform a personal drawing between, say, red and black with probabilities of, say, 51 and 49 percent. If red comes up, he gives the true answer. If black is the outcome, he lies. No personal information can be drawn—even by the tax authorities—concerning possible tax evasion. But for large populations, the small difference between 51 and 49 is enough to get reliable statistics. The statistical rationality did not convince the public. Those who were asked to take part in such polls did not believe in the method.

Here is another example of the lack of understanding of decisions based on randomness. In elections to Israel's parliament, the Knesset, voting takes place by placing a chit with the name of the party the voter chooses into an envelope that he then seals. Placing more than one chit in the envelope, even if they are for the same party, let alone if they are for different parties, leads to that envelope being disqualified and the vote is a spoiled vote. Yet sometimes, for reasons we do not need to detail here, it is not easy for an Israeli voter to decide which of the parties is worthy of his vote. Moreover, if the system was such that every voter put five chits into the envelope, some voters would put five chits into the envelope all for the same party, others might put three for one party and one each for two other parties, in line with their opinions and leanings. This would be logical also from a conceptual viewpoint. The elections determine the composition of the Knesset, and most of the voters do not put all their faith in one party, or they would prefer to give equal power to two parties. Such splitting of votes, however, is not allowed. A few elections ago I proposed the following procedure to those who would have liked to split their votes. Assume that if you could split your vote you would give two-thirds to one party and one-third to another. Take two chits of your preferred party and one chit of the party of your second choice, choose one of the three randomly, say mixing them up and selecting one behind your back, and put it in the envelope without looking at it. Then throw the other two away, also without looking at them. That way you not only divided your vote in the proportion you wanted, but the only information you have at the end of the process is about the division you chose. If you are asked which party you voted for, you can answer only with that division, that is, the division of the probabilities. Subjectively, you divided your vote, and the subjective aspect is what is important, because to a great extent the reason for going to the polling booth is the subjective feeling. If you were to measure the chance that your vote has any effect against the trouble of going to cast your vote, you probably would not bother to go at all. A talented journalist named Yivsam Azgad heard about my idea and published an article on it in the *Ha'aretz* newspaper. The day before the elections I was actually invited to a television studio for a short interview, in which I explained and illustrated my idea (at that time politicians could not be interviewed on television on the day before the elections, so television had to make do with mathematicians and the like). The reactions were astonishing. On the positive side I received compliments on the proposed system, and many of my friends and also people I did not know beforehand told me that they had adopted my method. On the other hand, there were also many who opposed it. A listener on a radio program phoned the studio and complained in anger, "And what if I draw the chit of the party I hate?" (he even gave the name of that party). He clearly did not quite understand the proposed system, in particular that you, the voter, determine the weights according to which you divide your choices, and you would certainly not include in the drawing the chit of the party you loathe. For that caller, a lottery is a lottery, and you never know what will come up in a lottery. Another acquaintance, an activist in a political party, was also opposed to the system and angrily accused me of "wanting to let a lottery decide who our leaders will be." For her, any decision based on randomness is unsuitable.

50. EVOLUTIONARY RATIONALITY

We will now expand the analysis of how people make decisions not necessarily in situations of randomness. Here too we see the effect of evolution on the way we decide and act and on how our brains think and analyze. The decisions are not always rational, but we can recognize evolutionary rationality. In this section we will again benefit from the contributions of

Amos Tversky and Daniel Kahneman and others, who identified the structures according to which the human mind works.

Here is a description of a tendency that Tversky and Kahneman called *anchoring*. A roulette wheel, with the numbers from 1 to 100, is spun in front of a group of people, and it is clear to everyone that the number the ball falls on is completely random. Say it falls on 80. The people are then asked to estimate whether the number of people, in millions, living in a certain geographic area, say West Java Province of Indonesia, is greater than or smaller than the number where the ball fell on the roulette wheel. In other words, is the population of West Java Province greater than or smaller than 80 million? We chose somewhere that the people may have heard of, but would certainly have no idea of the correct answer to the question. Their answers are then based on sensible estimates, or guesses, each with his or her own approach. At the second stage, the same people are asked to give a figure of what they thought the population of that province was. Here again the answer can be only an intelligent guesstimate. One surprising fact emerges. The numbers given in answer to the second question are affected by the random number that the people were exposed to in the spin of the roulette wheel. In other words, if the number that came up on the wheel was high, say 80 as in our example, the guesstimates of the population of West Java will be higher than if the number that came up was a low number, say 2. That is so despite the fact that all the participants saw and could confirm that the number that came up was completely random. The very fact that they were exposed to a particular number when first asked about the population resulted in that number affecting the figures they gave as their estimates of the population.

The anchoring effect may be viewed as irrational. The axiom that an irrelevant factor is not relevant, that is, it is not meant to have an effect, would be accepted by any reasonable person. It is not always easy to recognize when a factor is relevant and when it is irrelevant, but the random outcome of a spin of a roulette wheel is definitely irrelevant to estimates of the population of anywhere. But the fact is that the roulette wheel outcome does have an influence. This is not really surprising. The effect reflects evolutionary rationality. People who have to make a decision do not have time to "waste" on an intelligent analysis of what is relevant and what is not and, after such analysis, to ignore the irrelevant data. We have already stated that the brain cannot think subject to conditions and axioms that are imposed or examine logically what is relevant and what is not. The brain looks for any data that arise in connection with a decision and decides according to those data, without checking and clarifying the details of the extent of their relevance. This system is apparently more efficient from the cost/benefit aspect in most cases where action or a decision is required, and it therefore became ingrained by evolution in human behavior.

The above example indicates one of the structures according to which the brain operates when action or a decision is called for. *The brain completes the picture of the situation with the information it already holds, without carefully checking the relevance and logical impact of the information.* We saw other examples of such analyses in section 42, when we examined answers given intuitively compared with those given following calculations based on Bayes's formula. Although the questions lacked information that was essential to solving the problem, those questioned were unaware of that when they answered intuitively. The brain completes the picture in such a way that it can provide an answer.

The following is another example (taken from linguistics) of the implications of the claim about what the brain already knows, without a careful logical examination. Try to rephrase the following:

There is no head injury too minor for you to ignore.

The great majority of people coming across this statement understand it to mean that every head injury, however minor, should be treated. A close examination of what it says will reveal that it claims the opposite, that no head injuries at all need to be treated! What causes the confusion? First, as we claimed back in section 5, the brain has difficulty in analyzing claims that contain a negative, as in sentences that start with "there is no" or that have the word "ignore," with its inherent negation. At the same time, the brain recognizes the content (not the meaning) of the claim, as it has often come across warnings about head injuries. It therefore combines the general content of the claim with what it already knows and skips the logical analysis. Evolutionary rationality dictates this type of reaction.

Another characteristic of decision making is also related to the structure of the brain. *The brain cannot analyze a problem without preconceptions* (not to be confused with prejudices). When someone is confronted with a problem or a question, a preconception is utilized immediately, sometimes based and dependent on how the question is worded or the information presented. Thus, the reaction of a sick person to the doctor's prognosis that he has an 80 percent chance of recovering is different than his reaction to the news that he has a 20 percent chance of remaining sick for the rest of his life. The rational mind would recognize that an 80 percent chance of success is the same as a 20 percent chance of failure. A mind that acts only in an evolutionarily rational way will not recognize the equivalence. That recognition is not worth the effort needed to compare and analyze the facts in every case, and we therefore cling to concepts and attitudes already present in our brains. The popular belief that we can examine a question without preconceptions is incorrect.

Some time ago, one of Israel's past ministers for internal security proposed that a criminal case passed from the police to the state prosecutor should no longer be accompanied by a recommendation from the police whether or not the suspect should be charged, as it always had been until then. The minister said that it was up to the state prosecutor's department to make its own decision on the basis of the material it received, without preconceptions. The honorable minister was clearly unaware that there is no such thing as "without preconceptions." If the police do not give their view regarding prosecution of the suspect, the discussion will start in the state prosecutor's department with a different preconception, almost certainly a less authoritative one, based for example on media reports.

Another behavioral pattern rooted in evolution is the default *to believe what you are told*. Most of us are familiar with the situation in which newspaper reports of events we ourselves witnessed are not exact or reliable. Yet if we read in the same newspaper a report of an incident whose accuracy we are unable to verify, we believe what is written. Again, the reason is evolutionary rationality. It is not efficient to doubt what we are told, even if that sometimes leads to absurdities.

Here is a description of an experiment that, although it could be considered a quaint tale, teaches an interesting pattern of behavior. A group of monkeys was placed in a cage, in which there was a pole with a banana at the top. When one of the monkeys started climbing the pole to take the banana, all the monkeys were given a small electric shock. The monkeys soon learned when it was they felt the shocks, and whenever a monkey started to climb the pole the others would stop him by blows and threats. Then one of the monkeys was replaced by another one not from the cage. That one immediately tried to climb the pole to get to the banana but was stopped by the others so that they would not get an electric shock. After a while another fresh monkey replaced one of the group, and the same happened. The new member of the group tried to get to the banana and was prevented from doing so by the others. Among those hitting him was the monkey that had been brought in previously. After this process of replacing monkeys had been repeated several times, there were no monkeys in the cage that had ever received an electric shock. Moreover, the electrical current was disconnected. Yet whenever a monkey showed signs of intending to climb the pole to take the banana, the others fell upon him and beat him vigorously.

It is easy to identify such behavior among all human societies and among many groups of animals. The property is the result of evolution. A child who devotes himself to checking every instruction and advice his parents and teachers give him will not get very far. Although science allows, and even supposedly requires, every scientific theory to be treated skeptically and critically and every result to be checked, such doubts and checks are not part of students' and researchers' practice, even in mathematics! That is evolutionary rationality that is ingrained in our genes, and it is of course consistent with the efficiency of our day-to-day lives. Such behavior, however, is responsible for many serious ills in human society. Apparently the evolutionary advantage of the strategy of believing what we are told outweighs the drawbacks resulting from that strategy.

Another pattern deep-rooted in our brains is the belief that *what was, will be*. The default is not to believe any prophecies, and in particular prophecies of doom, about future changes. The reason, again, derives from evolution. Devoting time to what may occur and preparing oneself accordingly takes valuable means that are required in the evolutionary struggle, as we have mentioned previously in the hypothetical possibility that the dinosaurs could have grown gills to preempt the disastrous effect of the meteoric dust that led to their extinction. Acting according to the default that what was, will be, explains the great surprise with which we react to every abrupt change, although when changes occur it is easy to see that they could have been foreseen.

Yet another behavioral pattern derived from evolution that deviates from mathematical rationality can be seen from an experiment performed by Güth, Schmittberger, and Schwarze, reported in 1982. Two decision makers have to divide a hundred dollars between them in the following way. The first, let us call him A, must decide what share of the hundred dollars he wants to be left with and what share he will offer to the other, B, with the proviso that he must leave B at least one dollar. B can either accept what he is offered or reject the offer. If he rejects the offer, neither of them will receive anything. It is explained to them that this is a one-off, non-repeated game, and furthermore, to emphasize the point, neither knows against whom he is playing, and both are assured that this anonymity will be maintained. What amount is it worthwhile for A to offer, and how ought B to react? The second part of that question has a very clear rational answer (assuming that B prefers to receive a positive number of dollars rather than nothing). Whatever positive amount A offers B, it is worthwhile for B to accept. Hence, assuming such rationality, it is worthwhile for A to offer B one dollar. The argument is clear and is based on rationality that says to choose what is best for yourself. But it is not surprising that the participants in the experiment behave differently. A generally offers B an amount between forty and fifty dollars, and B generally rejects any offer less than forty dollars. The explanations given by the participants were largely related to arguments of *justice and fairness*. The search for justice is inherent through evolution, and the sense of justice has been observed to exist even in day-old infants, as we mentioned at the end of section 3. The dependence on justice and fairness reflects evolutionary rationality and not logical rationality.

Another property that is sometimes not rational but reflects evolutionary rationality is the *protection of property*, that is, the tendency to keep property that has been acquired. This tendency certainly came about in the evolutionary process, when private property was essential to subsistence. Daniel Kahneman and his colleagues carried out a trial on this. Two groups of people are given the choice of a cup worth about ten dollars and a ten-dollar bill. The first group is given the choice between the cup and the bill, while the second group is given the cup as a gift when they come into the hall where the experiment is taking place, and they are then offered the ten-dollar bill in exchange for the cup. The fact that the cup is already in the possession of the second group had a strong effect on their choice. The subconscious desire to preserve their property, reflecting evolutionary rationality, overcomes simple logic and mathematical rationality. The reader will find more examples and broader discussion of this subject in the list of sources.

And at last we arrive at the question we asked at the beginning of the previous section. Actual human conduct often deviates from rational behavior in the sense that it contradicts generally accepted basic assumptions, accepted also by those behaving in that way. That said, can mathematical methods be used to describe human behavior? Is it possible to formulate a theory based on other basic assumptions through which, with the help of the usual mathematical rules of induction, it will be possible to analyze and understand human conduct? In my opinion, the answer is yes, but the basic axioms for such a theory have to be derived from characteristics embedded in humans in the evolutionary process.

Can people be taught to behave in a logically rational way, that is, to fulfill in practice those axioms on which they agree in theory? And if so, is it worth doing? My answer to those two questions is the same as the answer I gave to a similar question related to Bayes's law: in my opinion there is no possibility of instilling rational behavior into the brain when that contradicts behavior that is consistent with evolutionary rationality, and it is also not worth doing. The advantages of behavior that developed during evolution still exist. Nevertheless, there are cases when it is important to reach a rational decision, when the results of a decision are fateful and could cause immense harm if the decision follows evolutionary rationality but not logical rationality. In those cases it is worthwhile to invest the time to look for the logically rational solution and not to base the decision on "gut feeling."

Finally, there is serious debate about whether decision makers in key positions who make decisions with fateful outcomes rely on their gut feelings, or can we be confident that when all is said and done they arrive at their decisions based on mathematical analysis? For instance, can we rely on a leader who must decide whether to start a war to analyze the situation calmly, perhaps subjectively, but from our point of view, rationally? It seems that that is an open question.

CHAPTER VII
COMPUTATIONS AND COMPUTERS

Why is an abacus so called? • How did weaving looms help the computer world? • How many computers does the world need? • How can you win a million dollars through sudoku? • How should the winning numbers in the national lottery be encoded? • Can a computer think? • Can a computer impersonate a human? • What do genetics and mathematical calculations have in common?

51. MATHEMATICS FOR COMPUTATIONS

In Assyria, Babylon, and Egypt, mathematics dealt only with calculations. These calculations were needed in building, agriculture, trade, and so on. As already mentioned (in section 6), there is no evidence that the mathematicians in those days invested time or effort in formulating or documenting the methods they used to solve problems, but it is clear that they were familiar with them and could draw an analogy from the solution to one problem to solve a similar problem. They passed this knowledge on by recording the actual problems they solved, either on clay or on papyrus. Thousands of shards and a number of papyri have survived and testify to this enormous accumulation of knowledge. They show many instances of different types of mathematical calculations. For example, shards have been found with multiplication tables inscribed similar to those used today. Others contain lists of numbers broken down into their factors, and the like. Some give clear indications that they were used in a lesson in mathematics. Others were apparently used by builders or traders, or even by mathematicians carrying out calculations simply for their own amusement. Counting systems, including the choice of a base used to present the numbers, developed such that they simplified the writing of the numbers and also calculations, but there is no evidence of discussion of the methods themselves and the relation between the method and the calculations. To solve a new problem, the user had to analogize from among the collection of previously solved problems. This culture of calculation did not include abstract formulae or general methods that would help the user to solve new problems. Further evidence that mathematicians of that period did not deal with mathematics beyond the realm of calculations can be seen from the fact that there was no insistence on solutions being exact. Many of the Babylonian records have exercises in addition and factorizing numbers in which the results are just approximations of the correct solution. The approximation may have been good enough for the purposes for which the solution was used, but it would not meet standards of mathematical precision. We have commented previously that this need not surprise us. Generalization, per se, is not a characteristic that evolution has implanted in us, and hence it is not natural. Generalization and abstraction yield results after a long time, and the recognition of the importance of precision came about only in classical Greece.

The Greeks, while using mathematics as a means of expressing rules of nature and developing mathematical subjects with insistence on proofs, rules of inference, and so on, did not neglect the mathematics of calculations. They made great efforts to develop methods of calculation and also realized the importance of their formulation and documentation. Neither did they hesitate to develop new mathematics to help the efficiency of calculations. We will give two examples.

The first is trigonometry. The relation between the length of chords of a circle and its diameter and the use of these relations in geometric constructions are attributed to Thales of Miletus, since then the study of the geometry of circles and their chords has made great progress. Later, the use of trigonometric sizes for calculations developed, motivated by the interest in astronomy. The sine of the angle alpha was described then as the length of the chord a divided by twice the radius, *b* (see the diagram). (The equivalent definition that we learn at school, and which relates to the ratio of two sides of a right-angled triangle, was adopted only in the eighteenth century.)

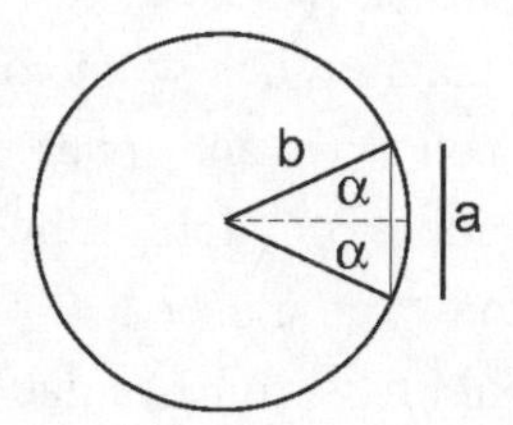

The impact of the study of chords of circles was also evident in methods of calculation. Ptolemy presented many trigonometric relations in his books, but they always related to right-angled triangles. Menelaus of Alexandria (70–140 CE) extended the rules of plane trigonometry to spherical trigonometry and used the methods he developed for calculations in astronomy. Menelaus, Ptolemy, and their colleagues relied on tables of trigonometric quantities. Apart from the fact that the lists in these tables related to lengths of chords of circles, the tables were similar in structure to those included in mathematics textbooks that were in use until quite recently. With the advent of computers it became more efficient to recalculate the required quantity than to look in the tables.

The source of another system of the mathematics specific to calculation purposes can be found in the dichotomy paradox of Zeno of Elea (490–430 BCE). This paradox, one of four paradoxes of motion, claims that anyone wanting to reach a certain place can never reach it. First he has to cover half of the distance, then another quarter of the distance, then an eighth, and so on, ad infinitum. Thus he can never arrive at his destination. Aristotle referred to the philosophical aspect of the claim and its implications for motion, but as far as the paradox itself was concerned, he stated that the time of each successive stage also gets smaller, and the target will be reached. However, he did not include a method for calculating the time required. Eudoxus's exhaustion method (see section 7) discussed a similar problem and showed how to calculate areas by dividing them into smaller and smaller subareas. Archimedes improved the method, as we shall see.

The system was used later by Newton and Leibniz as the basis for the development of calculus.

Archimedes's method was to enclose the shape whose area he wished to measure between two shapes whose areas are easy to calculate; for example, the circle in the diagram below enclosed by the two hexagons. When the difference between outer and inner enclosing shapes is small, the area of each is a good approximation of the area we wish to measure. To obtain an approximation even closer to the area of the circle in the diagram, the hexagons would have to be replaced by polygons with more sides. The more sides there are to the shapes, the closer the approximation.

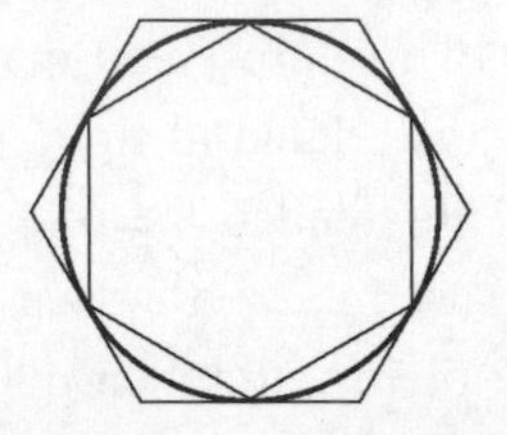

Archimedes also referred to the concept of the limit. The areas obtained via the series of approximations that converge closer and closer to the sought area is called the limit and, according to Archimedes, is the right area. Newton and Leibniz extended the system beyond the method of calculation and made it a mathematical theory, a tool both for describing nature and for mathematical analyses and calculations. Archimedes developed the system as an auxiliary instrument for calculations. He also presented the computational concept of an "approximation as close as we wish." There is no possibility of presenting the area of a circle with radius 1 as an exact decimal number (the area of a circle with radius r is πr^2 and π is an irrational number). We can, however, obtain as close an approximation as we like. That is, if you, the reader, wish to know the area of a circle with an accuracy of one millionth, I, Archimedes, can find it for you. If you wish to improve the accuracy to a level of one billionth, I can calculate that as well; in fact, I can do so to any degree of precision. Note that such a stipulation, that is, "for any degree of approximation I can find a solution with that degree of precision," is based on the logic developed by the

Greeks. Since then, the concept of an approximation as close as we wish has become the cornerstone of modern computations. As is the case with most contributions of the Greek scientists, no original writings of Archimedes have survived, and we must rely on copies of them accompanied in the spirit of that era by corrections and interpretations. One of the well-known interpreters of Archimedes was Eutocius, who was born in about 480 CE and worked in Ashkelon, a city in Israel.

Calculation methods continued to improve throughout the development of mathematics. Another contribution that made calculations easier was the invention, or discovery, of logarithms. The system was developed by John Napier (1550–1617), a Scottish mathematician who also dealt with physics and astronomy, and in keeping with the spirit of the time, also astrology. Napier developed logarithms to make calculations easier. It was only later that the ingredients of the system, that is, the logarithmic function and the exponential function, played a major role in the description of nature.

The logarithm of a positive number, say N, to a given base, say a, is the number b that satisfies $a^b = N$. This is denoted by $\log_a(N) = b$. Any number bigger than one can be chosen as a base. Napier himself defined a function very similar to the logarithmic function, whose values do not depend on the selected base. It was discovered very soon that calculations become much simpler if the number e is chosen as the base, called the natural logarithm. Its value is approximately 2.71828 (it is an irrational number, so it cannot be given exactly as a decimal number). The function is shown as $\ln(N) = b$, that is, the base e is omitted.

We will not go into the details of the calculations, but we will just note that the equality $\ln(NM) = \ln(N) + \ln(M)$ that derives from the laws of powers is the key to the efficiency of the system. This equality enables us to convert the multiplication of two numbers, say N and M, by finding the logarithms of N and M in a table of logarithms, adding them together (and the sum is the logarithm of NM), and then finding in the table the number whose logarithm is that sum. In that way we can convert the complicated calculation N multiplied by M to finding in a table of logarithms the values of $\ln(N)$ and $\ln(M)$, then adding the two numbers, then looking up in a table of logarithms the number whose logarithm is the result of the summation of the two numbers that were added. As it is much easier to add two numbers than to multiply them, especially if they are large numbers, the system of logarithms simplifies mathematical computations. The calculation depends on finding the appropriate tables, but these tables can be produced once and then used for any calculation. Napier presented such tables of logarithms, and even in his days they were improved, and they were used from then until quite recently. That method of computation was part of the compulsory syllabus when I was a student in secondary school and then at university, and the log tables were study aids. Nowadays it is easier to press the appropriate button, and the electronic calculator will provide the result much faster and with greater accuracy than can be achieved by using tables.

Gauss made an important contribution to computation methods, in addition to his other contributions. One of his contributions, mentioned above (section 39) related to the asteroid Ceres, which was discovered in 1801 by Giuseppe Piazzi, an Italian astronomer from Palermo. The discovery of the asteroid aroused great interest, among other reasons because its position and size were consistent with the possibility that it was in fact a new planet, a possibility that several astronomers had indicated prior to its discovery. Within a few months, however, Ceres approached so close to the Sun that it could not be identified, and the few measurements that had been taken up until then were insufficient to enable the calculation of when it would reappear on the other side of the Sun. The young Gauss, then aged twenty-four, took this task upon himself. To carry out the calculations Gauss developed a special method, based on an approximation of Ceres's elliptical orbit by summing cyclical orbits given by trigonometric functions and then computing their parameters.

The presentations as a sum of such orbits is very similar to what are today known as Fourier series. Joseph Fourier (1768–1830) described his method in France at about the same time as Gauss presented his, but it is almost certain that Gauss did not know about Fourier's work. The mathematical problem was a double one. First was the use of very little data. Second, rapid calculation was required. The new method met both these requirements, and Gauss predicted with impressive precision the location and time when Ceres would reappear. This success earned the young Gauss fame throughout Europe. He published his method in a short article that dealt with its use in astronomy and just mentioned that the method could be extended and used for other purposes, but he did not go into any further detail. In practice, no new uses were found, and the method merely gathered dust among other texts.

More than 150 years later, when computers were already in relatively wide use and Fourier's method of approximating functions by summing trigonometric functions was in common use in mathematics, there were several attempts to make these calculations more efficient. The efforts to streamline the method were spurred on by the many uses to which it was being put, including computerized tomography (for instance, for CT and MRI tests) and signal processing. In 1965 two American mathematicians, James Cooly of IBM, and John Tukey (1915–2000) of Princeton University, published an algorithm that calculated the coefficients of the Fourier approximation more efficiently. This was a great improvement, and the method, called the fast calculation of the Fourier transform, or FFT, caught on immediately. It did not take long until it was found that the method discovered by Cooly and Tukey was exactly the same as the one Gauss had used 150 years earlier.

The examples brought here are of course the tip of the iceberg relative to the wealth of methods developed during the history of mathematics to streamline computation processes.

52. FROM TABLES TO COMPUTERS

Alongside the mathematical improvements in the methods of performing calculations, mathematicians always managed to develop mechanical aids, and later electric and electronic devices, to help them carry out computations. The bones mention in section 6, found in the Belgian Congo and that were dated to about 20,000 years BCE, were in a sense a tool for carrying out simple arithmetic tasks.

Shards from the Assyrian and Babylonian periods dated to about 2600 BCE contain tables that, according to the accepted interpretations, served as the first primitive version of the abacus. Greek texts teach us that the abacus was used as a tool for calculations by the ancient Egyptians. It was also used for arithmetic calculations in Greece and Rome. Roman abaci, remarkably similar to those still used today, were found in archaeological excavations throughout the Roman Empire. Its name provides evidence regarding the antiquity of the abacus as an accessory for calculations. The name derives from the Hebrew word *avak*, meaning dust, which appears several times in the Bible. Apparently, the connection is the dust they used to pour onto stone slabs so that they could write numbers and calculations. Abaci are still in widespread use in Eastern cultures, including in Korea and in China, and similar devices have also been found in ancient American cultures such as the Maya and the Inca.

Immediately after Napier presented his logarithmic method for rapid calculation, mechanical aids for calculation were developed alongside printed tables. The British mathematician Edmund Gunter (1581–1626) presented a logarithmic scale, which translated the logarithmic function into geometric terms, following which another British mathematician, William Oughtred (1574–1660) invented the slide rule. By moving the parts of the rule and aligning the different scales marked on it, the values of the logarithmic and exponential functions can be found. Oughtred himself improved his slide rule to enable various operations and other functions to be calculated mechanically, including trigonometric functions. Slide rules became more sophisticated over time and were available in a multitude of forms: straight, triangular, circular, and so on. They became everyday work tools of engineers, who could not operate without them until relatively recently, when they were replaced by computers.

While still a youth, Blaise Pascal, whom we met in section 37, helped his father in his job as a tax collector. When he was sixteen years old he had the idea of simplifying and speeding up the calculations needed in collecting taxes by constructing a calculating machine. The machine consisted of a system of cog wheels connected in such a way that turning them in the required direction and by the right amount gave the correct answer for the amount of tax due. Obviously the machine could perform general mathematical, not just tax-related, calculations. His machine was called the Pascaline, and Pascal built a number of them and tried to sell them, but without commercial success. Several of his machines are in the Conservatoire National des Arts et Métiers (CNAM) museum in Paris.

Other famous mathematicians also tried to build calculating machines, including Leibniz, some of whose machines can be seen in museums in Germany, including at the Deutsches Museum in Munich. These too were based on cog wheels, and calculations could be made as a result of the ratio between the different wheels. To make the machine suitably efficient for calculations, Leibniz developed binary arithmetic, that is, representing numbers using two digits only, 0 and 1. In our daily lives we write numbers on the base of 10, and specifically we use ten symbols to represent all numbers. In earlier times other bases were used. The Babylonians, for example, mainly used the base 60. The choice of a convenient base clearly depends on uses and can be reached either by trial and error, or by intelligent analysis. That was how Leibniz chose the binary system as the base according to which the calculating machines would operate. That base has remained the most convenient for performing machine calculations still today, and it is particularly appropriate for the way computers work.

Pascal's and Leibniz's calculators and those of other contemporaries could not be programmed, and for each calculation the cog wheels had to be realigned. In the nineteenth century Charles Babbage (1791–1871), a British mathematician and an engineer, made a great stride forward. As a result of his contribution, Babbage is considered one of the pioneers of modern computers. He took his idea from weaving looms and the perforated paper tapes that were fed into them to determine the combination of shapes and colors in the material. Babbage used that as a basis for his construction of a calculating machine into which a perforated strip of paper was fed, as in the weaving industry, and the machine then produced the result. That was what led Babbage to call them *input* and *output*, terms that have remained in use in basic computer terminology still today. Babbage built several calculating machines that were large and complex, and that could perform fairly complicated calculations. They obtained their power from steam engines. (Electricity had not been discovered yet. If electric motors had not been developed, today we would probably need a small coal-powered steam engine alongside every personal computer.) Copies of Babbage's machines can be seen in the Science Museum in London.

With the development of electric engines, electric calculating machines also developed, including relatively small ones that could be placed on office desks. These were in use until quite recently, and some can still be found in use today. In addition to computation aids we have mentioned that were used for numeric calculations, different types of mathematical calculations were installed into machines that were used for specific purposes, such as speedometers of various sorts (the Romans already had such speedometers attached to horse-drawn chariots), compasses, and other engineering instruments.

The technology accompanying all the computational aids we have mentioned made calculations easier and faster but still within the limitations of what the human brain could follow and absorb. The different types of calculating machines—from tables, to slide rules, to mechanical machines—mimicked as far as possible the path that the mathematics of calculation had traced, that is, using complex operations to the extent that the technology allowed. The revolution occurred when electronics was enlisted for the benefit of calculations. Their speed was such that elementary calculations such as the addition and subtraction of two numbers were so fast that other results could be easily obtained by many repetitions of those operations.

There are at least two claimants to the title "Father of the Electronic Computer." One is the German engineer Konrad Zuse (1910–1995) who in the years 1935–1938 developed and built an electronic computer that became fully operational in 1941. A number of the computers he built are displayed in museums around Germany, including at the Technical University in Berlin and at the Deutsches Museum in Munich. Nazi activity in Germany and World War II resulted in Zuse working in isolation from what was happening in the rest of Europe and the United States, and his contribution had limited impact. Zuse's competitors as computer pioneers were John Atanasoff (1903–1995) and Clifford Berry (1918–1963) of Iowa State University, who started building an electronic computer in 1937, the final version of which was also completed in 1941. That computer is exhibited in the university where they worked. Zuse's and Atanasoff and Berry's computers were constructed along the lines of the calculating machines that preceded the electronic age. In other words, their ability to be programmed was limited, and the input and output were specific for each computational task. The great leap forward in electronic computers started with ENIAC (Electronic Numerical Integrator and Computer), the computer whose structure owes much to John von Neumann.

We met John von Neumann above, in section 47, in the context of game theory. His contribution in that field was only a small part of his mathematical activity, which included work in many areas. He contributed to the foundations of mathematics and set theory, in which he was helped and supported by the mathematician Abraham Halevi Fraenkel (1891–1965) to whom von Neumann had sent a rough, hardly legible draft of a paper. Fraenkel immediately realized its potential and guided von Neumann through the first stages of his career (we shall refer to Fraenkel again in section 60, on the foundations of mathematics). Von Neumann also worked on hydrodynamics and quantum theory, in which he introduced an axiomatic basis. He was born to a Jewish family in Budapest in 1903, where his father was a banker and a lawyer. John's father was elevated to the nobility for his service to the Austro-Hungarian Empire, and John inherited the title, so that *von* became part of his name. John's exceptional abilities were discovered very early and were expressed in more than just mathematical talent. At an early age he mastered several languages and showed interest in social issues, economics, and related subjects. He obtained his doctorate in mathematics in Budapest at the age of twenty-two, having already written a number of fundamental mathematical papers. He held a teaching post in Berlin and in 1930 was invited to visit Princeton University. In 1933 he was offered a post at the Institute for Advanced Study in Princeton, which was established partly to take in scientists who had to flee from Nazi Germany. Von Neumann was one of the first scientists there; others included Albert Einstein and the mathematicians Oswald Veblen, Hermann Weyl, and Kurt Gödel, whom we shall meet again in the next section.

John von Neumann was one of the most prominent of the developers of the atom bomb and one of the main proponents of using it against the Nazi enemy. ENIAC was built for the American army to calculate trajectories of shells and missiles. Von Neumann was not part of the original team building the new computer, but when he heard about the project he joined the team and suggested, among other things, a new structure and other improvements that turned the computer into a hitherto unknown type of machine. The official announcement that the construction of the computer was completed was made in 1946, although it had been put to use earlier.

Von Neumann was one of the first to realize the potential of the unimaginable speed of calculation of the electronic computer. It was he who suggested the structure of the computer as we know it today. He inserted elements of memory into the computer and showed that software could be considered the same as data, thus making the computer a multipurpose tool. Modern computers can perform calculations faster and can process and store more data than could the computers of the mid-twentieth century, but the basic structure remains that proposed by von Neumann. That is why he is considered to be the father of the modern computer.

When the possibilities offered by the computer became apparent, the concept of computation changed as well. Until then, mathematical computation related essentially to numerical calculations, the solution of mathematical equations, weather forecasts, and so on. Today they calculate travel routes, maintain public records, retrieve data, search dictionaries, help in translation, search the Internet for the nearest restaurant, and of course there are the social networks. All these result from computations.

Recognition of the potential embodied in the fantastic speed of calculation was slow. The assessment that the world would require at most five computers is attributed to Thomas Watson, president of IBM, although there is no direct evidence that he ever said it. Even if he did not say it, the fact is that there was widespread lack of recognition of the new opportunities. The first electronic computer in Israel was built in the Weizmann Institute of Science; it was named Weizac and can be seen in the Ziskind Building where it was built, which now houses the Faculty of Mathematics and Computer Science (see the photograph).

The construction of Weizac was completed in 1954, and it was in use until 1964. Albert Einstein and John von Neumann were members of the steering committee for the building of the computer, and a large share of the institute's total budget in those days was devoted to that project. Einstein, so it is reported, expressed doubt about it and asked why Israel needed its own computer. Von Neumann, who as stated above realized earlier than others the enormous potential of the electronic computer, answered that Chaim Pekeris, the founder and leader of mathematical activity in the

Weizmann Institute and a leading light in the construction of the computer, could himself keep the whole computer occupied full time. During part of this period the president of the institute was Abba Eban. I have a copy of a letter Eban wrote to Pekeris, saying more or less: "Dear Chaim, Would you agree to explain to the members of the Institute's Board of Directors what an electronic computer can be used for. I would suggest," wrote Eban, "that you explain that it could help in carrying out surveys in Africa." (Eban was interested in foreign policy and later became Israel's minister for foreign affairs.) Pekeris himself realized the possibilities that the computer could offer, and his desire to build a computer was motivated by scientific considerations. One of his objectives was a worldwide map of the ebb and flow of the tides. With the help of the computer he succeeded in identifying a location in the Atlantic Ocean in which there is no tidal movement, namely, a fixed point for tides. Eventually, the British Royal Navy carried out measurements that confirmed Pekeris's computed result.

As stated, the new element in the electronic computer is its high speed. We recall that one of the reasons for the rejection of the model in which the Earth revolves around the Sun was the difficulty in trying to imagine such a high speed for the Earth's movement. Intuition that says at such a speed people would fly off the face of the Earth prevailed over the mathematical model. Similarly, despite the fact that arithmetic operations are inherent in the human brain, and mathematical calculations can be easily accepted and understood, it is not easy to understand intuitively the speed at which the electronic computer operates. That speed cannot be grasped by the human brain, and I am not sure that even today the possibilities offered by the computer are fully realized.

53. THE MATHEMATICS OF COMPUTATIONS

The development of the electronic computer necessitated a change in the attitude toward mathematical calculations, and this resulted in the development of a new mathematical subject called *computability theory* or *compu-*

tational complexity. To understand the source of the change let us recall the method of logarithms discussed in section 51. The method was developed to simplify calculations. It converted multiplication into addition and reference to two tables. This conversion constitutes a great improvement for the human brain. What eases things for the human brain, however, is irrelevant for an electronic computer. For a fast computer, multiplication consists of repeated additions and is therefore a much more efficient operation than using logarithms. For a computer it is simpler and more efficient to carry out the multiplication directly rather than to calculate the logarithms and then add them. The fact that the process is based on elementary operations means that the crucial factor in determining the efficiency of a computation is the number of elementary operations required. To compare two methods of calculation by an electronic computer we need to compare the number of elementary operations that each method must perform to give the result of the calculation. But what is an elementary calculation? And how can we compare two methods or two programs developed for different purposes and different computers? It was the British mathematician Alan Turing who provided the appropriate framework for discussing these questions.

Alan Mathison Turing was born in London in 1912. His talent for and leaning toward mathematics were revealed when he was very young. He completed his first degree at King's College, London, stayed there as a fellow, and carried out research on the foundations of mathematics. He presented an alternative approach to Kurt Gödel's result regarding the theoretical limits of calculations, an approach based on a system that would later be called the Turing machine. In 1938 he received his doctorate from Princeton University, with Alonso Church (1903–1995) as his doctoral advisor. Church was a well-known mathematician whose research in mathematical logic dealt with similar topics but with a different approach. After returning to England, with the outbreak of World War II, Turing joined a team of scientists working for the Allies, trying to decipher the coded messages transmitted by the Germans. The team met at a place called Bletchley Park, which still today serves as a meeting place and as a museum of the history of computing. Within a short time Turing was in a central position in the deciphering efforts. With the help of an electromechanical computer available to the team, the Bletchley Park group succeeded in deciphering the German Enigma code, an achievement that played a significant part in the Allied victory. Turing had been declared a war hero by the wartime prime minister, Winston Churchill.

After the war Turing continued his research at Manchester University. He made an important contribution, which we will discuss later on, to a subject called artificial intelligence, attempted to build an electronic computer, and carried out research in different areas of biology. His work was cut short when he was accused of having relations with a young homosexual, which was then a criminal offence in Britain. Turing admitted to the charges and agreed to undergo chemical treatment to depress his sexual urge, offered as an alternative to a jail sentence. However, the social pressure together with the physical pressure from taking the treatment was apparently too much for him, and in 1954, two weeks before his forty-second birthday, he ended his life by cyanide poisoning. Many years later, in 2009, the British prime minister Gordon Brown apologized on behalf of the government for "the appalling way he [Turing] had been treated," and in December 2013 Queen Elizabeth II granted him a rare "mercy pardon." In 1966 the Association for Computing Machinery (ACM) instituted a prize in honor of Alan Turing, the prestigious Turing Award.

Before we present the main points of the theory of computability, we should state clearly that it does not presume to assess the efficiency of specific mathematical computing programs and thus to help the potential user to select a calculation method or program better than others. To do that the user needs to try the alternatives and to compare the practical results. Computability theory is a mathematical theory that tries to find the limitations of the capabilities of different algorithms. There is no reason to be alarmed by the terminology. An algorithm is simply a group or a list of instructions to be carried out. Thus, detailed directions on how to reach a destination B from point A constitutes an algorithm. A cookbook is a collection of algorithms whose objective is to prepare tasty dishes from the ingredients by following preset recipes. Operating manuals for different equipment are sometimes in the form of algorithms. How to add or multiply two numbers is generally presented as an algorithm.

A list of instructions as a way of performing mathematical calculations always existed, of course. The sieve of Erastothenes (mentioned in section 14) as a method of finding the prime numbers is an algorithm. The algorithm states that to find all the prime numbers from, say, 2 to 1,000, first delete all multiples of 2, that is, all even numbers greater than 2. From the remaining numbers delete the multiples of 3, that is, 9, 15, 21, and so on. Then, delete all multiples of 5, and so on, until all the non-prime numbers up to 1,000 have been deleted. At the end of the process, that is, when the algorithm has been implemented, only the prime numbers between 2 and 1,000 will remain. Thus, we have described an algorithm that finds the prime numbers between 2 and 1,000.

The word *algorithm* is a distortion of the name of the Arab mathematician Abu Ja'far ibn Musa Al-Khwarizmi, who lived in Baghdad from approximately 780 to 850 CE. He made many contributions to mathematics. He determined the digits for the numbers from 1 to 9 and was the first to designate 0 as a legitimate digit in the notation of numbers. He developed methods for solving algebraic equations. The word *algebra* is derived from the title of his book *hisāb al-ğabr wa'l-muqābala* (*On Calculation by Completion and Balancing*), in which he included methods for solving such equations. He made no less a contribution by translating and in the spirit of the time, corrected and improved, the writings of the Greek mathematicians, and he thereby deserves the credit for playing a major role in and being largely responsible for the preservation of those important texts.

We have already said that the efficiency of different algorithms is assessed by comparing the number of elementary operations that they implement. Yet the questions remain: What is an elementary operation, and how can algorithms with completely different purposes be compared? That was the purpose for which the Turing machine was devised. It is not a machine in the usual sense of the word, using electricity or other fuel, producing smoke, and requiring maintenance from time to time. The Turing machine is a virtual machine, meaning it is a mathematical framework (and there are several versions of this framework). The machine receives input, an ordered finite number of symbols, or an alphabet, marked on a tape. The tape is not limited in length. The machine can be in one of several predetermined states. The machine has a "reader" that can identify what is written in one of the cells on the tape. As a result of what it reads, the "program" of the machine causes a change in the machine's state according to a predetermined rule, a change in the letter in that cell to another letter, also in a predetermined manner, and a move to read the letter in the adjacent cell, again according to a predetermined rule. Such a program creates a series of moves from cell to cell, possible changes to the letters written in the cells, and possible changes in the state the machine is in. These operations are the elementary operations that will enable algorithms to be compared. The algorithm ends when the reader reaches a cell with a prescribed symbol. The result of the algorithm is defined by the state the machine is in when it stops and what is written on the tape at that time.

What is surprising is that every calculation imagined until now, whether a numerical calculation, finding the minimal distance for a journey between two towns, or finding the meaning of a word in a dictionary, all these can be translated into the mathematical framework of this hypothetical, virtual machine. In other words, a Turing machine can be programmed by its letters and the rules for moving from cell to cell to perform the calculation. The emphasis is on "an acceptable calculation," meaning every imaginable calculation can be translated into a calculation on a Turing machine. That is a "thesis," known as the *Church-Turing thesis*, and not a mathematical theorem. So far no discrepancies have been found in this thesis.

We have stated that the efficiency of algorithms is compared via the number of elementary operations they perform on a Turing machine to achieve a desired outcome. We will repeat our warning, however, that the comparison that interests the computability-theory experts is not between two methods of finding, say, the prime numbers between 2 and 1,000. With such a definite objective, it is incumbent on the high-speed computers to find the solution that will be to our satisfaction. The comparison of main concern between two algorithms will be between their performance when, in our example, the upper limit is not a fixed number, say, 1,000, but a general number, N, where N becomes larger and larger. Obviously the larger N becomes, the larger will be the number of steps the algorithm will have to perform to find the prime numbers. The comparison between the two algorithms will be between the *rate of increase* in the number of steps in each algorithm as N increases.

We will explain now what is meant by the term *their rate of increase as N increases*. If the expression describing the number of steps required to solve the problem is given by an expression of the following form,

$$N^2, \text{ or } 8N^{10}, \text{ or even } 120N^{100},$$

we say that the rate of increase is *polynomial*, that is, an expression that is a power of N. If the expression giving the number of steps required is of the form

$$2^N, \text{ or } 120^{4N}, \text{ or even } 10^{4000N},$$

the rate of increase is *exponential*. The names reflect how the parameter appears in the expression.

Despite the fact that for a given algorithm there is a difference between a rate of N^2 and N^3, computational complexity theory experts put them together in the polynomial rate category, and similarly with the exponential rate. The class of problems that can be solved by algorithms that reach the result of the calculation at a polynomial rate is called class P.

Here there is another interesting distinction, as follows. In many cases it is far easier to check whether a proposed solution to a problem is correct than to find the solution itself. For example, in sudoku there is a 9×9 square with numbers between 1 and 9 appearing in some of the smaller squares. The object is to complete the entire grid such that every row and every column and every 3×3 sub-square of the nine that make up the whole table contains all the digits from 1 to 9.

5	3			7				
6			1	9	5			
	9	8					6	
8				6				3
4			8		3			1
7				2				6
	6					2	8	
			4	1	9			5
				8			7	9

Image from Wikimedia Commons/Tim Stellmach.

The objective of finding the solution, that is, filling all the empty boxes so that each row, column, and so on contains all the digits from 1 to 9, is likely to be difficult and to require many attempts. Yet when someone suggests a solution, it is a simple matter to check whether it is correct. Here too the computability experts are not interested in the number of steps needed to discover whether the proposed solution to a 9×9 sudoku is correct or not. They are interested in the number of steps required to check a sudoku whose size keeps increasing, in other words, sudoku of size $N \times N$, when N keeps rising, and in the rate at which the number of steps increases. (A sudoku in which arbitrary N numbers have to be arranged still uses an input with a finite number of symbols, as every natural number can be represented, for example, as a decimal number that uses only ten symbols.)

We quoted sudoku purely as an example. The difference between finding a solution and checking a proposed solution arises in many algorithmic tasks; we have already mentioned (in section 36) the public competitions for solving equations. To solve equations is hard. To check if a solution is correct is, in most cases, easy. Also, in checking solutions, a distinction can be drawn between a polynomial rate and an exponential rate. The class of problems for which proposed solutions can be checked using algorithms that suffice with polynomial rates of increase in the number of steps required is called NP (the letter N is derived from the term *non-deterministic*, as the check can be performed simultaneously on all the proposed solutions).

Now here is a chance to win a million dollars. The experts in computability theory do not know if the two collections of problems, those in class P and those in class NP, are the same or different. (It can be proven that every problem in P is also in NP, in other words, if it can be solved at a polynomial rate, then the proposed solution can also be checked at the same rate.) At the beginning of the third millennium the Clay Mathematics Institute in Providence, Rhode Island, published seven unsolved problems in mathematics and offered a million dollars for the solution of each one. The question whether P = NP is one of those questions. sudoku, for example, is in the NP class. If you can prove that every algorithm for solving $N \times N$ sudoku, with N rising necessarily, requires a number of steps that depends exponentially on N, you will receive a million dollars (provided you do so before anyone else does).

The discussion of P and NP classes is just one example of the many questions that the mathematics of computation deals with. For example, computability experts showed that if you present a polynomial algorithm to a sudoku problem, then for every problem in the NP class there is a polynomial algorithm (in particular, if you do that you will have solved the problem of the equality of P and NP, and you will receive a million dollars). Such problems, the polynomial solution to one of which implies that every problem in the NP class has a polynomial solution, are called NP-complete. Research in computability has so far identified a very large number of NP-complete problems.

The research is also involved in seeking efficient solutions to specific problems, or in proving that there are no efficient solutions to other specific problems. For instance, just a few years ago, in 2002, a polynomial algorithm was presented for checking whether a given number is prime. We will expand on this question in the next section. Another question that has far-reaching implications that we will discuss in section 55 is whether there is an efficient algorithm for finding the prime factors of a number that is the product of two prime numbers. The efficiency characteristic becomes even more precise, for instance, in identifying problems whose solutions require a number of elementary operations that are not larger than a linear expression of the size of the input, that is, an expression of the type aN for a given number a, or an expression that is a product of the type $N\ln(N)$, that is, linear in N multiplied by the logarithm of N. The latter is a slower rate than quadratic, and hence preferable. Hundreds of problems have been examined and the degree of efficiency of their possible solutions has been identified. Many problems are still unsolved; in other words, the rate of increase in the number of steps required to solve them has not yet been identified.

The question arises whether the abstract discussion of the number of steps required to solve a problem with an input of size N with N increasing is relevant to finding a solution to real problems that interest the users of the computer. No one in his right mind intends to try to solve a sudoku of size one million by one million, let alone of size N, as N tends to infinity. The answer is surprising. There is no logical argument that relates the rate of increase in the number of steps required as N tends to infinity to the number of steps required to solve real problems. Moreover, there are problems in class P that theoretically can be solved efficiently but in practice cannot be solved by today's computers in a reasonable time. Nevertheless, many years of experience have shown that, in general, intuition gained from questions with a parameter that tends to infinity also applies to versions with a large but finite parameter.

This is an appropriate point to discuss the difficulty and validity of intuition rooted in various algorithms and in the way computers work. As an algorithm is a series of instructions, it is easy for the human brain to understand, implement, and even develop intuition about how the algorithm works. Associative thinking, which is the basis of the human thought process, is founded on the approach of deriving one thing from another, a process analogous to an algorithm. The same applies to the computer. It is easy for very young children and youngsters to master the handling of a computer, and they even relate to it intuitively. Not infrequently, if you ask for help related to the working of your computer from someone with computer experience (not necessarily an expert), the answer will be, "Let me sit at the keyboard and I'll find out what to do." Sitting at the keyboard activates the associative reactions, the intuition. It is often assumed that this is age-related. I do not think it is. To use the necessary intuition apparently requires you to overcome the psychological barrier, a barrier of fear perhaps, regarding the new machine, to operate it, which seems superficially to be an impossibility. Adults are challenged by this barrier, not youngsters who are growing up in proximity to computers. Nevertheless, at least two difficulties along the route to an intuitive understanding of the computer should be acknowledged. First, as mentioned, the incomprehensible speed of the computer clashes with human intuition, the latter of which developed in an environment where changes occur much more slowly (although it will be easier for the generation born in the computer age to develop such intuition). Second, the computer itself does not "think" associatively. The computer, at least the computer of today, does not think the way humans think. It carries out very precise instructions, coded in the software, and sometimes logical operations are incorporated in the software instructions. To develop intuition about logical arguments is a difficult or even impossible task. Which is why one will come across reactions from even the most experienced experts such as, "I don't know what the computer wants."

54. PROOFS WITH HIGH PROBABILITY

The title of this section may sound like an oxymoron. Since the Greeks laid the logical foundations of mathematics there has been a deep-rooted consensus that the proof of a mathematical contention is either right or wrong. The title refers to a proof that a mathematical claim or proposition is correct with a certain probability. This does not mean subjective probability reflecting the degree of faith in the correctness of the proof, but rather an absolute probability that can be calculated. This is a new element in the logic of mathematical proofs, an element umbilically related to the possibilities that the computer opens up before us.

In practice, people encounter proofs with certain probabilities routinely in their daily lives. The idea of changing from an absolute precise examination of a claim to a statistical examination has always been inbred in humans. The stallholder selling apples at the market assures you that all the apples are fresh and perfect. It is impractical to check every single apple, so you will check a few apples, and if you find that the ones you check are all fresh and perfect you will be convinced that the seller's claim is correct, or at least you will consider it highly probable that his claim is correct. It is important that the apples you choose to check are chosen at random, so the seller cannot trick you. The higher you want the probability to be (that his quality claim is correct), the greater the number of apples you must check. If you want to confirm his claim with mathematical certainty, you would have to check all the apples on the stall. Customs officials wanting to satisfy themselves that the contents of a container at the port are as stated on the shipping documents, or Ministry of Health inspectors trying to establish whether the shipment of pharmaceuticals meets the requisite standards, must for reasons of efficiency settle for examining only a sample.

The innovation in proofs with a high probability lies in their mathematical content. Why then would mathematicians agree to detract from the apparent purity of absolute proof? The answer lies in the efficiency obtained by accepting high probability proofs, efficiency in the sense we defined in the previous section. Even in mathematics it is sometimes possible to prove a claim correct efficiently if it is acceptable that the proof is correct only with a high degree of probability. We will now illustrate this and then present a derived concept, and that is how to convince someone that a proposition is correct without divulging anything of the proof itself.

The example relates to prime numbers. We have seen that these numbers played a central role in mathematics throughout the ages and continue to do so in modern mathematics and its uses. Mathematicians have always been seeking efficient ways of checking whether a given number is prime. No efficient way was discovered, however. The naive checks, from the sieve of Eratosthenes described above to more sophisticated checks, require an exponential number of steps as a function of the number being examined so that it is not practical to use these methods to check a number consisting of several hundred digits. In the terms introduced in the previous section, checking the primeness of numbers was not efficient.

In 2002 there was a theoretical leap regarding the efficiency of checking whether a number is prime. The scientists Neeraj Kayal, Nitin Saxena, and Manindra Agarwal of the Indian Institute of Technology, Kanpur, presented a polynomial method of checking whether numbers are prime. The method is known as KSA, after its discoverers. According to the theoretical definition, KSA is efficient, but the algorithm proposed in the system is far from practical. The number of steps used by the algorithm was given by a polynomial of power 8. Since then the system has been refined, and the power is now 6. This is still too high to make the system practical for everyday purposes. Thus, still today, no practical way of checking definitively whether a number is prime has been found.

Years before the KSA system was published, the mathematician Michael Rabin of the Hebrew University of Jerusalem proposed an algorithm whose result was a declaration that either a number was not prime, or it was prime with a probability that could be predetermined. The inference of the approximation proposed by Rabin is of the type that Archimedes proposed for approximate calculations. That is to say, you, the consumer, choose the level of probability you would like, and the computer will give you the answer at that level of probability. The computation will take longer the smaller you want the probability of error to be. The system is practical, however, also for imaginary levels of probability. For example, you can determine that a number shall not be declared prime if the probability that it is not prime is greater than $\frac{1}{2}$ to the power of 200. As we have noted, the human brain has difficulty in imagining an occurrence with such a low probability, and in everyday decisions people ignore such events. So if the computer tells you that a number is prime unless something with a probability of less than $\frac{1}{2}$ to the power 200 happens, for all purposes it is reasonable to accept that the number is prime. The system is not applicable if someone wants absolute mathematical certainty. But in that case there will generally be no reasonable way of providing a solution at all.

Rabin was the winner of the Turing Award in 1976 for his work on automata theory and later became the pioneer of probabilistic algorithms. He based his primeness algorithm on one proposed in 1976 by Gary Miller of the University of California, Berkeley, that checks deterministically whether a number is prime. The algorithm is known as the Miller-Rabin algorithm. For those interested, we will describe one instance of it in detail and then briefly explain the general case (skipping the paragraph will not detract from the clarity of the text that follows).

Assume we want to discover whether a number of the form $n = 2m + 1$ is prime, where m is not an even number. A well-known theorem in number theory states that for every integer k between 1 and n, the following holds. If n is prime, then either the remainder when k^m is divided by n is 1, or the remainder when k^{2m} is divided by n is $n - 1$. Hence, if we choose such k and the two remainders above are not upheld, we are certain that n is not prime. It could happen, however, that the number is not prime and both the above requirements are fulfilled, so that the fact that the requirements are satisfied does not enable the conclusion to be drawn that the number is prime. In such a situation another well-known mathematical result is used, and that is if the number is not prime, at least half of the numbers between 1 and n will not fulfill the remainder conditions. Thus, if k is chosen randomly and the two conditions are satisfied, this means that the probability that the number is not prime is less than a half. If we randomly choose another number between 1 and n and again the two conditions are satisfied, the probability that the number n is not prime reduces to a quarter. If this is repeated 200 times and the two remainder conditions are fulfilled each time, the probability that n is not prime is less than $\frac{1}{2}$ to the power 200. Thus, in no more than 200 trials, we either know with certainty that the number is not prime, or we declare that the number is prime, with the chance of an error no more than $\frac{1}{2}$ to the power 200. Performing two hundred trials is trivial for a computer. The general case is somewhat more complex, but not very much so. Every odd number can be written in the form $n = 2^a m + 1$, where m is an odd number. The mathematical theorem mentioned above states in the general case that for every integer k between 1 and n, either the remainder when k^m is divided by n is 1, or the remainder when k to the power $2^b m$ is divided by n is $n - 1$ for at least one number b between 1 and a. Hence the two remainder conditions in the previous case become $a + 1$ conditions in the general case. This too is a simple operation for a computer to perform.

Probabilistic algorithms led to another interesting conceptual development, and that is the integration of interactive proofs and zero-knowledge proofs. The idea was proposed and developed in an article written in 1982 by Shafi Goldwasser, Silvio Micali, and Charles Rackoff, all then at the Massachusetts Institute of Technology (MIT). Goldwasser, a faculty member of the Weizmann Institute, and Micali were jointly awarded the Turing Award in 2012. Also here the goal is reached subject to a probabilistic error, and here too the probability of error can be extremely small, and the estimate of the error is an absolute estimate defined mathematically. We will learn about the method from an example.

The United States is divided into about 3,000 counties (the exact number changes). Assume that you know how to color the map of the United States using only three colors so that every county will be in one color, and no two counties with a common border (not just a corner) will be in the same color. Assume further that you want to prove that you are capable of coloring the map of the United States in this way, but without giving the least hint of how you are going to do it. Can it be done? Even if you show the colors of two nonadjacent counties, you have divulged some information, because we will then know if you have colored them with the same color or different colors. Here is the interactive method we referred to. We decide on the three colors you are to use, say yellow, green, and red. You color the map of the United States but do not show us the result. We select two neighboring counties, say Los Angeles and San Bernardino. You show us the colors in which you have colored them, say yellow and green. As they are not the same color, we have not caught you in an error. This just means that we did not catch you coloring two contiguous counties with the same color, yet you still may not be able to color the whole map in accordance with the conditions. If, however, we chose the two adjacent counties randomly, we obtained a certain measure of confidence that you can in fact color the map as required, because the probability that we will find an error in the two counties you showed us (that is, they are in the same color) is greater than say one in 20,000 (as there are not more than 20,000 pairs of counties with common borders). At the same time, you have not revealed anything about the way you colored the map, as we knew in advance that two neighboring counties will be in different colors. In the second stage, you repeat the coloring of the map, but you change the colors. Again we randomly select two adjacent counties, and you show us the colors you have used for them. If they are not the same color, the degree of our confidence in your ability to carry out the task as required will rise. Yet again we have learned nothing about the method you used to color the map, as the colors were changed. For example, even if Los Angeles was drawn by chance again, this time you may have colored it red. We can repeat this procedure many times until we will have as much confidence as we want in your ability to complete the job properly, and you still will have divulged nothing about how you do it.

Let us review another example. You wish to convince me that you have completed the sudoku shown above, without giving me any indication of the solution. For each digit from 1 to 9 you substitute a color, without telling me what color represents what digit, and without showing me the completed, colored sudoku. I select one of the nine columns or one of the nine rows or one of the 3 × 3 sub-squares at random, and you then show me your colored version of the column, row, or square I have chosen. If you have not completed the sudoku properly, I may not see nine different colors in my selection, and that will have a probability of at least $\frac{1}{27}$. Repeating this procedure two hundred times, with your choosing different colors each time and with me seeing nine different colors in the column, row, or square I select, will lead me to the conclusion that the probability that you did not complete the sudoku properly is less than $\frac{1}{27}$ to the power of 200. Yet the solution is not complete. You have to convince me that in the coloring process you did not change the numbers revealed in the sudoku at the outset. That can be achieved if at any stage, at random, I can ask that you show me the numbers represented by the colors in the squares where the revealed numbers were situated. If the numbers match that constraint, the chance that you cheated remains as low as it was.

Clearly to carry out these procedures manually is not a practical proposition, but for a computer it is a simple matter. Furthermore, the task will remain efficient even if we need to check a map with many more than 3,000 counties or a sudoku much larger than 9 × 9. That is not the case if you want to calculate how to color the map with only three colors with the given constraints or how to solve a huge sudoku. In addition, it is not known whether the three-color problem is in the P class. We know that it is NP-complete. In particular, as with the sudoku problem, if you manage to find a polynomial algorithm that can check whether the map can be colored as required with three colors, you will have answered the question does P equal NP, and if you are the first, you will have won a million dollars.

55. ENCODING

Encoding information, or cryptology, has always fascinated mankind. We have mentioned the German code, Enigma, used in World War II; it was broken with the use of electromechanical calculators. The mathematics of computation enables practical methods of encrypting to be used, which as far as we know today, cannot be decrypted even by the fastest computers. We will present the basic idea, which is known as a one-way function.

Take for example the multiplication and factorization of two prime numbers. When given two prime numbers p and q, finding the product, $pq = n$ is a very simple task, even if the numbers are big, and certainly for a computer. If we are given only the product, that is, n, to find p and q is today an impractical task, even for the fastest computers. An exponential algorithm can be found to solve the problem, but it is not known whether a polynomial algorithm exists. Such a relation, which can be calculated easily in one direction but is difficult to calculate in the reverse direction, is called a one-way function. Thus, in the framework of the state of mathematical knowledge today, the relation of the product of two prime numbers to the factorizing of the product is a one-way function. If you find a polynomial algorithm for the factorizing of this product, the function will no longer be called one-way and, moreover, your discovery will have far-reaching implications. Some other one-way functions, not many it must be said, have been put forward in the professional literature. The product of prime numbers and factorizing the result is a function that has found its way into day-to-day use.

The idea of using a one-way function for encoding was presented by three American scientists. Two of them, Martin Hellman and his colleague Whitfield Diffie, published the idea of encryption using the so-called public-key cryptology in 1976, following a paper written by the third, Ralph Merkel, on a similar topic. The three are generally acknowledged as being the pioneers of the method. The idea was translated into a practical system by three scientists then at MIT, Ron Rivest, who is still there, Adi Shamir of the Weizmann Institute, and Leonard Adelman, currently at the University of Southern California. The three were joint recipients of the Turing Award in 2002; and their initials, RSA, provide the name by which the system—currently the most widely used encryption system—is known today. The method is based on the fact that the product of two prime numbers is a one-way function. Moreover, not even an efficient probabilistic algorithm is known today for finding the prime factors of a given number. If tomorrow you find an efficient algorithm for doing that, you will have neutralized the primary instrument for encoding currently in use. The mathematics underlying the use of a one-way function is as follows.

I want to convince you that I know the result of tomorrow's drawing in the national lottery without actually telling you what the winning numbers are. I choose two large prime numbers and insert the results of the next day's draw into the lower of the two numbers; for example, I identify the six winners in the drawing at the beginning of the lower prime number. I multiply the two prime numbers, show you the product, and tell you how the winning numbers appear in the lower prime factor without showing you the number itself. If you manage to factorize the product, you will know the numbers that I encoded. Multiplying the two numbers is a simple matter, as we have said. In contrast, to factorize the product is extremely difficult. Even if you were to work all night and all the next day, using the fastest computers, you would be unable to factorize the product and discover the winning numbers of the draw. After the draw and the announcement of the winning numbers, I reveal to you the factorization and the winning numbers written there. I cannot cheat. The product is known to you, and it is easy to check that that number is in fact the product of the two prime numbers.

The method can be used to send encrypted messages. I provide you beforehand a large prime number. When I want to send you a message, I embed it in a new prime number and send to you the product of it with the prime number you already have. You can decipher the message very easily. The product of the two numbers can be safely delivered over the telephone or via e-mail, and there is no concern that anyone who might intercept the e-mail message or eavesdrop on the phone call would be able to understand the message. Even if the method is widely publicized (this is the source of the term *public key*), no one but you will manage to decipher the code, that is, to calculate the prime factors of the number. This example illustrates the mathematical basis of the system. Its practical application requires several adjustments, which we have not described here, particularly if we wish to send encoded messages to many customers and to enable each one to open the message. The underlying idea of the system in this case is to give each customer one of the prime numbers and to send him the product of the encoded number and the prime number he knows. It is a simple matter to divide the product by the prime number he has been given, which enables him to decipher the message easily, while the message remains coded and inaccessible to anyone who does not have one of the prime numbers.

56. WHAT NEXT?

The ideas presented in the last two sections are specific to the era of very fast computers. Yet even the fastest computers today do not fulfill all our desires for rapid computation. Moreover, as computers become faster, our desires and aspirations for calculations become more intense, and we will never reach a situation in which the computational power available satisfies us. In addition, as we have seen, there are problems that theory indicates cannot be solved efficiently. On the other hand, some daily computational tasks are solved by the human brain satisfactorily, tasks that the fastest computers using the best programs known are incapable of handling. In like fashion, processes in nature can be seen that are essentially computational, whose implementation is incomparably superior to that of the fastest computers. Therefore, understanding nature's processes is likely to provide a clue as to how to arrive at more efficient computational processes. Thus nature does not only serve as the main source of objectives, it also provides inspiration for developments on which mathematicians and computer scientists are engaged in their efforts to advance computational ability. In this section we will describe several directions in which mathematics and technology are developing, with the intention of promoting computational possibilities and the possible uses of computers.

The first objective is related to the human brain. The human brain is capable of performing tasks that are out of reach to the fastest computers.

For example, facial recognition. Sometimes a quick glance at a large, tightly crowded group of people is sufficient for you to recognize a school friend whom you have not seen for some years. There is no computer program that can come anywhere remotely close to being able to do that. That gap is used for instance in entry to certain websites on the Internet, when you are asked to identify and copy a random alphanumeric series written in a deformed way. The human brain can identify the numbers or letters, while computer programs cannot. A second example is understanding a language and translating from one language to another. Even if you hear just part of a badly structured sentence with an unusual accent or pronunciation, you may well be able to understand what is being said. A computer will understand and react correctly only if the sentence falls within the framework that it has been programmed to recognize. There are programs for translating from one language to another, but their quality is acceptable only if the material is of a technical nature and is in a predetermined structure. Checking the meaning of what is written, particularly reading between the lines and translating accordingly, straightforward tasks for a human translator, are beyond the reach of any computer today. To go further, people sometimes mean the opposite of what they write or say. For instance, someone might want to say that every head injury, however serious, should be treated, and might write, "There is no head injury too serious for you to give up hope" (compare with the example in section 50). Someone else will correctly realize the writer's meaning and will comment and act accordingly, but not a computer. A computer would understand from the written text that one must give up on all head injuries—the correct grammatical meaning. No current translation program can correctly analyze the intention of what is written, which humans do relatively simply. The research area that develops computer programs that can imitate human abilities is called artificial intelligence.

We have spoken of Alan Turing as the man who laid the foundations of the mathematics of computation. He also recognized the gulf between the computational abilities of the human brain and those of the computer, and in an article in 1950 he presented a challenge to the science of artificial intelligence. It is known as *Turing's test* and is the following. When will we be able to say that a machine has human intelligence? A person can talk to another person or to a machine even when he cannot see his collocutor. When the speaker cannot decide whether his collocutor is a person or a machine, the machine has reached the threshold of human intelligence. The test highlights the great difference between human thinking and the computational procedure. Turing's test does not relate to the extent of the knowledge or the speed of reaction. In these, every computer is vastly superior to humans. The test examines the extent to which the computer can imitate human thinking. We have stressed more than once that evolution has taught us to think intuitively, which is incompatible with mathematical logic. Our judgment is based on our own life experience and also on experience gained through millions of years of development, experience coded into our genes and transmitted by heredity.

Can the computer emulate that experience? The answer is unclear. Artificial intelligence has many achievements to its credit and has made impressive progress, but only time will tell whether programs based on logic can reach the achievements of thinking produced by evolution.

One of the ways of advancing a machine with artificial intelligence is to copy the way the brain itself carries out its thinking tasks. We do not know or understand all the elements of thought in the human brain, but we do have an idea of the brain's anatomy. Specifically, the brain consists of a huge number of neurons, forming a network of neurons, with each neuron connected to several of its neighbors and capable of transmitting an electrical signal to each one of them, according to the condition of the neuron and the environmental conditions. The thought process after receiving any input occurs by means of transfers of these electrical signals, whose number is inconceivable, and they take place simultaneously between different clusters of neurons. At the end of the thought process, the brain is in a new situation, where the output might be an instruction to one of the organs of the body to react in the appropriate way, or preserving this fact (the input) or another in the memory. One direction in which efforts are being made to construct a machine to achieve the same capabilities as the brain is focusing on attempts to copy this structure, called neural networks.

That is an attempt to build a mathematical model that takes advantage of the structure created by nature to perform efficient computations. At this stage the target is still far away, partly because the implementation of the mathematical model uses those logical computers that lack human intuition. If indeed intuition and the exercise of human discretion are just the outcome of a huge number of neurons in the brain, computers and computer programs that imitate neuron networks have a chance of passing Turing's test. It may well be, however, that within human thought there is another structure as yet unknown to us.

There is another impressively effective computational process in nature, demonstrated by evolution. The debate between the proponents and opponents of the theory of evolution, with the latter favoring the concept of intelligent design, is not purely science based. The opponents of the theory of evolution do however put forward a scientific argument, that such an ordered and successful structure cannot be the result of a random, unprogrammed process. It should be noted that although randomness does play a role in evolution, the process itself is defined deterministically. Among the random mutations, the one to survive will be the one best adapted to its given environment. The mutations occur in the process of gene reproduction. It is difficult to grasp how efficient the process is. It is hard to imagine the development via mutation and selection from one protein molecule to the vast array of animal life on Earth. That very success serves the opponents of the theory of evolution, who claim that the outcome does not stand to reason. Yet the evidence that that is indeed the way all living things developed keeps accumulating. Such evidence includes computer simulations of the process of evolution, which show that the process is feasible. Over and above computers' ability to simulate evolution itself, they are capable of imitating the principles of the process of evolution, that is, they can construct algorithms based on mutation and selection to achieve other computational tasks efficiently.

Such algorithms are called *genetic algorithms*. It is difficult to identify the first stages of the system, as so many scientists contributed to its development in different directions. One well-ordered formulation that incorporated basic assumptions and rules governing its progress, which made the subject very popular, was proposed by John Holland of the University of Michigan in a book published in 1975. We will illustrate the principles of the system by means of an example.

Assume that we have to build a program for takeoffs and landings at a very busy airport. The aim is to produce a timetable that will meet the safety requirements, constraints regarding the time period between landing and takeoff, airlines' requirements of weekly landings and takeoffs, and other such requirements, while minimizing the time a plane is unnecessarily on the ground at the airport. The problem can be stated in the form of complex mathematical optimization equations, but there is no way of solving such problems completely even with the most advanced computers. The genetic algorithm method offers a different approach. First, construct a timetable, which may be far from optimal, that satisfies the various constraints and requirements. Next, try various mutations, small changes to the timetable, while still fulfilling all the requirements. For each mutation, measure whether the optimization index has improved, and if so, by how much. A computer can easily perform hundreds of such changes and checks. Then select the ten results with the best optimization outcomes. For each of these, repeat the process, that is, introduce further changes, and again select the best ten of these second-generation results. Keep repeating the process until there is no further significant improvement in the results. The final result may depend on the initial timetable proposed, but a computer can go over the process with a different starting timetable. Several such attempts will produce a satisfactory timetable for the airport.

Widespread use is made of the above method in different areas of programming and engineering. This development may be classified as a stage in mathematical developments prompted by the speed of computers. Recently uses of the method in new areas have been published.

The purpose of one of these is to identify the equations, or laws of nature if you will, underlying experimental data. This is not really new in mathematics. Gauss used the measurements that were available regarding the asteroid Ceres and, using the least-squares method on the parameters of the equations of the motion of the asteroid, predicted when Ceres would reappear from behind the Sun. Gauss's starting point was the information that the equations of Ceres's motion were Newton's differential equations. Economics provides another example. The current practice in applied economics research is to assume a general form of dependence between two macroeconomic variables, say input and output, and to use mathematical calculations to find the parameters that provide the best correlation. In such and similar cases, the first proposal, whether Newton's equations or the relation between input and output, is based on human understanding.

The computer scientist Hod Lipson of Cornell University is trying to go one step further. When it is given a collection of data, say data on the motion of a particular body, the computer starts with a general equation and measures to what extent it describes the body's motion. Then the computer tries "mutations," that is, small changes in the equation itself, and from the thousands of possibilities, it selects say the ten that best describe the body's movement, and so on, advancing by means of mutations. This genetic algorithm ends with the equation that best describes the body's motion. Of course the computer is restricted by the collection of mutations that the program allows, but the incredible speed of the calculations enables efficient examination of a wide range of equations. We could by way of caricature say that if Newton's laws of motion were unknown to us today, it would be enough to know that the physical law is a differential equation connecting location, speed (the first derivative), and acceleration (the second derivative), and the genetic algorithm would reveal the equations from the empirical data. This does not mean that Newton was superfluous; he introduced the differential equations to formulate his laws, and such a step is beyond the capabilities of a computer (at least today's computers). However, with regard to many engineering uses, it may be easier to use the algorithms that Lipson and his colleagues are developing than to try to use analytical logic to find the exact equations that describe motion.

Another even bolder development is the attempt to use a genetic algorithm to develop computer programs. Moshe Sipper of Ben-Gurion University of the Negev is building such algorithms. Assume that we wish to write a computer program that will perform a particular task efficiently. Let us start with a simple program that more or less carries out the task. We let the computer examine the programs that it obtains by introducing small changes to our initial program. From these mutative programs, we select say ten that best perform the set task. We perform another round of mutations on these and continue thus until the program performs the task we set. Of course that is the idea. Its implementation is much harder. It is difficult to estimate today how far these developments, based on a method in which evolution works, will reach and with what measure of success.

We will briefly mention two other directions of development, which in effect are trying to replace electronic computers with computers of a different type. The first is molecular computation. The immune system can be seen as an extremely efficient computation process. The input is the invasion of bacteria into the blood stream. The computer, in this case the immune system, identifies the bacteria and sets in motion the creation of white blood cells of the type needed to kill the bacteria. In most cases the output is the dead bacteria. The computation is performed by the meetings in the blood between the huge number of bacteria and the huge number of blood cells. Biomolecular scientists can draw conclusions regarding the process from the outcome, which is a certain protein. The biomolecular knowledge can be used to calculate mathematical results. From the result of mixing several types of molecules we can learn about the components that went into the mixture. These components can represent mathematical elements, and the mixing element can represent mathematical operations, for instance, on numbers, and thus mathematical computations can be performed. The initiator of this system is one of the pioneers of the RSA encryption system, Leonard Adelman, of whom we spoke in the previous section. The molecular calculations are still in their infancy, and we cannot know where these experiments will lead

Another direction is quantum computers. At this stage the idea is almost completely theoretical. The computational structure of the quantum computer is the same as that of the electronic computer, thus we are back again within the framework of Turing machines. Every quantum computer can be described by a Turing machine, but the speed of a quantum computer will be much greater. The reason is that the states of the quantum computer are based on waves that describe the states of the electrons. The wave characteristics of quantum superposition make situations that are like an assembly of elementary states possible, that is, a combination of the chances of being in one of those states. In that way the number of possible states can be increased. The result is that the states are not a series of binary numbers, 1 or 0, but a row of numbers from among more values. Specifically, a calculation that, in a normal Turing machine requires an exponential number of steps, is likely to require only a polynomial or even linear number of steps in a quantum computer. In the context of the subjects dealt with in this book, it is not necessary to understand the physics underlying the quantum computer. Progress is essentially a matter of engineering, and is also far from implementation. If a quantum computer is realized, the world image regarding the computational difficulty we described in section 53 will change. Problems that currently require an exponential number of steps will immediately come into the P class; in other words, they will be capable of being solved efficiently. An efficient quantum algorithm for solving the problem of factorizing the product of prime numbers already exists, but the mathematical solution is waiting for the construction of the quantum computer.

We will conclude with something from a different direction with the potential for a revolution in the way mathematics itself advances. Harnessing the computer for mathematical proofs is nothing new. The most famous example is the proof of the four-color theorem. In section 54 we described the problem: given a two-dimensional map, can it be colored with only four colors? It is easy to draw maps that cannot be colored with only three colors (see the picture). The mathematical problem of whether every two-dimensional map can be colored using four colors was an open question for many years. The problem was already described in the middle of the nineteenth century. Toward the end of that century, in 1879, a solution to the problem was published, but within a few years an error was found. Almost a hundred years passed, and in 1976 two mathematicians from the University of Illinois, Kenneth Appel and Wolfgang Haken presented a full, correct proof, but it was based on calculations performed by a computer. They had to examine many possibilities, about two thousand, and that was done by computer. In this case the computer's contribution was the great saving in time, but it also added to the reliability of the proof. If they would have relied on humans to carry out all the checks, the chance of error would have been much higher. Such computer assistance to mathematics is still technical (and yet some mathematicians are not prepared to accept such technical help as part of a mathematical proof).

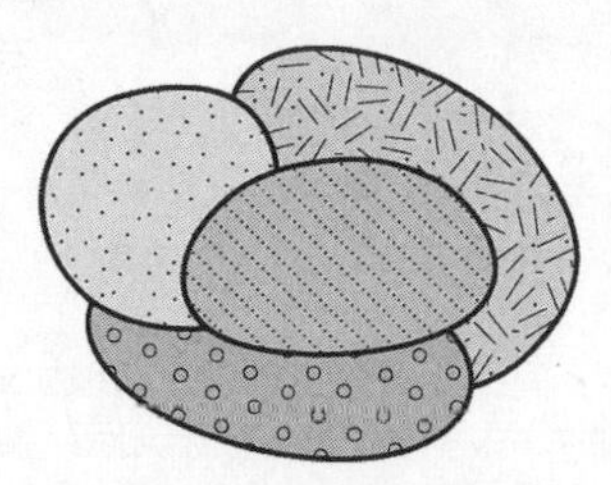

Can computers be used to prove mathematical theorems beyond being a computational aid? Doron Zeilberger of Rutgers University in New Jersey claims that the answer is yes. Moreover, he claims, the computer can reveal mathematical facts outside human reach. The type of theorems Zeilberger is working on are mathematical identities. The equality, or identity, $(a + b)(a - b) = a^2 - b^2$, was familiar to the Greeks in its abstract state. More complex identities came to light throughout the years of mathematical development, and they do not relate only to numerical identities but also to identities between other mathematical objects. Computer programs that operate on symbolic expressions have existed for many years. Zeilberger used these programs to prove important identities in algebra and used a computer to reveal new identities. He valued the computer's contribution so highly that he added the computer as a coauthor of some of his scientific papers. The computer is named Shalosh B. Ekhad. Clearly Shalosh B. Ekhad coauthored only those papers to which it contributed. On the other hand, it also cooperated with other mathematicians, and it is sole author of several papers. (At the time of writing these lines, there are twenty-three papers listed in Shalosh B. Ekhad's list of publications, and it has cooperated with thirteen authors.) Beyond healthy humor, I think there is something basic in this approach. Zeilberger claims, and that claim cannot be ignored, that the day will come when computers will reveal mathematical theorems that will be difficult for humans to understand.

CHAPTER VIII
IS THERE REALLY NO DOUBT?

Is a mathematician someone who doesn't know what he is talking about and doesn't care if what he says is true? • Is set theory the crowning achievement of the human race? • Is set theory a sickness that mathematics will overcome? • Can you believe someone who says, "I am lying"? • How many pages of the book *Principia Mathematica* do you need to read before you reach the proof of the equality 1 + 1 = 2?

57. MATHEMATICS WITHOUT AXIOMS

Since the Greeks made mathematics into a discipline based on axioms and logical rules of inference, mathematicians have viewed those as the solid foundation on which mathematics is based. Initially the axioms reflected physical truth, or ideal mathematical truth, which cannot be disputed. Later, the concept of an axiom was interpreted more widely, and lists of competing axioms were examined. Modern mathematics also allows mathematical developments under different lists of axioms that contradict each other with the intention of comparing the mathematical conclusions deriving from the various axioms. Be that as it may, the generally accepted assumption was that developments that fulfilled the criteria of the logical rules of inference could be relied upon. At the same time, as intuition that developed in the course of evolution is inconsistent with a significant part of the logical elements used in mathematical analysis, in many mathematical developments throughout the ages mathematicians ignored the strictness demanded by the logical method so that in day-to-day research both definitions and proofs relied only on intuition or formalism, without relating to the framework of logic. Here are few examples.

The Greeks had come across negative numbers, but they rejected them as being unacceptable. The mathematician Diophantus of Alexandria, who lived in the third century CE, published methods of solving algebraic equations, and in a formal way found solutions that were negative numbers. However, he did not accept them as real solutions to the equations. The reason was integral to the approach that led the Greeks to a definition of numbers that was geometry based: every number reflects a length of a section that can be related to a length of one unit. A rational number, say $\frac{5}{3}$, is the length of a section that is five times as long as a third of a section of unit length. An irrational number, say $\sqrt{2}$, is the length of the diagonal of a square with sides of length one unit. All these are intuitive geometrical definitions based on a system of axioms that the Greeks considered perfect. In such a system, there is no room for negative numbers.

Later on, the Indians permitted the use of negative numbers for purposes of solving problems. For example, a number can be written as negative on one side of an equation and can be transferred to the other side, but they too did not recognize negative numbers as independent mathematical entities. The Arab mathematician Abu Ja'far Al-Khwarizmi, whom we mentioned in connection with the discussion of algorithms, also allowed the use of negative numbers as a means of carrying out calculations, but he denied negative numbers the right to their own existence. Over time, many others used negative numbers to solve problems, and the method became embedded in mathematical practice, but those numbers were not accepted as legitimate entities. The debate over what they were also occupied mathematicians in the seventeenth century, who, although they used negative numbers, did not agree among themselves about various arithmetic operations involving those numbers. Leibniz pondered over the result of multiplying two negative numbers, or dividing 1 by negative 1. Euler proposed the solution $(-1) \times (-1) = 1$ to preserve the rules of multiplication but with no logical or intuitive basis. This disagreement persisted until the eighteenth century and beginning of the nineteenth century.

Augustus De Morgan (1806–1871), one of the best-known mathematicians of his time, used the following exercise to explain why negative numbers are of no significance. A fifty-six-year-old man has a son aged twenty-nine. When will the father be twice as old as his son? Simple arithmetic that any schoolchild today can do easily shows that the answer is –2, and we understand that to mean that two years ago the father was twice as old as his son. In the nineteenth century, however, De Morgan's conclusion was that the answer –2 was absurd. At the same time he argued that the question could be put differently and the correct answer obtained. That is, the question should be how many years ago was the father's age twice that of the son, and then the answer is 2. The approach accepted today, that the numbers are ranged along an infinite axis, with the negative numbers measuring the distance along what is called the negative direction, is a relatively modern one. Thus, we see that the negative numbers were developed independently in different places around the world, and they were used as an efficient tool for solving mathematical problems for a long time before they were recognized as representing a mathematical basis worthy of being defined by axioms.

Another example of a concept that preceded a logic-based system is that of complex numbers. These are numbers in the form of $a + ib$, where a and b are real numbers, and i is the square root of negative 1, that is, $i = \sqrt{-1}$. These also appeared as formal expressions in solutions of algebraic equations. For example, the equation $x^2 + 1 = 0$ has no real-number solution, but both i and $-i$ are solutions to the equation. Today this number system is well known and is used for various purposes, including the description of natural phenomena. There are also axioms that describe the system of complex numbers, but mathematicians were using complex numbers long before the system of axioms was constructed. One of the first to use them was Gerolamo Cardano, of whom we wrote in section 36, who was famous for solving polynomial equations and proposed complex numbers as a solution without explaining what those numbers are and without referring to axioms. Others developed the system further, but as in the case of negative numbers, many members of the scientific community opposed the adoption of the concept that complex numbers were legitimate numbers. Descartes, for example, was among the opponents, saying that such a solution had no significance, and he called *i* an imaginary number, intending that as a term of denigration. The name stuck, and is still used, but without any negative connotations. Neither did Newton, later on, accept those numbers as legitimate. Today complex numbers are used as a convenient description of mechanical systems that Newton himself analyzed and developed, a possibility that did not occur to him. Later, the arithmetic of complex numbers was developed formally, and many additional uses were discovered. The logic of complex numbers, together with the axioms that define them, was determined only in the nineteenth century, but that did not prevent its use much earlier.

Infinitesimal calculus as developed by Newton and Leibniz also lacked a base in logic accompanied by appropriate axioms. We have spoken of the article by Bayes that defended Newton's theory against the bitter attacks made by Bishop Berkeley. The latter, and others, denied infinitesimal calculus, as it was constructed without a basis in logic. The attack was given some support because of the fact that infinitesimal calculus was developed without regard to the normal rigorous mathematical approach, as a result of which many blatant errors were made and revealed. Newton and Leibniz defined the derivative of a function at a given point as the rate of change of the function at that point. The geometric expression of that rate of change is the slope of the tangent to the function at that point. They used the term *the tangent*, with the definite article *the*. They applied that definition to every function and at every point, ignoring the question whether a tangent exists. It is easy to find functions that do not have tangents at particular points, for instance, the function that ascribes to a number its absolute value (this is generally written as $|x|$) does not have a tangent at 0 and therefore does not have a derivative at that point. Perhaps we could come to terms with this example by ignoring the single point 0, but it is more difficult to satisfy intuition regarding derivatives in the case of the example given by Karl Weierstrass (1815–1897), one of the most highly respected German mathematicians. He presented a continuous function that did not have even a single tangent. The statement by Émile Picard (1856–1941) is quoted with good reason: If Newton and Leibniz had known of Weierstrass's example, infinitesimal calculus would not have come into being.

Thus, more than a hundred years passed before mathematicians undertook the task of proposing a more rigorous framework for infinitesimal calculus founded on accepted logic.

58. RIGOROUS DEVELOPMENT WITHOUT GEOMETRY

It was not until the nineteenth century that the mathematical community began to become involved in a reexamination of the logical basis of mathematics and the degree of rigor in its development. The words *not until* are used advisedly, as for thousands of years from the development of mathematics by the Greeks mathematicians relied on their intuition and were certain that it was consistent with the system of axioms determined by Euclid. This does not mean that the status of the axioms was totally ignored. There were sporadic attempts to check the compatibility of the axioms or to replace some of them with others, but these attempts focused on specific subjects, such as the discussion of the parallel-lines postulate discussed in section 27. Mathematicians in general did not consider it necessary to perform an overall examination of the foundations of mathematics.

Two interrelated factors brought about the realization that the concepts that mathematics had used for a long time and the axioms underlying mathematics should be reexamined. One was the growing number of definitions that were not sufficiently clear, definitions that even caused errors that were discovered in the use of infinitesimal calculus. Newton's formulation, specifically the use of fluxion as the basis for the description of the change of direction of a function, may have been appropriate for his own intuition, but it was thwarted many others. Leibniz's formulation, the division of infinitesimal quantities of the form $\frac{dy}{dx}$ was convenient to use, but it did not relate to the essence of the infinitesimal quantities, and mathematicians developed intuitions about them that often resulted in errors. Both approaches, that of Newton and that of Leibniz, and in their wake all the following developments of the theory, were based on geometry.

Then the second factor appeared, highlighting the need for renewed study of basics. The logical foundations of geometry throughout the generations were considered stable and thus not in need of examination and checking. And then, as described in section 27, questions and doubts arose at the beginning of the nineteenth century regarding the validity of the axioms that constituted the basis of geometry. The questions asked related to the absolute correctness of axioms that describe the geometry of nature, as well as to their logical completeness. For example, Euclid relates to the situation in which two points lie one on each side of a straight line. When we encounter such a description, a clear picture immediately comes into our minds of a straight line with a point on either side. But what are the two sides of the straight line? Their existence does not derive from the axioms. Imagine a long tube, and imagine a straight line drawn along its length. Does that straight line actually have two sides? Such doubts led mathematicians, with the leading French mathematician Augustine-Louis Cauchy (1789–1857) at their head, to redevelop differential and integral calculus, but this time based on the system of numbers and not on geometry.

We will not delve here into the details of the developments but will just illustrate one concept. As stated, for Newton, Leibniz, and their followers the derivative was defined as the slope of the tangent of a function at a point. A tangent and its slope are clearly defined in geometry. To define them using numbers alone, Cauchy used as a basis a precise formulation of the concept of the limit of a sequence of numbers. Archimedes had already related explicitly to the concept of a limit, but without defining it. Cauchy defined a limit as follows:

The number z is the limit of a sequence of numbers x_n if for every number ε greater than zero there is an index m such that for every index n greater than m the distance between x_n and z is smaller than ε.

Does that sound complicated? Indeed, the definition is a complicated one. We have already made the point that if a formulation has many quantifiers, and here we have at least three and the order in which they appear is also important, we cannot grasp it intuitively. If there are among our readers some who completed a course in differential calculus, they may have encountered this definition at that stage, but they will surely also remember the difficulties, or even nightmares, that this and similar definitions caused them and their colleagues.

When the concept of a limit is clear, the concept of the derivative of a function $f(x)$ at point x_0 can be defined as follows:

The derivative is the limit of the numbers

$$\frac{f(x_0 + h_n) - f(x_0)}{h_n}$$

for every sequence of numbers h_n that are different from 0 whose limit is 0.

Does that sound complicated? Indeed. Yet note, the definition is based on only numbers and is independent of geometry. The motivation for the definition, namely, the slope of the tangent, is geometric, but the definition itself does not use geometry.

The point should be made, and it will be repeated in connection with other developments, that the purpose of this rigorous development was not to give a better understanding of the concepts. We will go further and say that for a better understanding of the concepts they should be illustrated by a geometric drawing. The incentive behind the development was similar to that driving the Greeks, an attempt to prevent errors deriving from geometric illusions.

Basing infinitesimal calculus on numbers avoided the need to rely directly on geometric axioms, but it did not avoid indirect dependence on geometry, because the definition of the numbers was itself geometric. An example we have quoted previously is the definition of irrational numbers such as √2, defined as the length of the diagonal of a square with sides of length 1. With this realization, attempts began to provide a non-geometric base for irrational numbers themselves. Two of Germany's leading mathematicians at that time, Karl Weierstrass, mentioned in the previous section, and Bernhard Bolzano (1781–1848) based the concept of an irrational number on the limits of numbers. Thus, √2 would be defined as the limit of positive rational numbers, say r_n, that satisfy the requirement that the series $(r_n)^2$ itself has a limit, and that is the whole number 2. Later, the German mathematician Richard Dedekind (1831–1916) proposed a different definition of irrational numbers. His definition uses what is named after him, that is, Dedekind cuts, and is the definition that is taught still today in mathematics classes in universities.

The number $\sqrt{2}$, for example, is defined as the pair of sets of rational numbers, say (R_1,R_2), where R_1 is the set of rational numbers in which each one is smaller than a rational number the square of which is smaller than 2, and R_2 is the set of rational numbers in which each one is larger than a rational positive number the square of which is greater than 2. Other irrational numbers are similarly defined as pairs of sets of rational numbers.

To the reader who has not personally experienced this type of definition in the past, this will seem strange. A single number, whose meaning is clear and has been clear for thousands of years, is now defined as a pair of sets of rational numbers. Yet this is the price to be paid for the aspiration to avoid geometry. The definition of irrational numbers by means of Dedekind cuts was not intended to make it any easier to understand what an irrational number is. No one thinks he is clarifying what an irrational number is by presenting it as a pair of sets of numbers. The geometric definition is in fact simpler and more comprehensible. The reason for this development was to avoid using geometric language, even if by so doing it greatly complicated the concepts.

Defining irrational numbers without resorting to geometry did not entirely remove geometry from the picture, as rational numbers, sets of which are used to define irrational numbers, are also defined geometrically, based on the plane axioms. Again, the need arose to define the rational numbers without geometry. I will now present a development that I learned in my first class in university. It will be relevant to the last part of this book, but the details can be omitted without losing the main message.

Let us agree that we know what the natural numbers are, that is, 1, 2, 3, and so on, and we also know how to add and multiply the natural numbers. We now define the positive rational numbers.

First, we look at pairs of natural numbers (a, b) (the explanation, which helps us understand but which may not be used in the definition, is that (a, b) is the rational number $\frac{a}{b}$). We define equivalent pairs thus: the pair (a, b) is equivalent to the pair (c, d) if $ad = bc$ (according to our explanation, the equivalence does ensure the equality $\frac{c}{d} = \frac{a}{b}$, in other words, that it is the same rational number).

Once we have understood what equivalence means, we can define a positive rational number as follows: a positive rational number is a collection of pairs, with the numbers in each pair equivalent to each other. In addition, addition and multiplication of the rational numbers must also be defined. We define the addition of (a, b) and (c, d) as the collection of pairs equivalent to $(ad + cb, bd)$, and their product as the collection of pairs equivalent to (ac, bd) (we suggest that the reader check the operations in light of the interpretation).

These definitions reflect what we understand by the term *rational numbers*, and they are entirely free of dependence on geometry. It should be repeated and stressed that it is difficult to understand the definitions without relating to the intuitive grasp of rational numbers, and the only reason for this somewhat strange presentation of the definitions of quantities that we all essentially understand, is the desire to avoid reliance on geometry. Thus, as we have seen, irrational numbers, and hence the real number line, can be defined without resort to geometry.

Note that these definitions, and others that we have not included here, use the natural numbers and also the concept of a set. The sets were used in the definition of rational numbers, via equivalence of sets, and also in the definition of irrational numbers, via Dedekind cuts.

59. NUMBERS AS SETS, LOGIC AS SETS

The reliance on sets in the development of the foundations of mathematics, and in particular the redefinition of numbers on the straight line, the definition of limits and hence calculus, was in general accepted and greatly welcomed by the mathematics community. With regard to applications, it can be shown that all results previously proved geometrically were correct also on the basis of set theory. For example, the equality $\sqrt{2}\sqrt{3} = \sqrt{6}$ can be proven by geometry but can be proven more simply using Dedekind cuts (to do this we must define and understand the product of two cuts; we will spare the reader that step). It was even more satisfying to view this from the aspect of the foundations of mathematics. Indeed, as George Boole showed, there is full parallelism between sets and the natural operations that can be performed on them on the one hand, and logical arguments on the other (Boole developed this believing that was the way to analyze probabilistic events, as mentioned above in section 41). Some examples follow.

The parallelism between sets and logical propositions can be seen when we consider a claim to mean the set of all the possibilities of satisfying it. For example, we will take the claim "it is raining" to mean the set of all the situations in which it is raining, and the claim that "the sky is blue" will refer to all the situations in which the sky is blue.

We will examine the relation between two claims, say P and Q, and compare it to the relation between two sets, say A and B. The statement "either P holds (i.e., is true) or Q holds" is equivalent to taking the union of the two sets, that is, all the elements contained in A or B. Likewise, "it is raining or the sky is blue" is equivalent to the union of the sets of situations in which it is raining and those when the sky is blue. The proposition "both P and Q hold" is equivalent to the intersection of the two sets, that is, the elements contained both in A and in B. The intersection is denoted $A \cap B$. The claim "P does not hold" is equivalent to taking the complement of the set A, in other words, all the elements not in A. In this way, every logical argument can be stated in terms of sets. Thus, the statement "P and Q cannot both hold" can be translated into the statement that $A \cap B$ is an empty set. Such sets are also known as disjoint sets. The proposition "it is not possible that is raining and the sky is blue" can be translated into the language of sets thus: "the intersection of the 'rain' set of situations and the 'blue sky' set of situations is an empty set," that is, they are disjoint sets. Numbers can also be based on sets, by counting the elements in the sets (this will be described shortly).

This enables us to base the whole of mathematics on sets and operations on them, including numbers and all the definitions and conclusions drawn from them, and the resulting mathematical theorems. The following is the way to establish the natural numbers from sets. The formal description of the method will be accompanied by an intuitive explanation, but we emphasize that the explanation is not part of the actual process.

We begin with the assumption that an empty set, that is, a set that has no members, exists. We select a mathematical symbol for the empty set, and the symbol normally used is $\emptyset$. The explanation is that the set $\emptyset$ corresponds to the number 0. The set that corresponds to the number 1 is the set that includes the empty set, and only the empty set. We denote this set $\{\emptyset\}$. It is customary to write the elements of the set between the braces. The reason for considering the new set as corresponding to the number 1 is that it contains only one element, and that is the empty set. The next set will be

{Ø,{Ø}}, which has two elements, the empty set, and the set that includes the empty set. This set corresponds to the number 2. The set corresponding to the number 3 is {Ø,{Ø}, {Ø,{Ø}}}, and so forth. This formulation is much more complex than simply saying 1, 2, 3, and so forth.

The sole advantage of this type of construction is that it does not use numbers at all, it is formulated entirely in terms of sets. Hence, from here we can take a further step forward and identify other sets that have a finite number of elements equivalent to those we have already constructed by matching the number of elements in those sets to those in our sets, where matching reflects counting. The next step is to define the addition of numbers by means of sets, and that too can be done by combining two sets (the union of two sets) that have no elements in common. That is how small children would act. They calculate three plus four by counting the elements in the union of a set with three members with a set with four members. The same method is used for multiplication and other operations.

We will repeat what we have already mentioned several times. These structures were developed purely and simply to show that mathematics can be based on sets and operations on sets, thus establishing a geometry-free logical basis for mathematics. Nobody thought that this would yield a better intuitive understanding. Some mathematicians opposed this type of development on principle. Leopold Kronecker (1823–1891), for example, is quoted as saying, "God made natural numbers; all else is the work of man." In other words, there is no need to justify the existence of the natural numbers. Poincaré also saw no need for such structures. Most mathematicians of the period, however, accepted these developments enthusiastically.

The reliance on sets resulted in renewed interest in the concept of infinity. The sets used in defining the natural numbers include a finite number of members, but the sets required to form more complex foundations, such as irrational numbers, contain an infinite number of elements. The question arises naturally, is it possible to implement the arithmetic operations that are translated into logical claims on sets with infinite numbers of elements? As noted previously, evolution did not equip the human brain with the tools to develop intuition about the concept of infinity. Throughout the thousands of years of the development of mathematics in the Babylonian and Assyrian Empires and in Egypt, the concept of infinity was not considered. The term *infinity* itself was mentioned in the context of very large quantities or large numbers that could not be counted, meaning that it was difficult or impossible to count to such high numbers. Reference to God was also sometimes a reference to the infinite, meaning that God's wisdom and strength were so great that they could not be described. The Greeks were the first to relate to the mathematical infinity, for instance, counting that goes on and on, or lines that did not end. Underlying the interest in the infinite were the questions of whether the world had always existed and if it would exist forever. The Greek's solution to the inability to analyze infinite sets was to make a distinction between potential infinity and an actual infinite collection, a distinction based on Aristotle's methodology. They simply did not think of an infinite collection as a legitimate mathematical entity eligible for consideration. Potential, not actual, infinity—for example, constantly increasing collections of numbers, or finite lines that keep lengthening, or a world that will exist for a constantly increasing length of time—the Greeks considered to be collections of finite sets.

This doctrine of the Greeks persisted, seemingly, but as most mathematical developments were not based on axioms, mathematicians did not hesitate to use the concept of infinity even in the non-potential sense. For example, they referred to infinite lines in researching plane geometry, and thereby even reformulated the axioms, ignoring the distinction between potential and "ordinary" infinity. For many generations there was no rediscussion of the concept of infinity itself, apart from a contribution by Galileo. He noted that although there are "more" natural numbers than squares of natural numbers, there was a one-to-one relation between the natural numbers and their squares, thus:

$$1, 2, 3, 4, \ldots$$
$$1, 4, 9, 16, \ldots$$

This was also the correspondence that Galileo found between time and the distance that a body falls, as we have seen, and apparently that was the research that brought him to consider infinity. However, this discovery of Galileo's did not go beyond the statement that infinity has strange properties, a saying that was not accompanied by further study. Now, with the increasing dependence on infinite sets as a basis of mathematics, the time had come to explore that strangeness. That step was taken by Cantor.

Georg Cantor was born in St. Petersburg, Russia, in 1845 to a Christian family of merchants and musicians, a family apparently with Jewish roots. When Georg was eleven, the family moved to Germany, where he excelled in his studies. After studying at the University of Zurich, he returned to Germany and completed his doctorate at the University of Berlin. He studied under Leopold Kronecker and Karl Weierstrass. These two were bitter opponents, and their rivalry also held implications for Cantor himself. On completing his studies he hoped to obtain a post in Berlin or in another major city in Germany, but his path was blocked, apparently by Kronecker. Cantor settled for a post in the less-prestigious University of Halle, Germany, about a hundred miles (160 kilometers) from Berlin, and there he developed the mathematics of infinity. Kronecker was strongly opposed to this new mathematics and, among other things, also blocked Cantor's attempts to publish his papers in professional journals. These rejections, both regarding the posts he had hoped to be appointed to and of his findings, had a lasting effect on the young Cantor and apparently contributed to his mental crises. Cantor spent a large part of his time in the sanatorium in Halle, where he died in 1918. Nevertheless, he saw his theory accepted by the mathematics community, even with the logical difficulties and paradoxes it brought to the foundations of mathematics.

We will now give a short description of Cantor's theory. Its starting point is the same analysis of Galileo's that we spoke of above. Cantor suggested that we agree to state that both sets have "the same number" of elements if the elements of one can be matched one-to-one (bijection) with the elements of the second. Thus the set of natural numbers and the set of their squares have the same number of elements. Similarly, he showed that the set of rational numbers and the set of natural numbers have the same number of elements, although the rational numbers are spread tightly all along the real number line, while the natural numbers have empty spaces between them.

The next question was, do all infinite sets have the same number of elements? Here Cantor made a surprising discovery. He proved that the set of real numbers and the set of rational numbers do not have the same number of elements. Every rational number can be represented by a point on the straight line, but there is no such match for all the real numbers. From that he derived that the latter set had "more members." He denoted the "number" of members in the set of natural numbers as *aleph-null*, denoted $\aleph_0$, and sets with a number of elements like the set of natural numbers he called countable infinite sets. Cantor called the indication of the size of the set its power, or cardinality. Other powers of sets greater than that of the natural numbers he denoted $\aleph_1$, $\aleph_2$, and so on. It is not clear why Cantor chose to use the Hebrew letter aleph for this mathematical notation, and some relate that choice to his family's possible Jewish roots. Others note that as the Bible, including Hebrew writing, was studied by religious Christian groups in Germany, Cantor was familiar with Hebrew, and that was why he chose the first letter in the Hebrew alphabet, aleph. He denoted the power of the set of the real numbers C, the first letter of the Latin word *continuum*. Cantor went on to prove that the power of C was the same as that of the set of subsets of the natural numbers. Therefore, just as the number 2^n measures the number of subsets of the set with n elements (we recommend the reader to check this), Cantor denoted the power of C by $2^{\aleph_0}$.

Cantor also developed the arithmetic of powers. For example, it can be seen that the equality $\aleph_0 + \aleph_0 = \aleph_0$ holds when the sum is defined in the same way as we defined the sum of numbers by means of sets. That is, the sum of powers is the power of the union of disjoint sets that have the corresponding powers. And indeed, both even numbers and odd numbers are of power $\aleph_0$, and that is also the power of their union, that is, all the natural numbers. Cantor proved generally that the collection of subsets of a non-empty set is of a greater power than that of the set itself, thus ensuring that, as with the numbers themselves, the possible powers of sets can increase boundlessly.

During the development of this elegant theory, some troubling questions arose. For example, is there a power greater than that of the natural numbers and smaller than that of the continuum? In mathematical notation the question is, does $C = \aleph_1$? This question greatly perturbed Cantor himself and generations of mathematician who came after him until the answer was found in 1964, as we will describe below. Another question was, what happens to the set of all sets? It was always clear that there was no natural number that was the largest of all numbers. With any number, 1 can be added to it, and a larger number will be obtained. When examining sets, however, the question can be asked, what is the power of the set whose elements are all the sets in the world? Such a property would ensure that the power of that set is the greatest possible. On the other hand, Cantor proved that for every set that is not empty, and hence also the set of all sets, the power of the collection of its subsets is greater than the power of the set itself. We seem to have arrived at a contradiction. It was solved by the decision that not all sets are "legitimate" from the aspect of the new mathematics. Thus, the set of all sets, although we have called it a set, is not a set in the sense that we can apply the new mathematics to it, just as the arithmetic of the natural numbers cannot be applied to infinity.

That was the situation toward the end of the nineteenth century and at the beginning of the twentieth. Natural numbers were defined by sets. Then, as we have seen, the positive rational numbers could be defined, followed by the negative rational numbers (we spared the reader the detailed description of that stage). Next, the irrational numbers could be defined by means of Dedekind cuts, which were themselves defined as sets of power $\aleph_0$ and were thus permitted infinite sets. From there it was possible to continue to derivatives, integrals, and the rest of mathematics. Operations in logic, and in particular the laws of inference, could also be explained in terms of sets. At that time it seemed that the solid foundations of mathematics had been found that would replace the shaky foundations set by the Greeks.

In this context another important new development took place in understanding the concept of an axiom. For the Greeks, axioms expressed agreed properties of entities in nature. As such, the axioms related to familiar concepts, such as points and straight lines, which the mathematicians defined rigorously. The problem is that the definition itself uses concepts that have not themselves been defined. Here the question reappears: To what extent are the defined entities self-explanatory? The answer, on which there was a consensus in the nineteenth century, was that the axioms relate also to abstract entities, which do not need to pertain to anything recognized and defined but just need to be denoted, say, by letters, x, y, A, B, and so on. When we want to apply abstract mathematics, we must ascribe to undefined quantities a reference to known entities. If the explanation conforms to the axioms, the mathematics to be developed in line with the axioms will indeed describe reality. Unlike with the Greeks, however, in the mathematics of the nineteenth century, the elements dealt with by the axioms could have nothing to do with nature or with other uses. Speaking of elements that at the outset have no explanation, the English mathematician-philosopher Bertrand Russell (1872–1970) described a mathematician as a person who does not know what he is talking about, nor does he care whether what he is saying is true. Not knowing what we are talking about refers to entities that mathematics deals with that from the outset have no explanation or use. Not knowing whether what we are saying is true refers to truth for a particular purpose, for instance, in nature. In other words, the mathematician can engage in mathematics without any explanation and without being at all interested in an explanation of the entities he is analyzing. Despite Russell's humoristic note, and despite the agreement that axioms relate to abstract entities that are not represented in nature, I do not know mathematicians who can discuss and analyze what can be derived from the axioms without having in mind some sort of representation of a system that fulfills the axioms, except in limited and extreme cases. As we have seen, the human brain is incapable of relating intuitively to logical systems that are completely abstract.

The attitude of the leading mathematicians of the time, at the turn of the century (nineteenth to twentieth), to the basing of mathematics on set theory is interesting. Of particular interest is the reaction of the most famous mathematicians of that period, the German David Hilbert, and the Frenchman Henri Poincaré.

Hilbert (1862–1943) was born in and studied in Königsberg, Prussia (today the Russian city Kaliningrad), then moved to Göttingen, where he remained until the end of his life. In his lifetime he saw regimes in Europe change several times, and he died while the Nazis were in power. He was not one of their supporters, and after 1933 he tried to help persecuted Jewish mathematicians and physicists, even though at that time he was no longer at his prime. A leading figure in the Nazi regime turned to Hilbert in the course of an official dinner and said, "Herr Hilbert, at last we have rid ourselves of the Jews who contaminated German mathematics." Hilbert's reply was, "Yes, sir, but since the Jews have left, mathematics has ceased to exist in Germany." He made many contributions to mathematics in various spheres. He was responsible for fundamental developments, in particular the abstraction of concepts and methods, and was interested in logic and the foundations of mathematics. He made a major impact on mathematics worldwide. When he was invited to deliver the keynote lecture at the Second International Congress of Mathematicians in 1900, instead of presenting his own achievements, as usually happened at such congresses, Hilbert chose to present a list of unsolved problems in mathematics, which he predicted would become the central mathematical problems of the twentieth century. Those problems did indeed feature in research in mathematics throughout the twentieth century; some were solved relatively quickly, and others are as yet unsolved and await solutions in the twenty-first century.

Henri Poincaré, who featured in our discussion of the events that led to the development of the theory of relativity, actually came from the field of engineering. He studied in the École des Mines (a mining, or engineering, school), which was and remains a highly prestigious school in France. His abilities were discovered when he was very young, and he was elected to the French Academy, taught at the Sorbonne, and was possibly the most influential French mathematician of his time. One of his nonacademic activities was engaging in the defense of Alfred Dreyfus at the appeal stage of the famous trial. He and his colleagues, the mathematicians Paul Appel and Jean Gaston Darboux, examined the evidence and declared in a written report submitted to the court that probability theory showed that the charges did not hold up under serious scientific examination. In his private life Poincaré exhibited great courage in other situations too. In mathematics he was active in mathematical physics and dynamics. His career surged when in the context of a competition announced by the king of Sweden he promoted the understanding of the three-body problem, that is, the dynamics of three or more bodies in space, for instance, the Sun and the planets. In his research he discovered and characterized the dynamic behavior that results in what is studied today in the context of chaos theory.

Hilbert enthusiastically adopted the new developments in understanding the axioms and their connection to logic. He himself drew up a series of geometry axioms, which refined Euclid's axioms, and succeeded in showing that they do not depend on unreliable intuition, nor do they contain internal contradictions. Poincaré also adopted the dependence on logic with enthusiasm and actually said that logic is the material that disinfects mathematics by inhibiting errors. Nevertheless, with regard to the use of sets as the basis of mathematics, opinions were divided. Hilbert declared that set theory is the crowning achievement of man's creativity. Poincaré is reputed to have declared that set theory is a sickness that mathematics will be cured of. We will see the effects of this dispute on the status of set theory when we discuss matters related to the teaching of mathematics, in the last chapter of the book.

60. A MAJOR CRISIS

With the exception of a relatively small group of mathematicians, the mathematics community welcomed the reliance on set theory with open arms. Gottlob Frege (1848–1925), a leading German mathematician at the beginning of the twentieth century, decided to put in writing the set theory basis of mathematics. He published the first tome and was in the latter stages of the second when the crisis erupted. Bertrand Russell wrote him a dramatic letter and set out the famous paradox he had discovered in Frege's text, a paradox that led Frege to halt the publication of the second part of his book. For a while Frege tried to correct the theory he had developed but eventually decided to abandon the project he had undertaken.

Bertrand Russell was then a young mathematician whose astuteness had already earned him a reputation in Britain. He was later acknowledged as one of the founders of analytic philosophy and was known for his radical social and political views. He was a pacifist, a conscientious objector in the First World War, and bitterly critical of totalitarian regimes throughout the world. In 1950 he was awarded the Nobel Prize in Literature for his writings in which he championed humanitarian ideals and freedom of thought and for his famous bestseller *A History of Western Philosophy*.

Russell's paradox was in fact a variation on a paradox known already in the time of the Greeks, the liar's paradox. A person says of himself, "I am a liar." Can we believe him? If his statement is untrue, he is telling the truth, so he is a liar. If it is true, then he is a liar, so we cannot believe him and he not a liar. We have a paradox. Likewise with regard to sets, claimed Russell; define a set such that it contains all sets that do not include themselves as one of the members of the set. Does such a set include itself as one of its elements? If not, it is itself a member of the set and therefore includes itself. If it includes itself, then by the definition of the set it is not one of its elements. A paradox.

Russell's paradox could have been solved in the same way as the Greeks solved the paradox of the liar and similar paradoxes. The solution is to determine that a statement in natural language relating to itself is not acceptable, and it is not legitimate to analyze it by mathematical means. This rule can also be adopted with regard to sets. Just as the set of all sets was excluded from being a set that can be analyzed mathematically, it can be decided that a set whose definition relates to itself is not a "legitimate" set, and the set that Russell used in the paradox was such a set. However, Russell's paradox raised a more fundamental problem, one that even the Greeks were unaware of. We will examine again one of the basic laws of inference, the law of excluded middle.

For every claim P, either P holds or P does not hold.

As this rule of inference relates to itself, among other things, it is not acceptable! The removal of this rule of inference from the area of mathematics is too great a blow for the mathematics community to bear. One reason is that the method of proof via contradiction is based on this rule. Abolishing all the claims that were based on this system means going right back to square one, including, for example, placing a question mark by the discovery in Pythagoras's time that the number √2 is not rational (see section 7) and also doubting a large part of mathematics developed since then. It was clear that this proposed method of solving the paradox would not work, so there was an urgent need to get to the source of the problem and to reexamine the foundations of mathematics.

The attempts to reconstruct the foundations of mathematics focused on three main avenues.

The first was proposed by Bertrand Russell himself, together with his colleague the well-known English mathematician and philosopher Alfred North Whitehead (1861–1947). They realized that elimination of sets or logical claims would not yield the desired results. Instead, they decided to define what is a permitted system of logic, and they started with a very delicate classification of permitted logical claims, constructing the permitted structure of logic "from below." They called the permitted bases "types," and constructed a theory of types from which all mathematics could be developed. Whitehead and Russell began to write their theory and published the first parts in a weighty tome, from which it would eventually be possible to develop mathematics. Its title was *Principia Mathematica*. Their monumental project was never completed, because the approach was too complex. For example, the proof of the equality 1 + 1 = 2 did not appear before page 362. Clearly such a system could not play a future role in a vibrant mathematics.

A second approach, called intuitionism, was developed by a group of mathematicians led by the Dutchman Luitzen Brouwer (1881–1966). The mathematics that this approach allowed was limited to concrete constructive operations. For example, if you wish to show that a geometric body that has certain properties exists, you must point to it directly. Inferring that the body exists from indirect evidence is not acceptable as proof. In particular, according to this approach the method of proof by contradiction is not acceptable. Brouwer and his colleagues managed to reconstruct a large part of mathematics in accordance with their approach, but the awkwardness in mathematical practice deriving from that method, together with the need to relinquish many results in the existing mathematics, led to the non-adoption of the method by the mathematics community. Hilbert himself was strongly opposed to intuitionism, repeating and emphasizing that proof by contradiction is at the core of mathematics.

The third avenue pursued was the one accepted by mathematicians as a whole. The idea of basing the structure on sets remained, but like Whitehead and Russell's approach of constructing logic from the foundations, instead of declaring which sets are not permissible and risking future encounters with other paradoxes, here the construction is "from the very core." We start with permitted sets and, via specific building axioms, we show which sets can be formed from those already in existence. Only sets that can be constructed via the axioms are "legitimate" from the aspect of mathematical analysis. It was Ernst Zermelo, of whom we wrote with regard to his contribution to game theory, who presented these axioms, and they were completed later by Abraham Halevi Fraenkel.

Zermelo was a German mathematician who studied in Berlin, worked for some years in Zurich, returned to Germany, to the University of Freiburg, but resigned in 1936 in protest against the Nazi regime's treatment of the Jews. After World War II he was reinstated to his honorary professorship in Freiburg. Abraham Halevi Fraenkel was also born in Germany, where he published his work on the foundations of set theory and reached the position of professor. He was an active Zionist, and in 1929 he immigrated to pre-State Israel (then Palestine), joined the Hebrew University of Jerusalem, and worked there for the rest of his life.

The system of axioms developed by Zermelo and completed by Fraenkel were named the Zermelo-Fraenkel axioms. The axioms themselves are too technical to be of interest to the general public, and we will not present them here, but once they were put forward and tried on a range of problems, it seemed that hope had been reawakened that mathematics could be based on set theory. In addition to the set-theory axioms, other specific systems of axioms were also being examined, for example, the axioms of the natural numbers developed by the Italian mathematician Giuseppe Peano (1858–1932). Those axioms are quite simple and incorporate some self-evident statements, such as: the rule that the number 1 exists, the rule stating every natural number is followed by a number that is larger than it by 1, the rules of how to add and multiply and how to use induction, which is actually an independent axiom. The system is simple, and its purpose is to show that mathematics can be based on simple axioms, and those axioms can also be translated into terms related to sets. We mentioned above that Euclid's geometric axioms also underwent a reexamination that was led by Hilbert himself. It seemed that the new versions are free from lack of clarity and the errors of the mathematics of Euclid and his followers.

Alongside the efforts to improve the axiom systems, emphasis was also placed on understanding the properties that a system of axioms should have in order to make it acceptable. As mentioned, for the Greeks, axioms reflected absolute unassailable truths. The more modern approach permitted alternative systems of axioms and even systems that contradict each other. It was therefore important for all of that to clarify what is expected of a system of axioms itself for it to be accepted as reliable. Here are two basic requirements:

Consistency: Mathematical inferences from the use of axioms should not result in contradictions; in other words, mathematical conclusions deriving from the axioms should not contradict each other.

Completeness: Every mathematical claim about the system that the axioms describe can be proven or disproven by using the axioms themselves.

The consistency requirement is self-evident. In many day-to-day events the human reaction to an encounter with logical contradictions is not one of great upheaval because we are constantly exposed to situations that we do not examine in depth or even to apparent contradictions that we are prepared to accept as part of our daily lives. Mathematics, however, cannot allow itself internal contradictions. That is, a mathematical system cannot allow a conclusion that is both true and untrue.

The completeness requirement is more intricate. The basic idea is that when we formulate a hypothesis, let us say about numbers, the system of axioms that describes the numbers must be rich enough for us to be able to decide whether the hypothesis is or is not correct within the system of axioms. If that is not so, there could be, say, number systems that contradict each other that satisfy the system of axioms and we will not know which is correct and which is not. This does not mean that a system that is incomplete is useless. The completeness property, however, ensures that in principle we can conclude whether a claim is correct without having to resort to additional axioms.

When the Zermelo-Fraenkel system of axioms was formulated and the first successful steps were taken to corroborate its consistency and completeness, the mathematics community was spiritually uplifted. The axioms seemed reasonable and were formulated meticulously. Although their consistency and completeness have not been completely proven, the intuition and care with which the axioms were constructed appears flawless, and the first steps to establish their consistency and completeness are promising. Hilbert himself published the grand program: a full formulation of the system of axioms for mathematics that would be consistent and complete. He declared in a lecture in 1930 on the occasion of his retirement

We have to know, and we will know.

And that is what is inscribed, in German, on his tombstone.

61. ANOTHER MAJOR CRISIS

The "evil" came from Austria. Kurt Gödel was born in Brünn (now Brno), then a Czech town in the Austro-Hungarian Empire. When that empire broke up, Gödel automatically became a Czech citizen, but he considered himself Austrian and studied at the University of Vienna. There he completed his doctorate studies and two years later, in 1931, published his famous incompleteness theorem, which we will describe below. While in Vienna, when the Nazis came to power in Germany, he was severely affected by the brutality and anti-Semitism of Nazi groups in Austria and by the murder of Moritz Schlick on the steps of the university. Schlick was a member of the university's academic staff and of the logical-philosophical Vienna Circle, of which Gödel was also a member. Although Gödel was not Jewish, he became paranoiac and remained so all his life. In the 1930s Gödel received several invitations to go to the United States, and he stayed at the Institute for Advanced Study in Princeton, alongside Einstein, who became a close friend. Gödel's yearnings for Vienna grew stronger, and despite his fears he returned to Austria. He was in Vienna in the beginning of the war as a German citizen, citizenship he had received with the German annexation of Austria. But the pressure grew and in 1940 he moved back to the United States, accepting an invitation to become a permanent member of the Institute for Advanced Study in Princeton, where he remained until his death in 1978.

Gödel's interest in the foundations of mathematics and the link between logic and the developing set theory began in the time of his doctoral studies. His doctoral thesis presented a result that merged well with the program that Hilbert envisioned. Gödel proved that with a consistent and finite list of axioms, if every system that fulfilled the axioms had a certain property, it could be proven from the axioms themselves. This was a promising step forward. What was now needed to complete Hilbert's program was "only" to prove that the Zermelo-Fraenkel system was consistent and that every property of a system that satisfies the axioms exists in every system that satisfies the axioms. Two years later, however, Gödel himself presented a result that completely negated Hilbert's program, and that was *Gödel's incompleteness theorem*.

The theorem showed that in every finite system of axioms (or even infinite system, if it was created by algorithmic computation) that is rich enough to include the natural numbers, there will always be theorems that cannot be proven or disproven. In other words, such a system cannot be complete!

This result was a slap in the face for all those, like Hilbert, who believed that there is nothing in mathematics that we cannot know; in other words, those who believed that the logical approach based on axioms that can be stated clearly will always result in the resolution of the question whether a particular mathematical claim is correct or not. Gödel's result also relates to systems such as the system of natural numbers constructed by Peano, the system of types developed by Whitehead and Russell, and the system of axioms of Zermelo and Fraenkel. With regard to such systems of axioms he showed that it is impossible to prove that the system does not have contradictions by using the axioms themselves. That is, if we rely only on the axioms of the system, then either one day we will discover a contradiction, or we will never know whether the system contains a contradiction. (It may be possible to prove consistency by using a broader theory, but such a theory has not yet been discovered.) The inconsistency theorem did not affect the intuitionist approach, but neither did it breathe renewed life into it. There is a strong desire to operate in an environment without potential contradictions, but not at the price of significantly limiting the range within which mathematics can develop.

Beyond the philosophical aspect and the questions relating to the foundations of mathematics, Gödel's result had a direct effect on the practice of research in mathematics. Throughout the development of mathematics, when a mathematician set out to prove a theorem, he faced two possibilities. One was that the theorem was correct, and he had to find a proof. The other was the theorem was not correct, and he had to disprove it, either by bringing a counterexample or by finding a contradiction between the theorem and other results. The incompleteness theorem gives rise to a third possibility, that the theorem cannot be proven or disproven.

Let us take as an example Fermat's last theorem, which states that for any four natural numbers X, Y, Z and n, if n is greater than 2, the sum $X^n + Y^n$ cannot equal Z^n. For more than three hundred years, mathematicians tried to prove or disprove the theorem. With the publication of Gödel's incompleteness theorem a third possibility was added: Fermat's theorem could be one of those that cannot be proven or disproven. In 1995 Andrew Wiles published a proof of the theorem, and the cloud of incompleteness around Fermat's theorem was dispelled.

Goldbach's conjecture, formulated by Christian Goldbach in 1742, is also simple: Every even number is the sum of two prime numbers. Despite the simplicity of the formulation, still today no one has proven or disproven the conjecture, and here too nobody knows whether it is one of the theorems that Gödel's theorem applies to.

Now for a third hypothesis, Georg Cantor's continuum hypothesis, that is, the question referred to in section 59: Does $C = \aleph_1$? Cantor invested much effort in trying to prove or disprove this claim, using the naive approach to set theory, the approach in which Russell found the contradiction. Following the formulation of the Zermelo-Fraenkel axioms, the question became: Is the equality $C = \aleph_1$ correct or incorrect in the context of those axioms?

Another axiom also played a major role in the attempts to answer the question, and that is the axiom of choice, a highly intuitive claim when referring to a world about which human beings developed intuition. It says that given a collection of non-empty sets, a new set can be formed by selecting one member from each set in the collection. Such a selection is simple if the collection of sets is finite, but we have already seen that the concept of infinity can be deceptive. The axiom of choice was indeed recognized as an axiom, and the question of what would happen if it was added to the Zermelo-Fraenkel axioms remained unanswered.

Gödel himself contributed to the answer to this question and proved that if the Zermelo-Fraenkel system is consistent, that is, it does not contain contradictions (which has not been proven), adding the axiom of choice to the system will leave it consistent. Then Gödel added a slightly confusing finding: if the Zermelo-Fraenkel system is consistent, even if we add the axiom of choice to it, it will be impossible to prove that $C = \aleph_1$. Prior to the incompleteness theorem, this result would have ended the search for the truth: if it is impossible to prove the correctness of a particular claim, and if everything can be proven or disproven, then the claim is incorrect. In the world after the incompleteness theorem, however, there is another possibility, that the theorem is one of those the truth of which is impossible to resolve.

In 1964 Paul Cohen (1934–2007), a young mathematician from

Stanford University, showed that the continuum hypothesis was in that category. Moreover, he showed that if the Zermelo-Fraenkel system is consistent, the claim $C = \aleph_1$ or its negation can be added to the system without affecting its consistency. The question of whether the Zermelo-Fraenkel system is consistent, meaning, does not contain contradictions, is still unanswered today.

Research along these avenues, that is, the attempts to find a system of logic without contradictions on which mathematics can be based without doubts or uncertainties, continues. On the one hand, research into Gödel's theorem shows that the phenomenon is very general. For instance, Gödel's proof of the incompleteness theorem was based on the liar's paradox, that is, on a claim that relates to itself. Since then proofs have been found that do not use claims that relate to themselves so that we cannot dispose of incompleteness merely by not allowing claims that relate to themselves. On the other hand, the various attempts to find ever-more complete systems sometimes yielded strange results. For example, as the axiom of choice seems intuitive, we can define other axioms, no less intuitive, that contradict the axiom of choice but whose status regarding the basic axioms of set theory is the same as that of the axiom of choice. That is to say, even if we add them to the Zermelo-Fraenkel system we will not find a contradiction (provided, of course, that the Zermelo-Fraenkel system is consistent).

So what is the correct mathematics? Does mathematics that is correct beyond all doubt exist? The answer is clear: we do not know. At this stage there is room for faith. Some believe that in the world, the physical world and the Platonic ideal world, there is a correct mathematics, but we just have not found it yet. In that mathematics, for example, Goldbach's conjecture is definitely correct or incorrect. Similarly, in that mathematics, either the equality $C = \aleph_1$ is fulfilled or it is not. Others believe that there is no such absolute truth. Mathematics rests on the axioms that define it, and different systems of axioms yield different types of mathematics, possibly even contradictory ones, and we must learn to live with that dichotomy. But most mathematicians simply do not care. The many years of involvement in mathematics have resulted in faith that there is no basic contradiction in mathematics, and we can continue to work. If the logicians manage to arrive at absolute mathematical truth, whether via axioms or some other way, all the better. But even if not, even without any logical "proof beyond doubt," the mathematics we produce is correct beyond any doubt.

CHAPTER IX
THE NATURE OF RESEARCH IN MATHEMATICS

What does a mathematician do when he gets to the office in the morning? • How does sleep help to solve problems in mathematics? • Does creativity in mathematics decline with age? • Why would a mathematician refuse to accept a million dollars? • Does pure mathematics exist? • Why did the engines of steamships start exploding? • Can aliens do the sum 2 + 2?

62. HOW DOES A MATHEMATICIAN THINK?

We will start with the bottom line: there is no difference between the thought process of a mathematician and those in other disciplines. Before we clarify this issue, we must explain what we mean when we use the word *think*. Thinking means activating the brain to analyze situations, reach conclusions, and propose courses of action or solutions. With regard to these functions, there are various levels of brain activity, and we will concentrate on two that are related but different. The first type of thinking takes place in situations in which you have to carry out an action about which you have received previous guidance on how to proceed. For example, you have been given a cake recipe, and now you have to make that cake. Or, someone explained to you how to use a road map, and now you have to find the route from one place to another. Or, you have been taught how to use a paintbrush and oil paints to paint a still life, and now you wish to paint flowers in a vase. Or, you have learned how to design a car engine, and now you have to design one. Or, you have been taught how to solve a certain type of exercise in algebra, and in an examination you have to solve an exercise of that sort. All of the above need thought, and the thought process is matching the required process to those learned or tried in one way or another. We will call such thinking *thinking by comparison.*

The second type of thinking occurs when we use our brain to react to an unfamiliar situation in which we have not learned how to operate, or in which we consciously want to deviate from the normal course of action. For example, you are on a desert island, you have to put together a meal from the vegetation you can find, and you do not even know if the various plants are edible. Or, you have reached an unfamiliar location and have to find your way without a road map. Or, you have artists' materials in front of you, and you have decided to paint a picture in a completely new, unfamiliar style. Or, you have to design a vehicle that will be able to move on an asteroid, the nature of whose surface is unknown. Or, you are trying to find a mathematical characteristic of an unknown system that no one has researched. In such situations, creative thinking is needed.

The two types of thinking are not detached from each other. Even when comparative thinking is called for, in general there are differences between the situation previously encountered and the current one, and a measure of creativity will be necessary to match the solution properly to the new situation. Where new situations are encountered and creative thinking is required, creativity is not ex nihilo, and elements of comparative thinking will play a part. Comparative thinking focuses on searching and matching, which are essentially routine, sometimes automatic, operations and can even be performed by a computer. The other type of thinking, in contrast, is based mainly on intuition, feeling, hunches. When a person has to deal with a problem, the brain will "decide" whether to activate comparative thinking or to enlist creative intuition according to the person's familiarity with that or similar problems. The brain's "decision" is not generally a conscious one.

Thinking by comparison cannot be taught or learned in isolation from the subject; in other words, this method of thinking cannot be learned in the abstract. In every subject, from cooking, through engineering, to mathematics, more and more can be learned to a level such that additional problems will be within a person's knowledge so that he can make do with comparative thinking and shorten the route to an appropriate solution. Unlike comparative thinking, creative thinking cannot be taught or learned at all. Creative, intuitive thinking can only be encouraged and stimulated. Trying to teach creative thinking via examples is a positive step, but the effect of the examples is simply to enrich the collection of problems that can be solved by comparative thinking. With regard to intuition itself, the more you know, the more you will be able to widen the sphere in which you can exercise creativity, but we do not know enough about how human intuition works to be able to teach people how to develop their creative intuition.

Once we have understood what thinking is, we can repeat: there is no difference between the thinking of mathematicians and the thinking of those in other professions. Obviously they think about different subjects. Cooking, finding your way somewhere, painting, engineering, and mathematics are highly dissimilar, but the types of thinking and the methods of thinking are the same in all of them. The myth that to be a mathematician you need a special type of brain is just as correct, or incorrect, as saying that to be a chef you need a special type of brain, and the same applies to the ability to navigate or to introduce novel ways of painting. Research in mathematics for the most part deals with new developments and is more closely related to the creative type of thinking, but the same could be said of creative research in any subject. One difference between research in mathematics and other research is that the intuition required in mathematical activity needs more thinking time. That is because research in mathematics deals mainly with logical aspects, and these are less at home among the brain's abilities as formed by evolution. The expression "Wait a moment, give me time to think about it" comes up more often in mathematical discussions than in conversations about cooking or navigating a route. Therefore, for most of their research time, mathematicians are immersed in reflection and intuitive thinking. Only after they have reached a solution intuitively do they start work on recording the solution in the acceptable language of logic. Henri Poincaré is credited with having said, "It is by logic that we prove, but by intuition that we discover." The proof stage itself, however, belongs essentially in the first, more technical, type of thinking.

This distinction between the logical formulation of a mathematical result and the intuitive thinking carried out before the formulation of the result is also reflected in the way information is transmitted between mathematicians. A conversation on a mathematical subject is unlike mathematical writing that we recognize from our studies, books, or academic articles. We are often asked about the utility of flying for several hours to meet a colleague in a research project or to participate in a mathematics conference in some distant land. Would it not be enough, in this age of electronic communications, to read articles or correspond via e-mail? The answer is that in the intuitive thinking stage there is no substitute for face-to-face meetings. Information, which in a certain sense is subconscious and as such is difficult even to define, can be transmitted only in such a meeting. This is not confined to mathematics alone but applies to other areas of research too. My reason for stressing this aspect of mathematics is that it is a common misconception that logical mathematics will be less in need of the intuitive, subconscious level of thinking. The opposite is the case: most mathematical research is based on that intuition.

The differences we have described are responsible for the fact that unexplained events related to thinking, which also exist in other disciplines, occur frequently among mathematicians. It is not unusual that a colleague asks me to help him check a mathematical development he has conceived and insists on showing me the problem, despite my objections that I do not understand anything about the subject. He starts covering the blackboard in my office with the development in question, and after a while, without my having uttered a word, he stops, thinks, and says, "Thank you very much. Now I understand what has happened. You have helped me a great deal." And I have indeed helped him a lot, just as I have been helped in a similar manner by colleagues. The very fact that we have tried to explain our findings to someone who is likely to have at least some idea of what we are trying to achieve is of great help.

Another well-known fact is that intuitive thinking to a large extent takes place subconsciously, sometimes even during sleep. Both Poincaré, whom we have discussed previously, and Jacques Hadamard, another leading French mathematician, wrote in their books about mathematical problems that they tried to solve over a long period, and then, in the course of an event completely unrelated to mathematics, the complete solution suddenly dawned on them.

This phenomenon is known at all levels of the profession. Just recently I was deeply engrossed in trying to solve a problem but made hardly any significant progress in my daily work. One morning I woke up with the complete solution to the problem, or so I thought. Checking it in the office, I realized that the solution was incomplete. The complete solution required another night's sleep. The conclusion to be drawn is not to have a sound sleep every time you encounter a mathematical problem that you do not manage to solve. Investing effort and concentration are certainly essential. After you have invested time and much effort, having a break does not mean that your brain stops trying to solve the problem. Moreover, the break is likely to be beneficial.

Another aspect of subconscious thinking is the context in which the result of such thinking comes to light. I used to live about fifty minutes' drive from my office in the Weizmann Institute. With a frequency that I cannot explain, at a certain point where the tree-lined road curved as I drove home, I would realize that I had made a mistake in my work that day. It would take a day or two to correct the error, but then, at the same spot on my journey home, another error would spring into my mind (obviously, when I moved closer to the office, the number of mistakes I had to correct in my articles declined significantly).

Colleagues tell me that quite often the solution to a problem, or a new idea leading to the solution, comes to mind while they are watching television (even today my wife finds it hard to believe that when I am watching a sitcom, I am actually working!). It may be that there is no significance in and no statistical explanation for these occasions, and the fact that we attribute importance to them is a result of some mental illusion. Such an illusion could derive from the fact that exceptional events, like discovering an error while driving past trees, are more memorable than more normal occurrences, such as discovering a mistake while working in the office. Yet it may also be that something in our brain causes solutions to spring to mind in just such unlikely situations.

Another surprising factor is the different degree of difficulty between solving a mathematical problem if we know that someone else has already solved it, say, as an exercise in a class, and solving a new problem that no one has solved yet. The latter are called open problems. The story is told of the mathematician John Milnor, that when he was a student he had dozed off in a class when the lecturer wrote on the board an open problem in mathematics. When he awoke from his nap and saw what was on the board, he thought that it was an assignment to be done for the following week. The next week he arrived with the solution. John Milnor is a top mathematician. In 1962, at the age of thirty-one, he won the Fields Medal (officially the International Medal for Outstanding Discoveries in Mathematics), awarded to mathematicians under the age of forty; in 1989 he won the Wolf Prize; and in 2011 he was the Abel Prize laureate. Of course you have to be a genius like John Milnor to solve an open problem as if it were simply an exercise to prepare for the next class, but this sort of thing is known to occur at all levels of research: you can work on an open problem for many months, and only after you have found the solution do you realize that the problem could have been solved much more simply, but that is being wise with the benefit of hindsight.

The fact that research problems, that is, open problems so far unsolved, are more difficult than other problems set in class frustrates many students setting out along the path of research. After investing great effort in solving an open problem, they discover that they could have solved it more simply, just as they had solved other problems in the various courses they took. Many of them draw the conclusion that they are not suited to research, which may be an incorrect conclusion. It is not clear why it is that the fact that a problem is an open problem makes it more difficult to solve. In my opinion, a reasonable explanation is that thinking by comparison and creative thinking may take place in different parts of the brain. The creative section of the brain works less efficiently than does the comparative part. When the brain has to solve a problem, it channels the problem to what it considers to be the appropriate section for handling the problem, and that is the cause for the difference in the level of difficulty in solving a problem presented as an exercise and one introduced as an open problem. I would even go so far as to say that the capability of creative thinking is what differentiates man from the rest of the animal world, but that of course is pure speculation.

Does creative thinking in mathematics decline with age? This is the place to dispel a myth about the relation between mathematical research and age. According to the myth, after a certain age, some claim as early as thirty, creative ability in mathematics declines and eventually disappears. That is not so, and I will explain why. Physicists, and I have heard this from top physicists, agree that creative ability in physics declines from the age of thirty-something. That does not mean that after that age physicists do not contribute to their profession, but rather that the breakthroughs and innovative developments are achieved by the younger physicists. Yuval Ne'eman claimed that this does not refer to chronological age but to the length of time in the discipline (Ne'eman himself completed his doctorate at a relatively late age), and the reason for the reduction in the ability to introduce innovation is that as people accumulate experience and become accustomed to certain truths and to a certain method of research, it gets harder for them to challenge accepted norms and practices and to disprove or change them. And indeed, the beginning of most breakthroughs in physics was the refutation of deep-rooted beliefs. That is not the case in mathematics. The fundamentals laid down by the Greeks are still applicable. Important discoveries in mathematics that contradicted previous basic approaches were few and far between. Knowledge expanded enormously, new areas of research were added, unexpected applications were discovered, but today's work method is the same as that of the Greeks. Moreover, the results achieved by the Greeks and their successors throughout the generations are still relevant. No other discipline in the natural sciences can claim to have such a stable, cumulative nature. Thus, knowledge and experience play a much more important role in mathematical research than they do in other sciences. That is why we see mathematicians creating and making new discoveries at a more advanced age, as long as their enthusiasm for the subject is maintained and their health allows.

So what does a mathematician do when he arrives at his office in the morning? The answer is obvious: he drinks a cup of coffee. And then?

He drinks another cup of coffee. This description is not my idea. One of the more graphic mathematicians of the twentieth century, the Hungarian Paul Erdős (1913–1996), once defined a mathematician as a machine into which coffee is poured at one end, and at the other out come mathematical theorems. Most of a mathematician's research time is spent in thinking how to solve the problem he or she is working on. How to activate intuition varies from one mathematician to another. Some prefer to think via interaction with colleagues, others have to be alone in a perfectly silent room, others favor working while listening to classical music, and others think best while strolling around. One well-known example is that of Steve Smale of the University of California, Berkeley, a leading mathematician of the twentieth century. His many awards and prizes include the Fields Medal in 1966 and the Wolf Prize in 2007. He was given a grant so that he could devote his summer to research. The authorities discovered that he spent the summer lying on a beach in Rio de Janeiro and asked him to return the grant money he had received. Smale claimed that he was in fact working while lying on the beach, and indeed that was when he had some of his best mathematical ideas. He was able to prove his claim and convinced the committee investigating the case, which ruled in his favor. This was not just an excuse on Smale's part. I once participated in a conference at Luminy, near Marseilles in France. Steve Smale was among the participants, and he insisted that the conference timetable leave time for those who wished to do so to get inspiration on the beautiful Mediterranean beaches. Indeed, time spent on the beach did improve the quality of the lectures at the conference.

63. ON RESEARCH IN MATHEMATICS

In this and the next section I will put forward a few comments and clarifications relating to the nature of mathematical research, including some based on personal experience regarding research topics and researchers.

First, mathematics is the result of developments by many researchers. Anyone reading only the bottom line of a mathematical research study may get the impression that mathematics was developed by a small group of geniuses (which may also be the impression gleaned from earlier chapters in this book). The truth of the matter is that the work of all those geniuses is supported by the contributions of many mathematicians, without whom they would not have made their important achievements; as time passes, however, the others are forgotten and only the leader's halo becomes brighter. Being awarded a prize or a medal also adds to the fame and glory of the individual, although in many cases the prize could also have been given to someone else equally deserving. This applies also to current research. The solution to each of the mathematical problems in the recent past, such as Fermat's theorem or Poincaré's conjecture, was the result of ongoing research carried out by many mathematicians. Despite the fact that the final stage gets all the headlines, and the mathematician who completes that stage naturally gains all the honor and praise due, the intermediate stages are often no less significant and important.

The fact that the mathematics of the Greeks is still relevant today, as is most of the mathematics developed since then, has direct implications for the subjects of research in mathematics and for mathematicians' work methods. This situation is unlike that in other scientific disciplines, in which a research topic of several decades is likely to have become irrelevant. That is apparently why research in every other natural science is concentrated in a few major directions, whereas the range of topics recognized as worthwhile subjects of research in mathematics is much wider. The range and diversity is such that mathematicians in different areas of specialization may have difficulty understanding each other. Which is why, so it is said, one of the characteristics essential for being a mathematician is the ability to sit in a lecture in which the lecturer is describing his latest results and nod your head as if everything is clear, whereas in fact very little of what is said is understood. Obviously that is an exaggeration, but it is correct to state that the audience's understanding of a mathematics lecture is generally on the intuitive rather than the technical level. You can generally get an idea of what the lecturer is trying to achieve, what he has achieved, more or less, and possibly also the methods he uses, all in a general and intuitive way. There is hardly any chance that you will understand the details in the lecture, unless you happen to be one of the very few mathematicians in the hall who is working in the same field. Thus, although the content of the lecture is generally presented on the logical level, the audience's understanding is usually just intuitive. A lecture to students in university, who are expected to understand the material they are studying at the logical level, is very different than a lecture in a seminar for researchers, at which most participants understand the details of only a small part of what is being said. This difference is a trap for many students beginning their research. They construe their difficulty in understanding the lecture as their weakness and find it hard to believe that the lack of technical understanding is the norm for senior faculty members.

Beyond the range of research topics that covers a vast number of possibilities, mathematical research also offers totally different models of research itself, models that require various types of expertise. For example, some mathematicians are known for their excellent problem-solving abilities. Give them a specific problem in mathematics, and they will solve it. Others excel in opening up new paths and constructing new mathematical theories. Some have the knack, in addition to their other talents, of asking the right questions. Open problems in mathematics played a central role in research in the past and still do so today. It is not easy to ask the right question, one that will be interesting and that has a reasonable chance of being solvable.

We have mentioned Hilbert's address to the Second International Congress of Mathematicians in 1900, in which he presented a list of mathematical problems that he considered would leave their mark on the mathematics of the twentieth century. The list appeared in the proceedings of the congress and included twenty-three problems. The first problem on the list included Hilbert's question about the foundations of mathematics—the program that Gödel showed could not be carried out—and the continuum hypothesis, solved by Paul Cohen in 1964. Another problem was Fermat's last theorem, which was solved by Andrew Wiles in 1995. Hilbert was right with regard to the position these problems would occupy in mathematics in the twentieth century, but it may well be that the fact that they were presented in such an illustrious forum contributed to their great impact. Most of Hilbert's problems have been solved by now, but some still remain unsolved.

Around the turn of the millennium several similar lists of problems were published. The best known is the list of seven problems published by the Clay Mathematics Institute (CMI), which also offered a prize of one million dollars to anyone who solved one of them. We described one of the problems in section 53, namely, whether the P class of problems is the same as the NP class. Another problem on the CMI list was Poincaré's conjecture, which was first introduced by Henri Poincaré as early as in 1904, and which was solved only in 2002. We will describe later the problem and the controversy that its solution sparked. Paul Erdős was known as a serial presenter of open problems in his area of research, number theory and combinatorics. To motivate potential problem solvers, Erdős would offer monetary prizes for the first person to solve any of his problems, with the amount of money, between five dollars and thousands of dollars, reflecting the difficulty of the problem. His purpose was to promote mathematics, and indeed, the challenge and the financial rewards led to the solution of many interesting problems, more than any single person could have solved. I was present at a lecture at which Erdős was asked how he could shoulder the financial burden imposed by the distribution of many payments. He answered that he paid by checks, signed by him, and he hoped that the recipient would not deposit the check but would prefer to keep it as a memento. He was surprised, he said, when one of the mathematicians who won a prize did deposit the check and then asked Erdős if he could have the paid check back as a memento!

Offering and solving open problems with different levels of difficulty is part of the practice of research in the mathematics community. At a conference I attended in Warsaw, the well-known Polish mathematician Czesław Olech presented an open problem and promised a bottle of vodka to anyone who solved it. I managed to solve it before the end of the conference and showed the solution to Olech on the way to the airport. A bottle of vodka duly arrived with a messenger some two months later. The solution appeared in an article I coauthored with another participant in the conference, who had also solved the problem. I won another bottle of vodka, also in Poland, after

Ron Stern, a mathematician from Montreal, promised it as a prize to anyone solving a problem he presented. The problem was solved in parallel by Piermarco Cannarsa, and each of us received our own bottle. I gave a tasty bottle of dessert wine to Felipe Pait, who produced an interesting example as a solution to a question I had posed at a conference held at Rutgers University in New Jersey. The example completed the results in an article I was writing at the time, and I included it in the article, giving due credit to Felipe for his contribution. The reader should not conclude from these instances that research in mathematics consists of drinking liquor that we win as prizes for solving problems at conferences. These are just a piquant, but very small, part of the work we do. In most cases no prize is offered for finding a solution to a problem, and whoever finds a solution must settle for professional esteem, and even more, the satisfaction derived from the solution itself.

Much research in mathematics, as in other disciplines, is the product of cooperation between several mathematicians. The special nature of mathematical research resulted in the development of types of cooperation that are different than those among our natural-scientist colleagues. Even when the research is in cooperation with other mathematicians, most of the thinking is carried out alone, either separately or with the coresearchers sitting together, staring at the blackboard in silence. That is also why, when trying to identify the contribution of each of the coauthors of a joint paper, the sum of the parts, whether in mathematics or in other fields, turns out to be greater than the whole. On more than one occasion my collaborators in a research project and I have come back to it the next day, or even just after a lunch break, and we have the same solution in mind. That is also why in mathematics it is the practice, though not an unbroken rule, to list the coauthors of a paper in alphabetical order, unlike in other disciplines that require experiments, where generally the first author listed will be the one who made the major contribution, and the last-mentioned will be the head of the laboratory. The list of authors of a mathematical paper is usually shorter than that of papers in biology or physics, subjects in which large-scale experiments are performed. That said, if we were to draw a chart of joint authors of papers in mathematics, we would obtain a denser network than we would have expected.

In the establishment of such a network too Paul Erdős has fundamental rights. He was one of the most prolific cooperators ever. He coauthored papers with more than five hundred mathematicians. We can obtain an interesting mapping if we draw a chart in which we denote everyone who coauthored an article with Erdős as "Erdős 1." We denote by "Erdős 2" everyone who coauthored a paper with someone denoted by Erdős 1 but who did not personally coauthor a paper with Erdős. Anyone who coauthored a paper with someone denoted by Erdős 2 but who is not himself Erdős 1 or Erdős 2, we denote by "Erdős 3," and so on (I currently am denoted Erdős 3). This network has some surprising properties, particularly in light of the way in which mathematical research advances. The number of Erdős 2 mathematicians at the end of 2010 was close to ten thousand, and of course the number of Erdős 3 mathematicians was far greater. There are some mathematicians who are not connected to Erdős via this chain of coauthored papers (we usually refer to these as having an infinite Erdős number), but of those who are linked with him, the average Erdős number is 4.5. With the tools available on the Internet today, it is easy to determine the chain of joint authorship that links any mathematician with Paul Erdős, and even the link between any two mathematicians. These chains are short, a surprising fact considering the range of topics covered by such an individualistic discipline.

Before mathematical papers are published in the professional literature, they are reviewed, in general anonymously, by professional colleagues (peer review). The referee decides whether to accept the paper for publication, basing the decision mainly on the degree of innovation in the paper and on being convinced that the results are correct. The term *convinced* is open to interpretation. The referee is not meant to perform a detailed a check to see whether there is an error in the paper. That is just too hard to do. The results are written in a logical, well-ordered manner, and as we have stated previously, following logical claims without the attendant intuition is an extremely difficult task for the human brain to perform. The cure might have been to attempt to present the underlying intuition in writing. Good authors try to do that, but it is a formidable undertaking. The difficulty in identifying errors is the reason for many errors coming to light later, when others, in addition to the authors and referees, become interested enough in the results to check the proof.

In contrast to the public image of mathematics, mathematical errors are commonplace, and although attempts are made to avoid them, success is only partial, at best. One of the best-known errors in the history of mathematics relates to the four-color problem, which we referred to in section 56. The problem was presented in the middle of the nineteenth century, and a solution was published in 1879 by Alfred Kempe (1849–1922). His solution was highly regarded by many in the mathematics community. It was not until eleven years later that Percy Heawood (1861–1955) found a flaw in the proof. As mentioned above, the four-color problem was fully solved only in 1976, almost one hundred years after the publication of the flawed proof that was initially accepted as correct.

Mathematicians are not very dismayed that mistakes are made. On more than one occasion I have seen comments by referees and reviewers that although an article was flawed, it was still very important, as it contained novel and fundamental ideas. Generally, in research in mathematics, the proof of a hypothesis is no less important, and may be even more important, than the correctness of theorem that it verifies. Many mathematical results have been confirmed by different proofs over the years, either to simplify the existing proof or by means of a new proof that gives a different and sometimes even deeper understanding of the issue.

Mathematical results are presented in an ordered, logical, and precise form, and that is also how mathematics is learned in school and university, creating the impression that with the exception of possible occasional errors, a mathematical proof is irrefutable. In the previous section we showed that no foundations have been discovered yet that raise mathematics in general to the level of complete certainty. However, even if we agree that the foundations accepted today, that is, the axioms of set theory and the use of logic, ensure contradiction-free mathematics, it would not be practical to base the proofs themselves directly on those foundations. Therefore, even in ordered, logical writing there is no choice but to rely on the reader's and the author's prior knowledge, knowledge that cannot be checked fundamentally. We have already emphasized that belief and trust in what you are told is a characteristic promoted by evolution, and the same applies in the practice of mathematics. As a result, there is no answer accepted by the whole mathematics community to the questions of what is permitted and what is forbidden in a mathematical proof, and what constitutes a complete proof. Every subgroup of mathematicians develops its own standards of what it considers to be a complete proof. This practice sometimes leads to serious disputes, an example of which we now describe.

Poincaré's conjecture relates to geometry and is simple enough for us to describe here in full and in intuitive terms. As mentioned previously, the problem was presented in 1904 in the context of attempts to understand the geometry of the world more fully, including the properties of geometric bodies. The following is a graphic description of the conjecture.

Consider the boundary (i.e., the surface) of an ordinary ball in three dimensions. We will compare this boundary with the boundary of an egg or of a cube. Even if these three objects look different, they are very similar. For example, we can match points on the boundary of the ball continuously, one by one, with those on the boundary of the cube. That is, points close to each other on the ball are matched with points close to each other on the cube, and vice versa. Mathematicians call this relation homeomorphism and say that two bodies are homeomorphic or topologically equivalent. Thus, the boundary of the ball is homeomorphic to that of the cube and to that of the egg, and also to many other bodies that may not be round. For example, if we knead and squeeze the boundary of the ball at will without tearing it or sticking parts of it together, we will still have a body homeomorphic to the boundary of the ball. There are other bodies, however, that are not homeomorphic to the boundary of the ball, for example, a ring or a pretzel shape. The boundary of a ball has another property that can be checked easily: if we think of a loop on the ball, we can shrink it to a single point without removing it from the surface. The same applies to the surface of the cube or egg, or to any other body that is homeomorphic to them. The boundary of a ring, however, does not have this property. A loop around a section of the ring facing the center (see the diagram below) will remain like that even if it is moved along and if we try to shrink it, unless we cut it. An interesting question arises: Is there a body such that any loop on its boundary can be shrunk to a point without breaking contact with the boundary and that is not homeomorphic to (i.e., topologically equivalent) the boundary of the ball? The answer is that there is no such body in our physical, three-dimensional space. In other words, the property of shrinking the loop to a point is characteristic of bodies that are homeomorphic to the boundary of the ball.

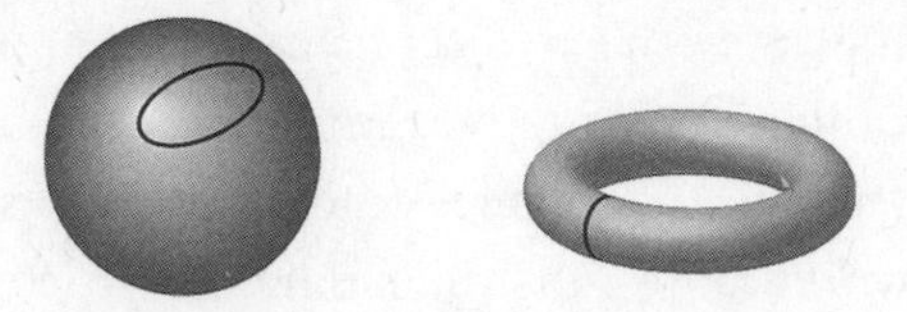

We now ask, does this also apply to spaces of higher dimensions, for example, the boundary of a ball in four-dimensional space? In section 32 we discussed the fact that physicists claim that physical space has more than three dimensions, but our senses cannot perceive them. It is difficult to develop a feeling or intuition about multidimensional spaces, but they can be defined quite simply in mathematical terms. If we follow Descartes and describe normal three-dimensional space by using coordinates (x, y, z), and the boundary of the ball is given by the equation $x^2 + y^2 + z^2 = 1$, four-dimensional space will be described by four coordinates, (w, x, y, z), and the boundary of a four-dimensional ball will be given by the equation $w^2 + x^2 + y^2 + z^2 = 1$, and likewise for spaces with more dimensions. The boundary of a ball in four-dimensional space will have three dimensions, just as the boundary of a ball in three-dimensional space has two dimensions (think of small squares or rectangles on the surface; each has two dimensions).

Poincaré asked: Is the boundary of a ball in a space with more than three dimensions, where the boundary of the ball has at least three dimensions, characterized by the property of the shrinkability of a loop on it? He himself did not suggest the possibility that the answer was positive. The intuitive reaction that this property of shrinkability is characteristic of the boundary of a ball emerged only after years of attempts to discover whether it was true. Indeed, it was proved, among other things, that the boundary of a ball in more than six dimensions is characterized by the property of a loop being shrinkable. It was Steve Smale who came up with the proof, for which he was awarded the Fields Medal. It was also proven that the property also applied if the boundary of the ball had four dimensions (the ball has then five dimensions). That was proven by Michael Freedman, for which he received the Fields Medal in 1986.

The question of the boundary of a three-dimensional ball (a ball in four dimensions) remained unanswered, despite many attempts to answer it. Richard Hamilton made a notable contribution by outlining a method for solving the problem, and in 2002 and 2003 Grigori Perelman, from St. Petersburg, wrote three papers that were considered preliminary papers, that is, before they had received confirmation from a professional journal that it would publish them, after they had been refereed (a process that can take years). Today there are Internet sites that publish papers at that stage, and that is accepted practice, especially if the author wishes to press ahead and preserve his rights as being the first to publish his findings ("first dibs"). The only reservation is that those papers have not passed the stage of being approved by referees. In his papers Perelman explained his solution of the Poincaré conjecture, basing it on the method outlined by Hamilton.

At this point we revert to the question of what is a proof. Two students of the Chinese American mathematician Shing-Tung Yau of Harvard University, named Zhu Xiping and Huai-Dong Cao, published a paper (after it had been approved by referees) setting forth a complete proof of the Poincaré conjecture. Their proof, as they themselves stated, was based on the Hamilton-Perelman method. Specifically, they claimed that Perelman's proof was incomplete and that the missing part was not trivial. Their claim was supported by their tutor, Yau, a famous mathematician who had won the Fields Medal in 1982 and the Wolf Prize in 2010.

The dispute was a professional one; that is, it centered on the question of what is a complete proof, although it is difficult to ignore the aspects of prestige and money involved in being the first ones to solve the problem. As one of the Clay Institute's millennium problems, the first solution of the problem would entitle its discoverer to one million dollars. The dispute was heated and intense, with accusations leveled in both directions. Perelman, who in an understatement could be described as a sensitive person, decided to abandon mathematics in reaction to the incident. Others did the work for him, completing what was missing from his proof, noting the fact that the missing parts were not significant. The mathematics community recognized Perelman's achievement, and eventually his Chinese adversaries did also. In 2006 it was decided to award him the Fields Medal, but he refused it. The Clay Mathematics Institute also recognized the partial publication and subsequent complementary sections published by other mathematicians as a whole publication and decided to award Perelman a million dollars, as promised. Once again Perelman refused to accept the prize and withdrew to his city of St. Petersburg, where he is currently absent from the scene of research in mathematics.

64. PURE MATHEMATICS VIS-À-VIS APPLIED MATHEMATICS

The title of this section appears to reflect a distinction between two types of mathematics, and, indeed, this distinction is acceptable to some mathematicians. I would like to put forward and justify the claim that this differentiation is artificial. I will also argue against the use of the term *pure mathematics* itself: Is mathematics that is not pure really impure or contaminated? When people use the term *pure mathematics*, they mean mathematics for its own sake, that is, mathematics that is not motivated by the applications for which it is intended. We will see that even when a mathematician undertaking research does not relate to the possible applications for his results, his findings are likely to be very useful. Here are some examples.

We have mentioned perfect Platonic bodies previously. The fact that only five such bodies existed was known already at the time of Plato. As far as we can reconstruct, the research that led to the discovery of perfect bodies and to the proof that there were only five such bodies was mathematical research for its own sake. Very soon, however, the result became extremely useful. The five perfect bodies became the basis of one of the first models of the structure of the world. Another attempt to use this "pure" mathematics result was made by Kepler, as we saw above (see mainly section 17). In the course of time, these uses lost their value as better mathematical models were discovered to describe the world. Conceptually, however, these were applications of mathematics just like the other uses made of mathematics to describe the world.

The research on the structure of bodies in a space has continued from the days of Plato until today, and among other things researchers in geometry tried to classify bodies that were semi-Platonic, that is, bodies whose boundaries consisted of parts of Platonic bodies. The research was completed some decades ago by Victor Zalgaller of St. Petersburg, who currently has affiliations with the Weizmann Institute in Israel. He proved that there are exactly ninety-two such bodies, and he found them all. Victor is an active mathematician in the fields of geometry and optimization. In December 2010 we celebrated his ninetieth birthday at the Weizmann Institute. He is also another example of the claim relating to age that we put forward in the previous section: in the last two years, he has published several original papers, including a number as sole author. This research of his, like the research that led to the discovery of all the perfect bodies, was carried out for its own sake, but I expect that it will also be found to have uses and applications.

Another example of research in mathematics for its own sake that turned out to have uses relates to tiling an area. A floor, say, can be tiled with square or rectangular tiles, used in most buildings, or diamond shaped (rhomboids) or triangular tiles, and so on. In older, historical buildings, such as palaces built by Islamic rulers in the Iberian Peninsula, the tiles are not even convex (i.e., the tiles may have, say, the shape of the letter L), yet they serve as flooring that is very pleasing to the eye, and the regular repeat of the pattern follows interesting rules of symmetry. Here too it was proved that the number of symmetrical patterns is finite, and examples of all these can be seen in the floors of palaces, such as the famous Alhambra in Granada, Spain.

Now here is a mathematical question asked for its own sake in the 1960s: Can a non-repeated tiling pattern be found with a small number of tile shapes? Non-repeated, or non-periodic, tiling means that there is no possibility of moving the tiled space in a certain direction so that the tiling after the move will merge with the tiling before the move. Roger Penrose of the University of Oxford, winner of the Wolf Prize in 1988, gave an interesting answer in the mid-1970s. He showed that it was possible to tile the area in a non-repeating pattern with only two kinds of tiles, both of diamond shape. Penrose's construction was mathematics, per se, although some places, including Texas A&M University, actually used the tiling proposed by Penrose in one of their halls.

In 1982 the crystallographer and computer scientist Alan Mackay of the University of London performed a computerized experiment to discover what the light diffraction pattern would look like if the atoms of a crystal were ordered in the Penrose tiling pattern. Such patterns are the main instrument used in the identification of crystallographic structure of crystals. Mackay's mathematical experiment can also be seen as mathematics for its own sake, as no one thought or believed that atoms in nature can arrange themselves in non-repeated patterns, as was written in all the textbooks.

However, at the same time, in 1982, the Israeli Dan Shechtman of the Technion, the Israel Institute of Technology, who was then on sabbatical at the United States National Bureau of Standards in Washington, DC, performed an experiment on the diffraction of light. He discovered, in contradiction to deep-rooted belief, a system that did not conform to the periodic structure. In the absence of a mathematical model the results of the experiment would have remained just a description of the findings, without an understanding or an explanation of the discovery. Therefore, Shechtman and a colleague from the Technion, Ilan Blech, published a model that explained the result. Within a short time two physicists who were then at the University of Pennsylvania, Paul Steinhardt and Dov Levine, found that the structure revealed by Shechtman was perfectly consistent with the results of Mackay's theoretical experiment, that is, Penrose's mathematical tiling pattern. Mathematics provided the explanation and the confirmation of the existence of what had been discovered. Researchers in crystallography laboratories searched for, and found, other instances of the phenomenon, but the leading theoretical crystallographers still persisted in expressing doubts and claiming that there might have been an error in Shechtman's experiment. They clung to the old mathematics that described repeated patterns and considered that to be the only mathematics that described crystals in nature. When many other crystals were discovered that did not have repeated patterns, the crystallography "establishment" also agreed to adopt the mathematics that described crystals in nature. Thus a door was opened to one of the most useful areas in material science. His breakthrough earned Shechtman the Nobel Prize in Chemistry in 2011.

In section 31 we referred to another example of applied mathematics: how group theory is a major tool in understanding the structure of elementary particles. The link between mathematics and particle physics was discovered in the 1960s, but group theory itself existed much earlier. The theory is formulated in textbooks in an abstract form. Here is its full version.

A group is a collection, which we will denote by G, of elements, which we will denote by the letters a, b, c, and so on. An operation takes place between the members of the group that we will indicate by the + sign (like the plus sign in the addition of numbers, although in our case we are not referring to addition, as the members of the collection are not necessarily numbers). The result of the operation $a + b$ is a new member of the group G. The operation has certain properties:

1. *Associativity*: $a + (b + c) = (a + b) + c$.
2. There is a member called the *zero*, or the *neutral, member*, which will denote 0, that fulfills $a + 0 = a$ for all members of the group.
3. For each member, there exists its *inverse member*, denoted by $-a$, that satisfies the equality $a + (-a) = 0$.

That is all. We reemphasize that although we use the symbol for addition, which we are familiar with from its use with numbers, and we use the symbol 0 to indicate the "neutral" member, here we are not dealing with numbers. These are abstract operations, and the usual symbols are used to help the brain absorb the abstract structure. Care must be taken not to confuse the dual use of the usual symbols. For example, the properties that we have listed *do not* imply that the equality $a + b = b + a$ holds for all two members of the group. In some groups, this equality holds. A group in which that does apply is known as a commutative group.

Of course the collection of integers is a group where the operation is addition. The natural numbers are not a group because it does not have the inverse property. The positive real numbers are also a group when the operation is multiplication, and the zero member is the number 1. There are also groups with a finite number of members, for instance when the members of the group are revolutions of the plane by multiples of 90 degrees, as described in section 31. Abstract group theory tries to reveal mathematical properties of groups relying solely on the simple definition of a group. As soon as such a property is proven, it applies to all examples of groups.

In addition to groups, mathematicians also define collections with other properties, for example, fields.

A field is a group on which, in addition to the operation of addition, another operation can be performed, and we will denote the additional operation by a point · like the multiplication point, and we will also call it "multiplication." The multiplication operation also has the associativity property, and it also has a neutral member that we will denote by the symbol 1 (again, not the number 1), and every element a that is not equal to 0 has an inverse that we will denote by a^{-1}. There is a connection between the addition and multiplication operations, and it is known as distributivity, that is, for all three elements in the field, the equation $a \cdot (b + c) = a \cdot b + a \cdot c$ holds. Here again the symbols are familiar to us from their use with numbers, and indeed the numbers are a field, but the purpose of the theory is to discover properties of fields without relating to any specific example.

For many years, group theory and field theory and similar theories served as the standard-bearers of "pure" mathematics, until their uses in describing nature were discovered. Group theory and abstract field theory started with the work of Évariste Galois (1811–1832), as part of his attempt to identify and characterize the solutions of polynomial equations, that is, to find when and how many solutions exist to the numerical equation $p(x) = 0$, when p is a polynomial. Finding solutions to such equations was considered then, and to a great extent also today, applied mathematics. The abstract concepts of a group and a field were defined many years after Galois's time. Thus, group theory is a clear example of mathematics that developed from applications, and its definitions are based on known examples, from which the notion of abstract, "pure" mathematics as a subject was derived. From there, as we saw in the section on groups of particles, applications, sometimes surprising ones, were found in totally different areas.

The day-by-day practice of mathematics sometimes stresses the theoretical aspect to the extent that it ignores what motivated the development of some mathematics and its possible applications. Influenced by developments in the study of the foundations of mathematics, a group of French mathematicians convened in the mid-1930s with the intention of systematically and rigorously recording and developing the mathematics that was known then, with the emphasis on mathematics for its own sake. They decided to publish a series of books, under the collective nom de plume Nicolas Bourbaki. The Bourbaki group was active for several decades and produced many books that had a significant impact on a sizable share of the mathematics community. No one questions the group's achievements in mathematics, but already in the course of its work, many disagreed with the approach that the group represented and championed, an approach that focused on mathematics for its own sake at the expense of its nonmathematical motivation and applications. The Bourbaki effect is still felt in certain circles, but today the approach that favors "mathematics shall dwell alone" is fading, and there is constantly growing recognition of the contribution of nonmathematical subjects to mathematics.

The following example describes mathematics that solved a very real problem and from which abstract mathematics developed along with largely applicable mathematics. The story starts with the accelerated development of steamships in the early nineteenth century. The engineering system used to stabilize the speed of the steamboats was what is still today known as the Watt regulator, named after the British scientist James Watt (1736–1819), whose name is also used for a unit of electrical energy. The steam engine would rotate, in addition to a propeller, a cylinder with two arms. When the engine was revolving rapidly, the centrifugal force raised the arms. When the revolutions were slow, the arms were in a low position. The arms were connected to a piston inside the cylinder, so that when the arms were raised, the piston blocked the flow of steam to the engine, and when the arms were low, this allowed a faster flow of steam to the engine. Thus, a faster speed raised the arms, reduced the flow of steam, and lowered the speed. A slower speed lowered the arms, increasing the supply of steam and thus the speed. The system worked well, up to the stage where the engineers improved the quality of the pistons and the cylinders, and then the boats' engines started to disintegrate.

The engineers did not manage to solve the problem and had no choice but to turn to James Clerk Maxwell, the famous scientist whose enormous contribution to science we have described previously (section 25), and he did indeed solve their problem. The first thing he did was to say to the engineers, and also to himself, that one must not trust human intuition. The action of the regulator that we described in the previous paragraph was based on the intuitive perception of the processes resulting from previous experience. In the absence of previous experience, intuition bases itself on understanding as fashioned by evolution. Steamships were not part of the process of evolution. Therefore, to check whether the intuition is right, mathematics must be used, just as the Greeks had taught the human race two thousand four hundred years earlier. Maxwell wrote down the differential equations that described the movement of the piston and showed where the intuition of the engineers had let them down. The failure lay in the fact that the mechanism described above related to the arms that regulated the steam being in a static state. The static intuition did not take into consideration the motion that the arms underwent in the transition from the low position to the high position and back again. The improvement of the piston made that movement faster, and that caused an overreaction and the destruction of the whole engine. The equations that Maxwell used identified the problem and led to the realization of how to solve it. He described his findings in a mathematical article published in 1868. Maxwell's applied research solved an engineering problem and, at the same time, opened up a new area of mathematics called control theory. Maxwell's colleagues, and prominent among them, his Cambridge associate Edward Routh, quickly entered the field and found mathematical criteria for the stability of general systems. Since then, the subject has been advancing in parallel in the areas of the applications of mathematics as well as of mathematics for its own sake. There are hardly any engineering systems in which control theory does not play a central role, and recently applications of the theory have been found in economics and finance. The mathematicians who deal with control theory contribute both to the engineering aspect, where the applications of the theory are of greater interest, and also to the mathematical aspect itself, where the mathematics for its own sake plays a greater part.

The next fascinating example shows on the one hand how a mental challenge can lead to important results, and, on the other, the difference between a quick and "elegant" solution, for example, with the help of a trick, and a useful solution. To provide the background to this distinction we recall that the purpose of a proof in mathematics goes far beyond persuading the reader that the result is correct. The function of the proof is to explain why the result is as it is and to learn from it how to act in similar situations.

The Greeks were, apparently, the first to have studied the question of planning the route a body should take to get from one place to another in the shortest time. The usual formulation of the problem, known as the brachistochrone problem (from the Greek for "the shortest time") is as follows.

Given points A and B in a vertical plane, as in the diagram, with A higher than B, plan the slide from A to B such that a ball falling without friction from A will reach B in the shortest time.

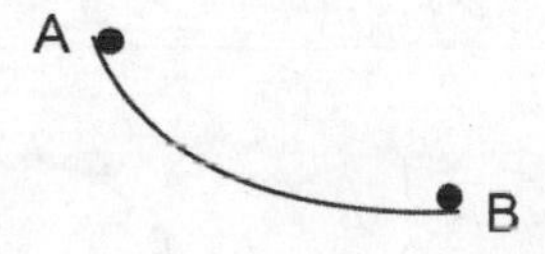

The Greeks thought that the optimal, that is, quickest, route was a straight line because that is the shortest distance between two points. That solution was rejected when the laws of motion were discovered, and in particular the law of falling bodies. Galileo proposed that the quickest route was a quadratic line, part of the circumference of a circle. He did not prove his claim completely, and the problem did not attract much further attention. In June 1696, Johann Bernoulli (1667–1748), brother of Jacob and father of Daniel, both of whom we have already spoken, published a letter in which he stated that he had a complete and attractive solution to the problem and challenged the mathematics community to propose a solution before January of the following year. The letter, published in the scientific journal *Acta Eruditorum*, attracted the attention of mathematicians. Leibniz even asked Bernoulli to extend the deadline for submission of a solution. The extension was given, and in May 1697 solutions were published in the journal, including that of Bernoulli himself, another proposed by his oldest brother, Jacob, a solution by Leibniz (that was actually published elsewhere), and another by Newton. Other solutions by the French mathematician Guillaume François Antoine, marquis de L'Hôpital (1661–1704) and Ehrenfried von Tschirnhaus (1651–1708) were also published there.

Johann's solution was indeed the shortest and most elegant. He suggested that although the problem relates to a falling ball, let us imagine that the path describes a beam of light that travels from A to B, passing through a medium in which the speed of light changes. The changes in its speed can be derived from the law of conservation of energy, with the potential energy of the difference in heights converted into kinetic energy, and with the conversion relative to the square of the speed. The beam of light, as was known in the days of Fermat, travels from one point to another in minimum time, and Snell's law (see section 21) defines the slope of the route along which the beam travels, which in our case is the slope of the path. As soon as the connection with Snell's law was established, prior knowledge could be employed, in particular, mathematics developed by Fermat and some of his contemporaries, to show that the path was cycloid, namely, the path traced by a point on the circumference of a circle, say, a round coin, that is rolling in a plane.

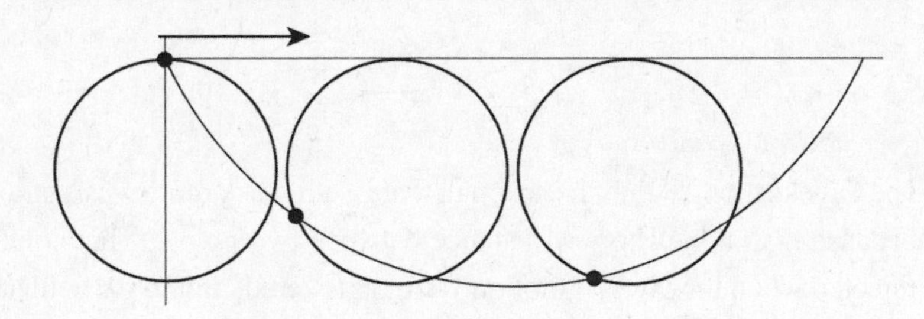

In contrast with Johann Bernoulli's trick of viewing the route as if it were the path of a beam of light, the solution proposed by his brother Jacob was long and complex and required laborious mathematical manipulations.

He started with a given line and asked what would happen if it were changed a little. If the original line is the solution, the change will result in the ball taking longer to reach its destination. If we calculate the difference in the times, we will obtain, using a method that today is relatively easy to describe but in those days was very complex, equations of a type that Jacob Bernoulli managed to solve, also relying on earlier studies. He also proved that the result was a cycloid.

The relations between the two brothers were not the best, to put it mildly; Johann even mocked his brother openly over his awkward solution. In response, Jacob used his method to solve other minimization problems that Johann could not solve using his trick or any other trick.

Johann Bernoulli's solution to the brachistochrone problem remained a one-off trick, attractive in form, but with almost no further use. In contrast, the clumsy and complicated method developed by Jacob Bernoulli grew into an extensive and important area of mathematics called the calculus of variations. This field of mathematics is implemented widely and has many applications, both within and outside mathematics, and it presents mathematical challenges that so far are in the area of mathematics for its own sake.

In the above examples, and in others we have not described, the connection can be seen between the abstract study of mathematics and its uses. Sometimes the practical use of mathematics leads to an abstract mathematical theory that develops into mathematics for its own sake, which often results in new real practical uses. On other occasions, research starts with questions that mathematicians ask themselves, prompted by purely mathematical curiosity, that apparently are of no interest for any practical purpose, yet whose solutions are found to have significant use outside the world of mathematics. Is this merely a coincidence? Or perhaps a statistical illusion? Or perhaps it is the inevitability of reality?

We may well be the victims of an illusion. For hundreds of years, mathematicians developed many and varied theories, most of which were assigned to oblivion because of their irrelevance. Theories that found their place within sciences other than mathematics have wider exposure and are more generally known than those that remain within the area of "pure" mathematics, so that the use of this "applied" mathematics seems to us more frequent than it really is. Similarly, when researchers in natural sciences look for a new mathematical theory to explain their findings, they search among theories already developed, including those developed as mathematics for its own sake, and that can lead to the illusion that mathematics for its own sake plays a central role in applications. There may be another explanation, however, that seems to me to be more reasonable: mathematics for its own sake was also developed by the human brain, a brain that learned to reveal patterns that appear in nature. That brain is not capable of revealing patterns that are totally detached from nature. When the mathematical pattern can be used by nature, it is not surprising that nature actually does use it.

65. THE BEAUTY, EFFICIENCY, AND UNIVERSALITY OF MATHEMATICS

It is generally accepted that beauty is a matter of taste, yet certain fundamentals of beauty and enjoyment are common to most human beings. Like other aspects of human conduct, the sources of enjoyment and the relation to beauty were formed by evolution. A prime example of the emotion inspired by beauty is that felt on recognizing patterns, symmetries, and so on. Recognizing a pattern or symmetry affords great pleasure, and, in surprising situations or places where one does not expect to find order, discovering them can give even greater delight. That is true in all areas of life, from natural scenery, to the plastic arts, to social and cultural environments, and so on. Patterns are also the basis of mathematical practice. Hence, every new theorem or new geometrical law that we encounter has the potential to give us pleasure. In most such cases, the patterns, and, hence, the enjoyment, are of the intellectual rather than visual sort. In many instances, before enjoying the mathematical pattern, we have to contend with terminology that is sometimes hard even for professionals to understand, learn special jargon, and also imbibe the mathematical content. These are some of the factors that keep much of the general public away from mathematics and its pleasures. I remember a cartoon showing three mathematicians in a jolly mood next to a blackboard covered with complex formulae, and one is saying to the others, "I knew you'd burst out laughing." There are indeed professional jokes and professional pleasures in mathematics just as in other disciplines. I will nevertheless repeat a statement presented at the beginning of this book: just as one can derive pleasure from music without being able to read a score, so can one derive pleasure from mathematical patterns without being able to read the mathematical notes, and one certainly can get enjoyment from knowing how mathematics fulfills its role as a tool for explaining nature. I hope that "non-mathematical" readers who have reached this section will agree with that statement.

Nonetheless, the general public is sometimes presented with other examples of the enjoyment and beauty in mathematics. Not long ago I received an e-mail with the following tower-shaped table, alongside other similar tables, under the title "The Beauty of Mathematics."

$$
\begin{gathered}
1 \times 8 + 1 = 9 \\
12 \times 8 + 2 = 98 \\
123 \times 8 + 3 = 987 \\
1234 \times 8 + 4 = 9876 \\
12345 \times 8 + 5 = 98765 \\
123456 \times 8 + 6 = 987654 \\
1234567 \times 8 + 7 = 9876543 \\
12345678 \times 8 + 8 = 98765432 \\
123456789 \times 8 + 9 = 987654321
\end{gathered}
$$

The table was presented in lovely colors, every digit in another color, and the colors accentuated the special pattern. Some of my acquaintances, of course including the person who had sent it to me, expressed their wonder at the beauty of this mathematics. I did not find anything interesting or special, in a mathematical sense, in this tower. It was pleasing to the eye, and its colors were attractive. The order that the arithmetic exercises reflected was also interesting and enjoyable, but it did not contain any marked element of surprise for me. I did not find any beauty in it from a mathematical aspect. I agree that the arithmetic reflected by the tower is aesthetic, but that is not the beauty of mathematics. If I were to delve deeper, or if someone were to show me that hidden within the structure is a special pattern, then I might find beauty in the equalities that constitute the tower. But I did not find it in the tower as it is.

Here is another example. One of Israel's daily newspapers published a diagram and puzzle under the heading "You Can Enjoy Geometry" (see my sketch of the puzzle below). The sketch shows four strips that could be considered bridges from one bank of an imaginary river to the other. The width of each band is the same and is constant along its whole length, but some are straight and some change direction, as in the diagram. The puzzle is: Which bridge would be the cheapest to paint, and which the most expensive?

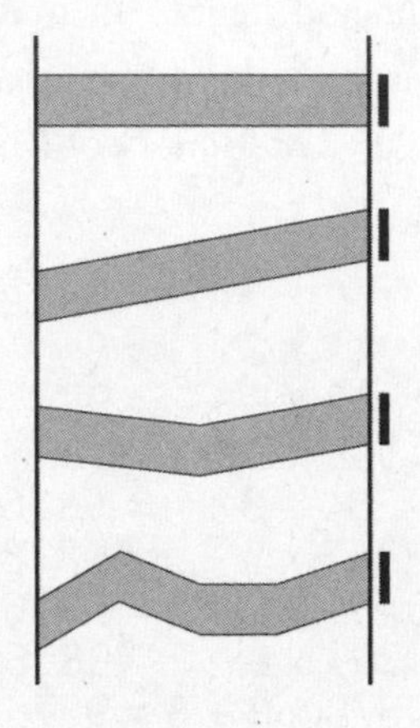

The answer is that there is no difference in the costs. All the bridges have the same area. The solution can be seen "clearly" (the word used by the paper; one could argue with the use of that term) if the picture is turned onto its side. Every bridge consists of parallelograms with the same sized bases, and the widths of the bridges stays the same from one bank of the river to the other, and the sum of the heights of the parallelograms is the same for all the bridges. Hence (as anyone who learned *and remembers* basic geometry will know), the areas of the bridges are all equal. And I ask, "What is enjoyable in this?" I did not get any pleasure from it, firstly because I did not find the solution myself. Moreover, whoever set the puzzle thought that I ought to see the answer immediately, and that is another reason why I did not enjoy it. I was also left with the feeling that I had been misled. The question was which bridge would be the cheapest to paint, and which the most expensive. To my mind, if the question is worded like that, it implies that one of the bridges has the largest area and one has the smallest, and if that is not the case, then I have been misled. Even if I derived a little enjoyment from the trick that I learned, to place the picture on its side, it was mixed with some disappointment that it was after all just a trick, let alone the fact that you do not need to place a parallelogram on its side to calculate its area; but that is how it is taught in schools. I have no objections to those who do enjoy such puzzles. To anyone like me who does not enjoy them I would say, don't think that you cannot get pleasure from mathematics. It is possible to get enjoyment from tricks, or to be frustrated that you didn't manage to find the trick yourself, but tricks of this type are not an essential part of mathematics.

A similar message is derived from this well-known story about John von Neumann. A friend asked him the following question: Two trains start moving toward each other at the same time from positions a hundred kilometers apart at a speed of fifty kilometers an hour. At the same time a bee starts flying at three hundred kilometers an hour from one train toward the other until it reaches it, and then starts flying back to the first train, until it reaches it, and then turns around again. This continues until the trains meet. What total distance does the bee fly? Before we give von Neumann's answer, we observe that there are two ways of solving the problem. The simple way is to note that the trains meet after exactly one hour, and in that time the bee will have flown three hundred kilometers. The second, more complicated, method is to calculate the distance that the bee will have flown until it first meets the second train (this requires solving a not-particularly difficult equation), then to calculate the distance it flies until it is back at the first train, and continue in this manner. This gives an infinite series that can be added relatively simply (for anyone who remembers the formula), and the result, if no mistakes are made, is three hundred kilometers. Now back to von Neumann. He thought briefly (he was known to be a very quick thinker) and then gave the right answer, three hundred kilometers. The friend confirmed that von Neumann's answer was indeed correct and then complimented him, saying, "One can see that you are an excellent mathematician. Some fools," he continued, "calculate the sum of the infinite series in order to get to the answer." Von Neumann looked at him in surprise and asked, "Is there another way?"

The message I am trying to convey is that mathematics is not a collection of tricks. On the contrary; the essence of mathematics is ordered and well-organized analysis of patterns already revealed or that still need to be found. We saw an example of the difference between a trick and a theory in the instance of the brachistochrone in the previous section. It is acceptable and possible to benefit from tricks, but the real beauty of mathematics derives from the patterns and rules found in mathematics, and the pleasure from, and sometimes amazement at, the link to nature and to applications that the patterns and rules open up for us. I believe that nearly all the material discussed in the book up until now reflects the beauty of that sort in mathematics.

Another aspect related to the beauty and pleasure of mathematics is the sometimes-wonderful efficiency of mathematics as an instrument for describing and explaining natural phenomena. The question why is it just mathematics that is so appropriate to the description of nature arose already in ancient times, and it reappeared even more pointedly following the revolution led by Maxwell, which led to the theory of relativity and quantum theory.

Until that revolution, which symbolizes the start of the modern era in science, the question was not why does mathematics describe nature so well, but why do natural phenomena occur according to any system of rules? Why does nature itself have to follow laws that we call laws of nature? Why are there such clear patterns in nature? Why is the law of gravity on Earth the same as the law of gravity on the Moon, and apparently the same as the law of gravity in other galaxies? Why is a small, finite number of particles the component that constitutes all known materials in nature? And why can we find ordered patterns, to the extent even of a structure of mathematical groups, that determine how those particles arrange themselves? We could extend this list of questions. There are no standard scientific answers to these questions, and interest and involvement in them is usually placed in what philosophers call the transcendental sphere. For example, the saying that nature is based on mathematics is attributed to Galileo Galilei, and even the Christian Church at the time of the Renaissance agreed that God in his great wisdom used mathematics to create such a wonderful world. That of course is not a satisfactory answer to the question of why mathematics is the appropriate instrument to be used in describing and analyzing nature. We should bear in mind that the very understanding that nature expresses clear sets of laws is not natural or self-explanatory. The ancient civilizations discovered that such laws existed but did not delve into them or try to fathom them. Also, the Greeks, who declared that the laws of nature could be described successfully through logic, needed hundreds of years to really absorb that understanding into their scientific view of the world. The fact that the earthly laws of motion were the same as the heavenly laws of motion was discovered, and then accepted, only a little more than three hundred years ago.

Throughout the generations mathematics was in competition with other approaches that proposed different understandings of how nature worked, including idol worship, astrology, and other strange theories, some of which are long forgotten. Today we view the mathematical structure of nature as obvious, but that was not the case in the not so distant past. What made it easier to instill the understanding that mathematics was the right instrument for describing the world was the fact that up to Maxwell's revolution, mathematics described mainly effects that people experienced and could measure. The Greeks saw that planets moved in the sky and enlisted geometry to describe the manifestation. Newton constructed infinitesimal calculus to describe the laws of motion, motion that we can observe. From the moment that mathematics was found to be an efficient instrument for describing events that we perceive with our senses, it is no surprise that it was developed, and is still developing, to elicit more and more from it.

What is more surprising is that the same instrument that serves to describe phenomena that we measure and experience also manages to foresee new ones, including phenomena of whose existence we are unaware and which we feel no need to try to understand. How did Maxwell's equations, created "only" to give uniform expression to the link between electricity and magnetism, foretell the existence of electromagnetic waves? How did the change in the variables proposed by Hendrik Lorentz to provide a formal explanation for the results of the Michelson-Morley experiment lead to the realization that everything is relative and that the geometry of the world is not what we see and experience? And how is the behavior of particles, which we also cannot perceive directly but whose effects we can measure, best described using solutions relating to waves, despite the fact that we cannot identify any medium in which these virtual waves move? Again, currently there are no answers to these questions that are not transcendental, that is, are not "beyond" the realm of scientific discussion.

We may be able to obtain partial understanding of these questions by recognizing that all the new phenomena that we discover are consistent with what evolution prepared us for. Even when mathematics, followed by experiments, leads us to the discovery of completely new phenomena, in order to understand them we translate them into a language and metaphors that we are familiar with. We speak of electromagnetic waves because we are familiar with waves in the sea and because we know what sound waves are. We describe the laws of relativity by means of geometry because that is something we know. Although we cannot perceive that the geometry of the world is not Euclidean, we have encountered non-Euclidean geometry in other experiences, such as the geometry of the surface of a ball. Thus, it may well be that nature seems to us to be subject to such elementary laws because that is what we are searching for, the same search and law-based system that are the products of evolution. Einstein said that the laws of nature are characterized by their simplicity, and a simple law must be preferable to a complex one. Could it be, however, that Einstein's statement expresses wishful thinking rather than a description of reality? Even when we discover manifestations that suggest a lack of clear laws, such as the phenomenon of mathematical chaos (in chaos theory), we nonetheless focus our efforts on finding order in the chaos. We will expand a little on this.

In 1961 Edward Lorenz (1917–2008), an MIT mathematician and a meteorologist, performed computerized experimental simulations of equations relating to weather forecasts. He found that although the equations were relatively simple, the results of the simulations were unpredictable. The reason was that minor changes in the data caused great changes in the results. This is critical in the context of computer simulations because in such calculations, perfect accuracy is never achieved. It soon became clear, however, that the deviations were beyond the range of mathematical calculations. Lorenz's discovery was related to the mathematical results that Poincaré had indicated previously, results that showed that the heavenly orbits, such as the Sun and its planets, were subject to significant irregularities. Poincaré's results also found expression in more-general equations of motion, for example, those by Jacques Hadamard and others relating to the movement of billiard balls. Another development related to these was made by Steve Smale, who showed that certain equations that satisfy relatively simple conditions (embodied in the function known as the Smale horseshoe) lead to the result that the dynamics represented by the equation is extremely intricate. An important step in that direction was taken by James Yorke and his student Tien-Yien Li of the University of Maryland. They found the simplest condition for the equation that results in the most complex dynamics, in which minute changes in the data result in huge changes in the dynamics. Li and Yorke also coined the term *chaos* for their result, and that word appeared in the title of their paper. An anecdote that also points to the nature of research and the nature of man relates that, not long after the publication of Li and Yorke's findings, it came to light that their mathematical result was a particular case of a far more advanced result published some years earlier by the Ukrainian Oleksandr Sharkovsky. The title of his paper, however, was not sufficiently eye catching to draw the attention that the article deserved. After some time, Yorke said jokingly that his own contribution to chaos theory consisted merely in providing the name for the theory. In truth, however, the important contribution was in drawing attention to the connection between the mathematical expressions and intuition about the complexity of the dynamics. The appearance of Li and Yorke's paper sparked off very extensive research on the chaos effect, research that spread into the spheres of philosophy and social science.

Here is another story relevant to the sometimes-strange interpretations given to mathematical results. To illustrate the dependence of a significant event on a very small change, the phrase "the butterfly effect" has been applied. According to this metaphor, the tiny movement of the wings of a butterfly in Southeast Asia could cause a hurricane in the Atlantic Ocean. I heard a television commentator say that butterflies in Asia cause hurricanes in the Gulf of Mexico, as if that were something well known and obvious. The commentator did not appreciate the difference between "could cause" and "causes." The commentator can be reassured: no hurricane in the Atlantic Ocean was ever caused by the fluttering of a butterfly's wings in Southeast Asia. The subject of chaos has become a wide-ranging and productive branch of mathematics, with many applications in physics and other sciences. However, most of that research itself focuses on finding the order within the appearances of chaos, either in the way chaos is created, or in the characteristics of the statistical rules governing its occurrences, that is to say, the same type of patterns that we generally seek. To repeat what has already been said, the essence of mathematical research is indeed the search for patterns, and we usually find them among those we already know.

We can even stretch the point and go further and ask, might there be laws of nature that we have not found because the human brain is limited to identifying patterns and rules of a certain type that are consistent with the way evolution molded our brains? In my opinion, the answer is yes, our brain is limited in that way. Indeed, Poincaré and Einstein both expressed the opinion that the laws of nature that we identify through mathematics are limited to the metaphors that our brains can create, and that in nature itself there exist phenomena that are beyond our capacity to understand. Those views are in line with the modern understanding of the way our brain, the product of evolution, perceives the world around us. The problem is that as long as research is carried out under the guidance of a human brain and is examined by a human brain, it is not clear how we can deal with that restriction.

And here we reach the last question in our current quest, the universal question of mathematics. Is mathematics that would have been developed (or that was developed) by various societies, either isolated human societies detached from our civilization or societies in another galaxy or another world, necessarily the same as our mathematics? The prevailing opinion is that the answer is yes. Mathematics developed independently and under different conditions may have different emphases, and certainly different symbols and a different language. However, the logical basis and the basic technical elements, such as the natural numbers and the operation of their addition, would be the same in all versions of mathematics. That is why when a spacecraft was sent into space in the hope that it would be taken by a foreign space civilization, it contained a board with signs for the numbers 1, 2, 3, . . . as if to announce, "We can count."

Nevertheless, I will allow myself to raise another possibility that, by its very nature, is no more than speculation, and which I can see no way to prove or disprove. Maybe one can question the belief that mathematics is uniform. The natural numbers are the result of the fact that our world is made up of items that we can count, and counting lends itself to the operation of addition. In a perfectly continuous world in which entities are not defined separately or as distinct units, there is no reason for the natural numbers to have any meaning (this insight is accredited to Sir Michael Atiyah, one of the most prominent mathematicians of our time). There is no reason that mathematicians in such worlds should be able to understand, say, the Peano axioms, let alone such concrete operations as $2 + 2 = 4$. In other worlds that developed differently than we did, there may be other rules of logic. Even if our brains cannot imagine the concept of another logic, there is no guarantee that our logic will be relevant to societies in other worlds. Even if we do not wish to go so far as to speculate about other worlds, we should bear in mind that the logic we use, including the elementary rules of inference, is the product of our brains. And that product is the outcome of experience accumulated over the course of evolution that formed our brains.

It is clear, nevertheless, that the fact that another logic might exist in another world with another mathematics in another society does not lead to the conclusion that the logic we use is faulty or that we must look for and try other methods, new and old. The mathematics we know and are continuing to develop has proved that it is correct, fine, and efficient.

CHAPTER X
WHY IS TEACHING AND LEARNING MATHEMATICS SO HARD?

Is it helpful to join a class in mathematical thinking? • How can you learn mathematics to a mother-tongue level? • How did prehistoric man discover mathematics? • What is the essence of a triangle? • Is there a link between a mathematical tree and a botanical one? • How does a centipede walk? • Do students have difficulty in understanding the parallel-lines axiom? • What are the chances that in a family with three children the third will be a boy?

66. WHY LEARN MATHEMATICS?

There is no doubt that mathematics is considered one of the hardest subjects in the education system, from elementary school to university, and achievements in the study of mathematics generally fall short of expectations. However, before we turn to the question of how to improve mathematics teaching, we should clarify what we expect from it. Once we agree on its aims, we may be able to examine the extent to which teaching practice achieves those objectives and improve the system to enhance its performance. The targets that we will formulate relate to students in elementary and secondary schools. We will also relate to higher education, from the aspect of teacher training.

The first objective is to provide the student with the *basic mathematical tools* needed to function in the modern world. Functioning clearly includes the ability to operate in a world driven by commerce, money, loans, investments, and the like. It is also advisable to be able to understand a world in which we are assailed by a mass of statistics, some of them important, and some unreliable. It is also important to possess the ability to estimate and calculate areas and volumes, that is, the ability to use basic geometric tools relevant to our environment. Lastly, in order to be able to function properly in the modern world it is worthwhile gaining at least an elementary mathematical understanding of the developing technological world.

The second objective is to become familiar with a *system of logic that rigorously checks claims*. This applies despite the fact that we have argued several times in this book that our natural tendency, resulting from evolution, is to believe what we are told. Yet when there is doubt whether the claims put to you have a firm basis, it is worthwhile being able to check them. No system can match mathematics in being able to clarify the difference between an assumption and a conclusion, between a conclusion drawn by deduction and a hypothesis reached via induction, and so on. Mathematics develops the ability to perform an ordered analysis using the tools of logic, to examine assumptions with minimal use of metaphors, and to query the quality of the information by using very precise language. It is true that people's day-to-day activity is based on intuition, and that is something that is impossible and unnecessary to change. Yet it is important that a graduate of the education system should have the ability to follow a logical analysis of claims, and even to perform such a logical analysis in cases where it is important.

The third objective of mathematics teaching is *to recognize mathematics as part of human culture*. Alongside history, literature, music, the plastic arts, social philosophy, and so on, mathematics and the sciences played a major role in human development. An educated person should know the part fulfilled and still being fulfilled by mathematics in the understanding and development of the sciences, the arts, technology, and society.

The last objective we shall mention here is *to open a window* for anyone interested in the subject and for future researchers in sciences and engineering in general, and in mathematics in particular. We are not claiming that secondary education should train scientists and engineers. The least that can be expected of the school is that it should arouse enough interest in its students and should show them how to learn and the possibilities of studying sciences and mathematics so that when they have to choose the path they wish to follow, they have the information necessary to make the right decision, in accordance with their aims and capabilities. The system should also impart to those who will continue to follow the scientific path the basic abilities to realize their potential.

Unfortunately the education system at present falls short in all of these objectives. The following sections will try to indicate some of the defects that lead to this sorry state of affairs. We will not presume to propose a detailed system of what and how to teach. That is too daunting a task that requires means, manpower, further development of didactic systems, and other similar serious issues and limitations. The range of reasons for the problem is beyond the scope of this book, but understanding some of the defects, particularly those related to the question of what is mathematics, may help to correct them.

67. MATHEMATICAL THINKING: THERE IS NO SUCH THING

The market is swamped with books, courses, advanced-study courses, and similar tempting offers that promise to teach and improve our and our children's mathematical thinking. All you need to do is to agree to pay the stated sum, and your thinking will improve and will become proper mathematical thinking. Are there any parents who would refuse to make the effort and pay the required amounts and prevent their children from benefiting from the opportunity to think mathematically? Without mathematical thinking it would be hard to function in the modern world. (Just to remove any possible doubt, the above is written in irony.) This went so far that one newspaper published an article about courses in mathematical thinking for two- and three-year-old children. I searched the Internet for the term *mathematical thinking* together with the word *classes*, and found that extracurricular classes for first graders and higher grades headed the list, with classes also for three-year-olds up to kindergarten age that included "mathematics as a mother tongue." Parents in one of the schools in the central part of Israel complained to the teachers and the headmistress that the school did not offer extracurricular courses in mathematical thinking. Our children are being left behind, they argued. The institute where I work and other institutes of higher learning offer courses for youngsters on mathematical thinking. The truth is, *there is no such thing as mathematical thinking*. If that is the case, you may ask, what is taught in all those classes and courses? The answer is, they simply teach certain mathematical subjects, generally the more logical parts of mathematics, as well as tricks for solving problems, most of which are called problems of logic.

At this point a number of questions come to mind, which we shall consider one by one. What does "there is no such thing as mathematical thinking" mean? What is wrong with calling the activity that takes place in these classes and courses mathematical thinking? Does participation in such classes cause any harm? Are they helpful in any way?

In various sections earlier in this book we discussed ways of thinking, in particular in section 62. We stated that thinking is mainly an activity of the brain intended to analyze situations and to make decisions. A rough-and-ready distinction divides that activity into two types of thinking. One is comparative thinking, in which people compare the situation they are facing to other similar situations with known solutions. The second type of thinking requires a new, more creative response to an unfamiliar situation. There too the brain uses the range of reactions that it knows, but it has to adapt and update them as necessary, and sometimes it will have to come up with a new and creative response. These two elements of thinking are *common to all disciplines* and are not confined to mathematics. These types of thinking cannot be taught. The more you know and the more experience you have in the subject you are thinking about, the more efficient your comparative thinking will be and the better your reactions. The more you know, the more efficient your creativity will be too. It is thus worthwhile to learn a variety of subjects in order to enrich your pool of knowledge and experience, and to help enhance your thinking. The same applies to mathematics. The more you learn and practice, the more you will succeed. This rule is valid for all subjects; you do not learn thinking, but content.

Friends who teach mathematics ask me why I am so pedantic. They say that the public calls these subjects, that is, exercises in logical problems, and so on, "mathematical thinking"; they say it is just a matter of what it is called, not of essence. Those who make this claim are the victims of the mathematics profession, and they are ignoring the effect of language on learning. Let me explain: it is the accepted norm in mathematics to give names, that is, to make up names, for new mathematical entities or specific operations. We have come across the terms group and field earlier in the book. Trees, matrices, manifolds, automata, and machines also appear in mathematics. The motive for giving a new and inclusive name to a characteristic, or to a certain collection, is clear. As soon as we have given a mathematical name to something, and we have absorbed what underlies that name, it can be referred to without having to explain from scratch what we are talking about. Giving a name is not arbitrary (a fact that is not clarified sufficiently in mathematics lessons, even in higher education). No authority can forbid you from giving any name you choose to a mathematical concept. Instead of a group, you could call it an elephant, or you could give the name Moses to what is generally called a mathematical tree, or you could concoct a name consisting of meaningless syllables. Nevertheless, that is not the practice in mathematics. A mathematical tree to some extent brings to mind a tree, and the word *group* was chosen because it refers to a collection of entities and the relations between them, and likewise with other names. Mental processes are associative. If we were to call a mathematical tree a steamboat, or replace the term *function* with the word *elephant*, it would disturb the users of the terms because the names would probably conjure up in their minds images of rivers or zoos. It is true that even if there is a good reason for choosing a particular name, it could still cause confusion. A mathematical tree is not a botanical tree. The term would not confuse a mathematics graduate but could well confuse someone not involved in mathematics. (The story is told that after the publication of an important paper by the mathematician Michael Rabin on automata on trees, he was invited to give a lecture on the subject by a faculty of agriculture!) Hence the importance of choosing a nonarbitrary name that will not mislead, that is, a name that would allow the relevant intuition regarding the concept to develop.

Those with higher degrees in mathematics are sufficiently well trained so that when certain subjects in mathematics are called "mathematical thinking" they will repress the intuitive meaning of the term "thinking." This apparently also applies to those who offer courses in mathematical thinking. To the general public, however, the concept of thinking means more than the learning of content. Using the term *mathematical thinking* for certain aspects of mathematics or logic is misleading, or at least an error, especially if there is a hint that the course offers more than just additional (and very partial) knowledge in mathematics. This is also a matter of raising unrealistic hopes. The result of the disappointment due to their nonrealization can cause real harm: if in the future children or their parents discover that they have difficulty in a certain topic in mathematics, they are likely to conclude that they are suffering from some disability. After all, they are graduates of a course that is not a vague general course in mathematics, but a course in mathematical thinking!

If we ignore the deception, we could ask whether there is something wrong in itself in learning some parts of mathematics, puzzles in logic, and various mathematical tricks. Of course there is nothing wrong in learning, per se; it depends on the parents' economic situation, the alternative cost of a babysitter, and the amount of pleasure the child (not the parents) gains from participating. There is something wrong, however, in presenting this mathematics as if it should guide daily conduct outside the classroom. Presenting "mathematical thinking" as a necessity, or as beneficial to life beyond mathematics and its applications, is likely to cause harm. Some time ago I met a teacher in one of the well-known secondary schools with a good reputation for its teaching of mathematics and known for preparing its outstanding students for the Mathematics Olympiad. That teacher was really angry with me when he heard the views that I am expressing here on mathematical thinking. He claimed that he was more than a mathematics teacher, rather, he prepared his students to think correctly in all areas of life. He argued that they must examine each step they take by means of mathematical thinking. Poor students, I thought to myself. Just like the centipede that was strolling along quite comfortably until it was asked: How do you manage to coordinate your steps so that your right leg number

23 and left leg number 12 are forward while left leg number 17 and right leg 19 are pointing backward? The creature stopped in its tracks and could not move any of its legs. In our daily lives there is no time for a logical analysis of every step we take. We have to use our intuition, and it is certainly acceptable, and sometimes even preferable, for mistakes to occur. Mathematics and logic do not help, and are likely even to hinder, if they are used as a tool to manage our lives.

Is there then any benefit in participating in classes for mathematical thinking? Again, that depends on the alternatives. If the choice is between that and watching a so-called reality show on television, the mathematics class would seem to be preferable. If the alternative is a theater group or literary circle, or an athletics or football team, the choice will depend on the child's preferences or what is better for his physical or mental wellbeing.

Such enrichment courses may actually cause indirect harm, as they are likely to give the impression that exercises of that type reflect the whole of mathematics, and they pressurize the outstanding students to advance as quickly as possible. This striving for excellence sometimes results in the student learning advanced material too soon, before he or she is ready for it. A child who excels in such groups is sometimes sent to university to participate in special courses for outstanding students. Some of the students are mature enough for such courses, but for others (the majority, in fact), the early excellence is in too narrow a field, and the push to advanced studies at that early stage is harmful. These students do not absorb the required material properly and form a misguided picture of mathematics not because they are less talented, but because they started before they possessed the necessary degree of maturity. Recently I participated in a number of interviews with candidates for graduate studies in the academic institute where I work; these candidates had obtained their bachelor degrees at a very early age in one of the special programs for outstanding students. For some of them, the programs had destroyed their chances for a real scientific career; more's the pity.

Here is another example illustrating the importance of accurate terminology. It was Galileo who came up with the saying that mathematics is the language of nature. That is a lovely analogy, but nothing more. "Language" in that sense is not a means of communication as we understand the word in the context of interpersonal expression. A committee appointed by the Ministry of Education in Israel headed by the scientist and educator Haim Harari used the metaphor of mathematics as a language when it recommended that mathematics teaching in schools needed greater emphasis and reinforcement. From that point it was apparently only a short step to the message from the principal of a well-known school to new students and parents that three foreign languages were taught in his school, English, French, and mathematics! Calling mathematics a foreign language is simply inappropriate. To me, that is similar to the promise given by the organizers of those courses I have spoken of, that they undertake to impart mathematics to children of three to six years of age to a mother-tongue level. That would be fine if all they meant was that the children would be a bit more familiar with mathematics, although I do not see the advantage of special lessons specifically in that subject. If, however, underlying the term *mother tongue* is a promise that the children would gain intuitive insight into mathematics in the way that we absorb a mother tongue and use it, then I am worried.

68. A TEACHER-PARENT MEETING

Here is a description of a parent-teacher meeting in which I took part, this time as a parent of a child in first grade. The description is quite accurate. The meeting took place many years ago, and the specific curriculum that we discussed in the meeting has changed since then, but the problems that arose in the meeting still apply. I should add that I did not play an active part in the discussion, nor did I reveal that I was a mathematician.

The parents had been invited by the teachers of that grade to talk about the new mathematics textbooks the children would be using that year. The teacher who opened the meeting explained that the material in the books was new, and even the teachers were not very familiar with it. Hence, a senior teacher from the Ministry of Education, named Batya, had been invited to present and to explain to the parents what their children would experience in the mathematics lessons during that year. After a few polite introductory words, greetings, and the like, the guest speaker started.

"It has recently been discovered how prehistoric man developed mathematics." At this one of the parents was heard to murmur, "I'd like to know just how it was discovered. Did they find a book left by prehistoric man?" It was said at a stage-whisper level loud enough for others to hear, but I do not know if Batya heard. In any event, she did not react.

She continued: "Once there was a cruel tribal chief who every morning would send one of the children of the tribe to herd the sheep. At the end of the day, when the young shepherd brought the flock back, the chief would beat him severely, claiming that he had lost some sheep. One of the children, called Ogbu" (here Batya clarified, apparently to add credibility to her story, that perhaps that wasn't really the child's name) "was very clever and found a way to avoid being beaten. Every morning, when he took charge of the sheep, he gave the chief a stone for each sheep. When he brought them back, he would take one stone back for each sheep. When the chief realized he had no stones left, he understood that all the sheep were back, and he didn't beat Ogbu."

My reaction to the story was that it was not very convincing. Every prehistoric tribal chief that I knew would have beaten Ogbu regardless of whether he was left with any stones or not. Moreover, I did not know a single tribal chief, cruel or kind, from prehistoric times until today, who would accept the idea put forward by a clever child (or a clever adult for that matter) for an innovative process of stones representing sheep to prove that no sheep had been lost. Ogbu would have been beaten just for his cheek in making the suggestion. I kept my thoughts to myself.

Batya went on: "Thus, Ogbu's clever idea led to the development over the years of the concept of a one-to-one correspondence (between stones and sheep in prehistoric times), and the correspondence between a set and a subset of another set, thereby laying the foundations for the natural numbers as we know them, and hence to addition and subtraction, and so on." I thought to myself, what a pity that the chief had enough stones to give back to Ogbu. If there had been a few short, that would clearly have led to the discovery of negative numbers thousands of years earlier than their actual discovery.

We will skip the questions parents asked. The approach and the material were new to them. Toward the end of her talk, Batya claimed that once it became clear how prehistoric man had developed the concept of numbers based on sets and matching them, we can use that today and teach the basic concepts better so that children will understand and imbibe the right concepts and will not just do arithmetic exercises mechanically. She went further and said, "The purpose of learning is to understand. It is not so important that the calculations come out exact. Understanding is the main thing."

One of the parents nevertheless asked how does one check understanding if the calculations are not exact. Batya did not have an answer. Another parent suggested, probably sarcastically, that if a child hesitates over the question how much is three plus four, he may well understand, while the child who immediately answers correctly is presumably answering automatically and does not really understand how to add. Another frustrated father banged the table and declared loudly that he hadn't understood much of what Batya had said (he said this in somewhat less diplomatic language), but what he wanted was that when he sent his daughter to buy a newspaper, she should come back with the right change, regardless of what she "understood." It took some time before a calm atmosphere was restored, and the meeting broke up with the general feeling "We will see what happens."

The children used the new textbooks throughout the year, and they and their parents were indeed deeply confused by them. The fun that the books offered the children when they were asked to match collections of items, generally sets of animals or flowers, was a marvelous opportunity to fill the pages with beautiful colors, each child according to his artistic ability. However, when the book tried to move on to develop natural numbers, several mental obstacles were encountered.

For example: one day my son came home with some friends and told me that the teacher had handed out blocks in the classroom, one block per child. Then the blocks were collected, and put into a bag, and the class went to visit the class next door. They gave the blocks out to the children in that class, and found that they did not have enough. The conclusion was, they explained, that there were more children in the second class than in their class. Obviously the purpose was to illustrate a one-to-one correspondence before the children learned the numbers themselves. In all innocence (or not) I asked them how they knew that there were more children in the second class. Perhaps some blocks had fallen out of the bag. "Oh, no," they answered together, "they have thirty-two children in their class, and we have only thirty-one!"

Another problem came to light toward the Hanukkah holiday in December, by which time the children had learned, needless to say on a fundamental level, the numbers from one to four. But on Hanukkah, the Festival of Lights, we are supposed to light eight candles. The creative solution proposed was that eight was a guest, a visiting number, who left immediately after the festival. Of course nearly all the children were familiar with and could use much larger numbers. But the teachers made them repress that knowledge because of the Ministry of Education indoctrination, according to which there was something defective in the children's intuitive understanding. It is clear there is a defect somewhere, but not with the children.

At a later stage, the teacher had to present the whole system of numbers, including the law of commutation, that is, the rule $a + b = b + a$ holds for all two numbers. But this property was already known, so why was it necessary to formulate a law? It took about two months to persuade the children that the commutative property was not self-evident and that there were other possibilities. For example, the order according to which first we put our socks on and then our shoes is important, and so that is not a commutative operation. When the whole class was convinced that it is not self-evident that $a + b = b + a$, then the punch line was delivered: nevertheless, the equality does hold. Why? The answer to that question was not given in the class. In the background material provided for the teacher, it states that it is an axiom. It also says there that the concept of equality is not obvious, and that it is also an axiom. We repeat: teachers are simply told that $a = a$ is an axiom, just like that; that is the "support" the teachers get from the Ministry of Education to help them understand the new teaching program.

These and similar instances made me realize how the following incident could occur. It happened before the school year had begun. My son was due to start first grade and had to meet the staff for them to assess him to be sure that he was at the right level. At home he would amuse himself (without encouragement from his parents) by counting forward and backward, 1, 2, 3, . . . and 9, 8, 7, and so on. The principal was the head of the assessment team, and one of her first questions was what number comes after six. My son answered straight away that it depends in which direction you are counting. It was clear from the expression on her face that she had doubts as to whether the child was ready for school, until I stepped in and explained the background and what my son meant. The principal (who knew my profession) said, "You mathematicians have to clarify for us the essence of the numbers." When I answered that numbers have no essence other than counting, she reacted as if I were belittling or mocking her. That was not the place or time to explain to her that the "mockery" was coming from another source.

That incident occurred a long time ago, but the strange system of confusing teachers by making them look for something in mathematics that is not there still continues. In a conference devoted to mathematics teaching, the lecturers complained recently that the students, very young children, did not understand the essence of a triangle. I confess, for me a triangle is a triangle, and I am not aware of any other hidden essence. When teachers and their instructors in teacher-training colleges try to achieve a situation where students will understand the essence of a triangle, they mean some sort of essence, a collection of properties, that they have decided upon arbitrarily. In fact, they want the children to take on the opinion of those who have made that arbitrary decision. Perhaps teachers should be required to impart to their young students the essence of a house or a bus? Why does the idea of "essence" arise, in the education of youngsters, only in mathematics?

I am pleased to say that some of the examples quoted here have been generally recognized and acknowledged as defects and are no longer part of the current curriculum. One-to-one correspondence, for example, has been taken out of most first-grade textbooks. Unfortunately the shortcomings that resulted in absurdities remain and have spread beyond the confines of elementary schools. One of the main failures is the result of the freedom that the curriculum developers take in defining mathematical content and approaches that they consider offer better content or are more instructive, regardless of material that the children or parents already know. One of the reasons is apparently that it is acceptable in mathematical research to present and use new definitions, and even new systems, as necessary. The difference is that in advanced research this is done carefully and in moderation, and in a manner appropriate to the researcher's target audience, while in school, teaching the target audience is not taken into consideration. The outcome is that parents cannot understand the textbooks of their six- and eight-year-old children. A book was published recently intended for mathematics teachers, with the purpose of placing the background to the pedagogic material on a sound basis. Professors of mathematics to whom I showed the first exercises in the book did not even understand the questions! To be able to answer them, one apparently had first to learn the terms that the authors themselves had invented. That is absurd. The damage caused by such alienation far outweighs any benefit of the new system, which in many areas is rather suspect.

At about the same time that I became aware of the above problems in teaching, a solution also came into my mind: maybe it is not worthwhile to teach mathematics formally in the low grades. I suggested to the supervisor of mathematics teaching in the Ministry of Education that he should try a system in which no formal lessons are given to the first three grades. The teachers could teach and give exercises in subjects like counting, addition, subtraction, multiplication, and the like, by means of contact with the real world, games, stories, and so on. This would rely on what the child had already encountered and recognized, and apparently absorbed, without compulsory lessons. The supervisor loved the idea and suggested it to a well-known school in Jerusalem. And then the idea was rejected by the parents. They were concerned that the delay in learning the basics would have an adverse effect on their children's future understanding of mathematics.

Several years later I learned that I did not pioneer that idea. In 1929 the superintendent of schools in Manchester, New Hampshire, Louis P. Bénézet, recorded a broad experiment covering mathematics teaching in the six grades of elementary school. The studies/lessons were conducted without any formal material, books, and the like. Instead, the teacher had to make the children aware of the connection to what they needed to learn, such as counting, numbers, estimating sizes, calculations, geometry, and so on, by means of games and trial and error. Occasionally the need would arise to present a mathematical symbol and the formal method of derivation, but then that was done only in the context of what had come up in practice, with a mixture of intuition and formalism. After seven years of the experimental method, comprehensive tests were carried out, and the achievements of the students who had participated were compared with those of schools that had continued with the old method. The unequivocal result was that students who had learned with the Bénézet method performed better. I do not know what happened following the experiment. In general, one can always question the validity of tests comparing the results of an innovative teaching method that attracts teachers' and supervisors' attention and boosts the efforts they invest with those of the commonplace system, which the teachers are already tired of using. The basic insight obtained, however, that understanding mathematics is not achieved via formal logic, is still valid today.

69. A LOGICAL STRUCTURE VIS-À-VIS A STRUCTURE FOR TEACHING

Many parts of mathematics consist of layer upon layer, that is, they are derived from previous developments, which leads to the belief that in mathematics, unlike in many other disciplines, you cannot advance to the next level without mastering the one you are on currently. That belief is true to some extent, but before drawing conclusions from it relevant to teaching, one must understand the stages on which mathematics is constructed. First, one must realize that there is a difference between the levels of mathematics as a complete structure of logic and the levels through which one understands mathematics, according to which it must be taught. The lack of realization of this difference is the cause of several of the difficulties in mathematics teaching.

One of the greatest stumbling blocks in mathematics instruction is the misunderstanding of the role of definitions and axioms. In presenting mathematics, generally the definitions come first, and then the axioms, and next come theorems and proofs. But it must be stated again: that is just a way of presenting the subject. The role of definitions and axioms is to clarify and to draw a precise outline of the limits of the discussion. For the definitions to clarify something, that something must be there, must be in the brain, even if it requires clarification. Mathematicians and teachers of mathematics have become accustomed to definitions coming before the subject itself, and they are prepared to accept it on the understanding that the subject will be presented thereafter. For a "trained" audience, it is actually easier to follow mathematical material when the presentation starts with definitions and axioms. Students of mathematics (generally the less-gifted students) sometimes even ask the lecturers for them. Once I was asked by a colleague in a non-mathematical field to explain what groups and fields are. I did the best I could, with little success. The same colleague sought help elsewhere. Some time later he asked me with some scorn in his voice, "Did you even know that a field is a triplet?" Many mathematics books do indeed begin the chapter on fields with the declaration that a field is defined as a set with two operations performed on it, and that is the triplet my colleague referred to. That same colleague, apparently in light of the explanation he had been given, ascribed the greatest importance to that part that indicates a field is a triplet to the extent that the fact that I had not stressed that aspect led him to doubt my skills. That is professional distortion. In a mathematics book, in which it is difficult to present intuition, there may be no choice but to begin with dry formalism. It must be borne in mind, however, that in the natural structure of mathematics, the definitions and axioms came after the intuitive understanding of the material had been absorbed. Try beginning an explanation to a non-mathematical audience, even engineers and scientists, with a list of definitions and axioms, and see their glazed eyes. The structure of definition, theorem, and proof exists only in the study of mathematics and is apparently necessary when reporting mathematical results to a professional audience. This structure used in schools, however, often causes harm and contributes to students' alienation from mathematics. The right way is to start with an intuitive discussion, and when the subject has been presented, at that point it can be shown that it is important to be clear and precise, and only then should the axioms be given. I am aware that presenting the purpose and content of a mathematical subject first on an intuitive level is more difficult than a logical presentation of it, but that is a challenge that mathematics teachers ignore on the grounds that mathematics is taught differently; that is a failure of mathematics teaching.

What position, then, should the foundations of mathematics, that is, the axiomatic and logical structure on which mathematics can be based, occupy in teaching? When it is necessary to present a complete picture of a certain topic in mathematics to experienced professional mathematicians, one can certainly start with definitions and axioms and only afterward deal with applications. If the subject is the general structure of mathematics known today, it may be worthwhile to start with set theory, follow that with natural numbers, and then rational and irrational numbers, as described in section 59. It should be borne in mind, however, that this theory was created simply to establish mathematics on solid fundamentals, and not so that we should better understand how to use mathematics, and certainly not in order for us to teach mathematics better.

For example, for every practical requirement and possible use, it is preferable to introduce the numbers directly, that is, 1, 2, 3, and so on, and to assume that they are something given and clear and do not need to be defined. Similarly, it is easy to explain what a real number is in geometrical terms of distance. The basis of arithmetic and basic fundamentals of geometry are already present in the human brain and should be used. Although comparing sizes, for example determining intuitively which is the larger set, was embedded in the human brain over the course of evolution, this intuitive comparison, which apparently predated counting, did not develop via one-to-one matching. The human brain works in such a way that it compares new concepts to familiar ones. If natural numbers are taught via sets, or if real numbers are taught by means of Dedekind cuts (see section 58), the mind must "erase" what it already knows about numbers and then absorb the new concept. That is no easy task and is generally not required. To teach a subject while ignoring what the student already knows is a didactic error. This principle is clearly relevant to the study of all subjects, especially learning a language, and is not confined to mathematics. Hence, if the aim is to teach simple arithmetic or calculus, it is worthwhile to use the knowledge the student already has. It would be a pedagogic error to try to teach grade 1 students to count by means of comparing sets; indeed, they already know what numbers are. It would be a mistake to start teaching calculus by presenting real numbers as Dedekind cuts; indeed, the students already know the numbers as means of measuring distances.

That said, there is a better reason for not using sets to describe the natural numbers as sets constructed on a basis of the empty set, or not teaching rational numbers as equivalence classes of pairs of natural numbers, or irrational numbers as Dedekind cuts. Those concepts came into being not to improve the understanding of numbers but to prevent the logical structure of numbers, and thereafter calculus, from being based on geometry. A structure not based on geometry is much more complex and less comprehensible than the classic geometry-based structures. For all the uses that will follow from the meeting of students with these concepts, a geometric structure is sufficient, is more efficient, and is certainly correct from the aspect of logic. When the purpose is to provide the student with mathematical tools to enable him to function in the technological world and also to advance in the world of mathematics, it is advisable that the concepts be based on geometry.

Why, then, did Israel's Ministry of Education decide to base the teaching of numbers on sets? How were teachers and parents persuaded that without sets their children would lack a vital stage in the understanding of mathematics? Why, in the first lecture I heard in university, did the lecturer define the rational numbers as equivalence classes of pairs of whole numbers, and irrational numbers as Dedekind cuts, without explaining the historical reasons for those structures? Is this lack of perspective of the profession the reason that in various institutions, and even in a book I saw recently intended for engineers, the irrational numbers are presented as Dedekind cuts? Those who teach in this way are apparently convinced that without these "foundations" their teaching would be deficient. They are wrong. Basing numbers on geometry is no less "complete" than basing them on sets.

On a similar subject, why does a college lecturer teach his students that the sequence a_1, a_2, a_3, and so on . . . of the real numbers is actually a function from the natural numbers to the real numbers? From the aspect of the foundation of mathematics, he is right. If you do not know or cannot understand what a sequence is (and for some reason you do know what a function from the natural to the real numbers is), that is indeed the way to define a sequence. I am not convinced, however, that the lecturer ascertained for himself whether it is at all necessary, and useful, to define a sequence instead of relying on the fact that we all understand intuitively what a sequence is.

Considering set theory as a step that cannot be skipped in trying to understand concepts constitutes a barrier to the teaching of mathematics. We showed this in the previous section when we discussed the teaching of numbers in elementary school. The failure goes deeper than that, however. Not long ago I participated in the entrance examinations for graduate students of mathematics teaching. A graduate of one of the teacher-training colleges said that she gained a great deal from set theory, which she learned in college. We asked her to explain, and she said that now she had a better understanding of the meaning that a child belongs to a particular class. In other words, in her opinion, or in the opinion of her teachers in college, the sentence "Benjamin is a student in Class 2A" is not clear enough. The sentence "Benjamin is a member of the set of children in Class 2A" would apparently clarify the situation. This is a distortion and misrepresentation of the role of mathematics and is harmful indoctrination. It is important to understand the logical structure of mathematics as part of human culture, but it is not right to start teaching by means of that structure. The need to base mathematics on sets arose as a result of the previous foundations being questioned and followed the lack of precision of the intuition used previously. But the study of the logical structure without understanding the need for it does not help and actually causes harm. The purpose of axioms and definitions is to check intuition and to prevent errors and illusions, even at the cost of complexity that is difficult to understand. They check intuition and do not render it superfluous.

That role is not always clear to teachers and to their teachers. The story of the parallel-lines postulate (section 27), for example, is fascinating as part of understanding human culture, and it also illustrates what mathematics is and the role in it of axioms. A committee appointed recently in Israel's Ministry of Education to review the mathematics curriculum reached the conclusion that the parallel-lines axiom was difficult for Israeli students to understand, as it dealt with infinite lines. The committee proposed replacing the parallel-lines axiom with another one, the rectangle axiom, the details of which do not concern us here. They went even further and put forward a detailed program of relevant geometric studies, in which the rectangle axiom takes the place of the parallel-lines axiom. Again, we are dealing with a failure. First, the difficulty that the members of the committee imagined that students encountered is artificial. "Does infinity exist?" and "If it does, what is it?" are questions for philosophers and mathematicians. Not all philosophical contemplation bothers people who are not philosophers. The average student, or even outstanding ones, will probably not identify a line that extends to infinity as a problem, unless it is explained to them that actually there is a problem of definition. A typical student will have no difficulty understanding the concept of a line that continues to infinity. If there is a problem, it is a philosophical, not intuitive, one. Moreover, the axioms are meant to express a property that is itself a natural, intuitive one, and in my view, reference to parallel lines is more natural than reference to a rectangle. Even if the members of the committee are right that the rectangle axiom is more natural than the parallel-lines one, it should not be adopted for teaching purposes. The story of the parallel-lines postulate is today part of human culture. It is related in countless (I almost wrote *an infinite number of* . . .) books, and it occupies a prominent position on the Internet. Teaching geometric axioms in schools by substituting another axiom in place of the parallel-lines axiom would result in the material to be learned being set apart from general culture.

70. WHAT IS HARD IN TEACHING MATHEMATICS?

We wrote in sections 4 and 5 that some subjects in mathematics are compatible with human intuition as it developed over millions of years of evolution, but others offered no advantage in the evolutionary struggle and are contrary to natural intuition. This distinction should be reflected in the methods of teaching and study. Unfortunately, that is not the case. Turning a blind eye to the problem causes conflicts and difficulties.

The following is an imaginary but definitely realistic exchange between a teacher trying to explain that the square root of two is irrational and a student (we discussed the proof in section 7).

Teacher: We will prove that $\sqrt{2}$ is irrational. We first assume that that it is rational.

Student: But how can we assume that it is rational if we want to prove that it is irrational?

Teacher: Wait a minute and you'll see. Assume that it is rational, and we will write it as $\sqrt{2} = \frac{a}{b}$, where a and b are positive integers. We can assume that one of them is odd, because if they were both even, we could divide them by 2 and would continue to do so until at least one, either the numerator or the denominator, is odd.

Student: I understand all that, so far.

Teacher: Now we square both sides of the equation, and we get $2 = \frac{a^2}{b^2}$, or $a^2 = 2b^2$. So a is even, and can be written as $2c$.

Student: Fine.

Teacher: Substitute $2c$ in the previous equation, and we get $4c^2 = 2b^2$. Dividing both sides by 2 leads to the conclusion that b is also even. But we started by justifying the assumption that either a or b must be odd. We have arrived at a contradiction.

Student: So what?

Teacher: The contradiction derived from the assumption that the square root of 2 is rational.

Student: I said at the beginning that we couldn't assume that.

The above was an imaginary conversation, but similar ones take place very often in the study of mathematics. The difficulty that it reflects derives from the fact that the teacher has become accustomed to the argument, may have taught the proof many times, and has simply forgotten the basic difficulty based on proof by contradiction. We should be aware of and remember the obstacles the proof had to overcome among the Pythagoreans, who concealed it for many years for reasons of their own, and it may well be that they did so because the proof is not easy to understand. We should also remember that even in the twentieth century there were those who argued against the method of proof by contradiction. The objections to the method would not have arisen if such proofs were obvious, that is, if they were compatible with intuition. I am not suggesting that young students be exposed to the tradition that views such proofs as problematic, but the recognition that such proofs can be memorized but are difficult to absorb should be reflected in the way they are taught. This difficulty cannot be completely overcome. The best way to deal with it is to separate the explanation of the principle underlying these claims from the proof itself, to be patient, and not to expect the student to arrive at such claims by himself.

The student will encounter similar difficulties in understanding claims containing logical quantifiers such as "all," "there exists" or "there does not exist." For example, if the teacher proves that a certain property, say of triangles, holds for all triangles, he cannot expect that henceforth the student will use such knowledge. The lack of accessibility to a fact that has already been proven is not due to the student possibly forgetting that property but is the result of the fact that the claim "all" does not sit naturally in the human brain. The student will face the same difficulty when the teacher uses the third of the classic rules of thought, the rule of the excluded middle ("either P [the proposition] is true or its negation is true"). The use of this rule is not intuitive. The only way to succeed in using it is to isolate it, and to draw attention to it whenever using it, and lessons must be planned accordingly.

One of the central difficulties deriving from the way the brain works is its failure to relate to stipulations or conditions. The mind does not think conditionally and generally does not discover that it lacks data required for an analysis of the situation. Evolution trained us to complete an incomplete picture somehow or other. A delay intended to enable the brain to discover the missing condition could have had dire results for the human race in the course of evolution, as a result of which, today we usually skip that stage. In mathematics the result is likely to lead to errors. I will illustrate with an example from probability studies.

We described the logical approach to probability and difficulties arising from it in sections 40 to 43. We will now see how experienced teachers are exposed to errors. It must be stressed, the example is not given to illustrate the error or to embarrass or belittle the writer of the article we shall analyze. We give the example in detail because we think that, in that way, we can identify the source of the error, which is the difficulty in identifying provisions and the completion of the picture by the brain. We will extend the discussion and go into detail in order to emphasize the refinement required in analyzing such problems.

The title of the article that appeared in a teacher's magazine some years ago was a promising one: "Conditional Probability as a Source of Paradoxes and Surprising Results." The introduction was in the same vein: "Mathematical thinking is an important means for discovering the world. . . . Phenomena that seem mysterious and paradoxical are explained rationally . . . ," and so on. The article itself gave examples of apparent paradoxes, which it then explains rationally, apparently. Here is one of the examples.

Boys and girls. We are given that the probability of the birth of a son is $\frac{1}{2}$. A certain family has three children. The example has two stages. The first is that outside their apartment we see two girls of that family. What is the probability that the other child is a boy? The second stage is that in addition to the two girls, we can see the outline of a baby, in the apartment, a younger brother or sister of the two girls we met. What then is the probability that the third child is a boy?

The approach of the article to the answer to the question is flawed. We will show the answer given in the article and will then point out the error. We reemphasize: it is not our purpose to belittle the author of the article. Errors are standard fare in mathematics. Our objective is to indicate the source of the error. First we will show the solution given in the article.

The writer solves the exercise only after stating that many students are convinced that the answer to the first part of the question is $\frac{1}{2}$. Their explanation, he claims, is that because of symmetry: the chance that the third child is a boy is equal to the chance that it is a girl. Then the author gives the "right" answer. He assigns the number 1 to a male child, and 0 to a female child. With the help of these symbols we can record eight three-digit permutations that describe all the possible situations for the three-child family:

$$\Omega = \{000, 100, 010, 001, 110, 101, 011, 111\},$$

where for example 011 indicates that the first child was a girl, followed by two boys. The article continues along the formal lines we described in section 41. We will use the symbol A for the event in which "the family has exactly one son," and the symbol B for the situation in which "the family has at least two girls." We must calculate the conditional probability $P(B \mid A)$, which is $\frac{P(A \cap B)}{P(B)}$. This is a fairly simple calculation. The article gives two methods, which give the same result. We will give the shorter one here: as it is known that there are two girls in the family, the relevant event, so claims the author, consists of four out of the eight possibilities listed above, and they are {000, 100, 010, 001}. In three of these there is a son in addition to the two daughters, so that the probability is $\frac{3}{4}$.

The article then addresses the second part of the question. It would seem, he warns, that in light of the above result, we would expect the probability to be $\frac{3}{4}$ as well. That is not correct (according to the article). Note: we know that the question relates to the child born last, so that the range of possibilities is just {001, 000}, and the probability is now $\frac{1}{2}$, and not $\frac{3}{4}$. As stated above, the writer considers this a surprising result, or a paradox, which mathematics clarifies.

The author's approach to the solution is incorrect, and the paradox, or apparent surprise, does not exist. The writer did not notice that we have insufficient data to reach an unequivocal answer to the question. First, the reason that we cannot accept the proposed solution is that the way the question is formulated, the solutions to the two parts contradict each other. If the answer is $\frac{1}{2}$ when we know that the child in the house is the youngest, then that would be the answer also if the child in the house is the middle child of the three, and that is also the answer if we know that the oldest of the three is in the house. These are three separate cases, and together they exhaust all the possibilities. And if the probability of each of these cases is $\frac{1}{2}$, the probability would also be $\frac{1}{2}$ if we do not know which of the three children is in the house, that is, the youngest, the middle one, or the oldest. Therefore, from the fact that we get $\frac{1}{2}$ as the correct answer to the second part of the question, we derive that $\frac{1}{2}$ is also the correct answer to the first part.

Where, then, is the author's mistake? I could make life easy for myself and claim that the writer was misled by the term *conditional probability* in the title, which is sometimes interpreted in the literature to mean the probability of A when it is known that B holds (see section 40). This is an incorrect interpretation of the concept of conditional probability, as it ignores the question of how it becomes known that B holds. Those who read section 40 will know that to solve problems like these we must use Bayes's thought process. If the author of the article had tried to apply Bayes's scheme, he would have seen that without more information it is impossible to arrive at an unequivocal solution. (We gave an example, the tale of the six competitors in a beauty contest, in section 42. A similar example was also "solved" incorrectly in the article.) As we have stated, if data are lacking, the brain of the person solving the problem supplies what is missing itself, generally unknowingly. The problem is that completing the picture in different ways yields different results.

We will put forward three different versions of how the picture can be completed, that is, how the missing information can be supplied, each of which yields a different answer. In the first, assume that in a family with three children, two, chosen at random, go outside to play. Then the solution to both parts of the question is $\frac{1}{2}$ (we skip the calculations). In the second version, assume that among the families in that neighborhood children always go out to play with another child of the same sex, meaning boys play with boys, and girls with girls, and if there are three boys or three girls in a family, the two older ones go out to play together. In that case, in the first part of the question the probability that the child who stayed in the house is a boy is $\frac{3}{4}$, and the probability in the second part of the question is $\frac{1}{2}$, as was claimed in the quoted article (again we skip the calculation, but note that here the calculation quoted in the article is correct). The third possibility is that two children of the same sex always go out to play together, and if there are three of the same sex, then the two who go outside to play are chosen randomly. In that case, in both parts of the question the probability that the child in the house is a boy is $\frac{3}{4}$. Thus we see that the information provided in the question can be completed in different ways, which all give different answers to the questions. What is the correct way to fill in the missing data? The formulation of the question does not provide an answer to that. The use that the writer made of the mathematical formula incorporates an unstated assumption about the question itself (for example, one of the assumptions we listed above), an assumption that does not appear in the formulation of the question. That is why I chose my words advisedly and said that the author's approach to the question was wrong, and not that the numerical answer in the article is wrong. (Although, as the writer related only to equations, it is hard to believe that he considered the various possibilities. Herein lies an important lesson for anyone using mathematics: do not use formulae before checking that they are relevant to the situation at hand.)

As we have said, the reason for the error is rooted in the fact that thinking under provisions is not a natural process for the human brain. So much so that the author of the article did not recognize it and thus also did not identify the source of the contradictory results he himself obtained. He preferred to think of the discomfort caused by intuition as a paradox, a paradox that formalism can explain. As we shall see, in this he is not alone.

Possibly the most famous instance of lack of clarity and unknowing completion of information is known from *Let's Make a Deal*, a television game show in which the competitors are given a mental challenge. The following is an exact formulation of the question:

You are a participant in the game. You are shown three closed doors. You are told that behind one door there is a big prize, while behind each of the other two is a goat. You are invited to select a door. Before it is opened to reveal the big prize or a goat, the host, who knows what is behind each door, opens one of the other doors, and reveals a goat. He now offers you a chance to change your mind and select the other door. Is it worth your while to change your selection, that is, to choose the door that the host did not open?

This problem has given rise to a huge volume of words, discussions, and arguments that turned into insults and abuse. The Internet is swamped with material on this subject. The problem also appears in textbooks, usually showing a solution, with the problem formulated similarly to our formulation, in most cases without indicating that the formulation lacks something. What is missing is the answer to the question whether the host was *obliged* to open one of the doors you did not select. If he is obliged to do so, it is not hard to show that you will not lose and may even gain by switching your choice of door (to know how much your chance of winning increases, you need to know how the host chooses which door to open). If he is not obliged to open a door, the right answer depends on the host's intentions, and these do not appear in the question. If, for example, he opens a door with a goat behind it only if you have chosen the door with the big prize, and he does that only to fool you, then it is not worthwhile for you to change your choice. (People approached Monty Hall, the host on the show, and asked if he was obliged to open a door, and his answer was that he doesn't remember.) Here the lack of clarity in ordinary language is revealed. Many people are convinced that the wording of the question shows that the host must open a door. Others disagree. Most books ignore this point, and the authors simply assume that the interpretation in their minds is the same as their readers' interpretation.

Teachers and their lecturers in teacher-training colleges ignore the fact that when we teach mathematics we use natural language, which is not subject to the laws of mathematical logic. One of the lecturers at a conference convened to determine what mathematics teachers must know was an expert, a professor of mathematics education. He complained about the lack of precision in the use of the language of mathematics among students and their teachers and claimed that a teacher must teach how to correct these errors. The many examples he brought included a quote of a definition from a textbook, which stated that an even number is a number whose units digit is one of the digits 0, 2, 4, 6, and 8. It was clear from the way he made this point that he thought everyone in the audience would spot the mistake, whereas I could not see any lack of precision in that definition. The professor went on to insist that the definition lacked precision because it would lead to classifying the number 26.5, say, as an even number. In presenting the issue, and in the discussion itself, he did not explain that when he referred to a number, he meant a real number or a decimal number. In that case, the definition clearly is not right. Yet when I heard the word *number*, what came to mind was whole numbers, so that I did not see an error in the definition. I do not know the background or framework of the book the professor was quoting from and criticizing for imprecision, or its target readership, but most of the examples he gave suffered from the same type of obscurity or lack of clarity. Apparently, he had not imbibed the fact that in describing mathematics we use natural language, and there is *no way* of escaping the lack of accuracy of spoken language.

Herein lies the difficulty in teaching logic-based subjects. Logical analysis does not allow lack of clarity, but living language is based to a large extent on intuition and structural imprecision deriving from the desire for efficiency. In our daily lives we deal with this one way or another, mainly by ignoring the meticulousness of logic. A father warning his son "If you don't eat the banana, I'll punish you," in some way promises that if his son does eat the banana, he will not be punished, although from the strictly logical aspect he has not made any such commitment. In mathematics lessons, logical precision cannot be ignored. How do we reconcile these two facts? We go back to probability. Some of my colleagues claim that the logic underlying probability theory is so far removed from intuition that it would be preferable to remove it totally from the secondary-school curriculum. I am more optimistic. We can teach, and it is important that we do teach, the logical foundations of mathematics, including the logic of probability theory. But it is vital to understand that it is a difficult subject, and it cannot be imparted intuitively. The teachers too, and their teachers in college, will be exposed to the risk of making mistakes if they do not first clarify the roots of the logical structure of every new exercise or problem in probability they encounter. Mathematics is a combination of an intuitive approach and logical considerations. Awareness of the logical aspects, and the fact that they must be treated differently than the subjects that are consistent with the wealth of material in our brains, is the first step in correct the teaching of mathematics. The lesson plan must be tailored in accordance with this inherent conflict.

Having said that, intuition should be used to teach and advance the appropriate parts of mathematics. For example, in section 4 we mentioned the sequence 4, 14, 23, 34, 42, 50, 59, . . . the natural extension of which is 72, as the numbers are the street numbers at which the subway in Manhattan has a station. The mathematician Morris Kline, whose critical book on mathematics teaching is very instructive, quoted this as an example of teaching without any logical basis, like the whole subject of finding the extension of a sequence. Certainly, the question is not appropriate or relevant to nonresidents of New York City, but searching for patterns is deeply embedded in human intuition, and this property served mankind in a way whose importance cannot be overstated. This property is a cornerstone of mathematics research itself and, as such, is respectable mathematics that should be encouraged and practiced, even if the extension of the sequence is not derived by pure logic. Similarly, we can and should take advantage of intuition relating to numbers that students can develop easily. A sense of numbers is innate in human nature, and it should be exploited, but we must be aware that there is no chance that schoolchildren, or indeed anyone, will develop a feeling or an intuition for logical operations, mathematical symbols, or other abstract systems without their being rooted in and backed by arithmetic or geometry. Teachers must also be alert to this, and lesson plans drawn up accordingly.

71. THE MANY FACETS OF MATHEMATICS

One of the aims of teaching mathematics, as we have said, is to arouse an interest in it among the students, firstly so that they should enjoy learning, but also so that those who want to carry on studying to qualify in professions requiring knowledge of mathematics, or even to continue to higher mathematics studies, will not be deterred from doing so. Here we can identify a failure related to the very perception of the profession. Like the elephant described by six blind people basing their description on the sense of touch and coming up with six totally different descriptions, mathematics is a huge elephant with many facets. If we present mathematics from one narrow aspect, we turn away all those not interested in that particular aspect, although they may find other sides of the "elephant" attractive.

First, we should know and describe the various aspects of mathematics. I have already confessed to my weakness in solving the type of problem that comes up in the Mathematics Olympiad, and the type whose solution requires the use of some sort of trick. At the same time I mentioned John von Neumann, one of the greatest mathematicians of the twentieth century, whose method of solving the problem he was facing would be belittled by every trainer of competitors in the Mathematics Olympiad. Mathematics does indeed have the aspect of solving problems by means of tricks, but it also has one of revealing patterns, and an aspect of constructing logical structures, and of course it plays a role in explaining natural phenomena, and in technological developments, and it also has a historical-philosophical aspect. All these should appear in the curriculum. Students should know that if they find difficulty or feel bored with a part of mathematics, they may well find another part very interesting. Someone who does not like classical music can still enjoy jazz.

The main element lacking in the mathematics teaching in schools is the broad perspective that encompasses the subject. Mathematics in schools has, to a great extent, become a presentation of a collection of solutions alongside a collection of questions. Confronted with a question, the student must learn how to find the connection between it and the right formula that will give the answer. That is indeed a natural way of thinking for the human brain, thinking by comparison or comparative thinking, but there is a great difference between a situation in which the brain constructs such a system of comparison for and by itself, and that in which the student must learn by heart a list given by the teacher. This is something that goes beyond the confines of the school, as can be seen from the following.

I recently taught in a higher-education course for mathematics teachers. As the time for the examination drew near, the question arose as to how the students could prepare themselves for the exam. One of them (like the others, a practicing teacher), seeing that my approach to teaching did not seem to fit the mold he was used to, suggested that I should give them an advance copy of the questions, but with different numbers in the questions! He was serious. Apparently that is the practice in secondary schools today. There are many reasons for the development of this practice. One of them might be that the assessment of success in teaching and of the quality of the teachers is carried out by means of standard examinations. As a result, the study focuses on the techniques for solving standard exercises, at the expense of developing a broad appreciation of the subject and introducing other interesting and important aspects of mathematics. We will not expand on this but will just note that this does a great disservice to mathematics and to the students and their future.

Clearly it will not be possible to broaden the range of mathematics teaching and turn it into an interesting field of study as long as the teachers themselves do not recognize the cognitive foundations of dealing with mathematics. For example, one may make a mistake! If a history student gives a wrong date for a certain event, or if a chemistry student misidentifies a certain chemical in a compound, the teachers do not come to the conclusion that the students do not understand history or chemistry. In mathematics, if a student does not answer a question correctly, it is taken to mean that he does not understand. This intolerance is harmful. I have an ongoing dispute with colleagues about questions that should be asked of applicants for entrance to the institution in which I work. Some of them give the applicants mathematical exercises and check the degree to which they manage to solve them. I strongly object to that and complain to my colleagues that they always ask questions to which they themselves know the answers. To succeed in completing exercises, especially in an examination, is a very small and nonessential part of mathematical capabilities.

Finally, the views I have expressed about mathematics teaching are the result of years of interest in the subject, following developments, and activity in the field. The defects I have described are only part, a small part, of the problems of the educational system. I have not referred to the difficult physical conditions, overcrowded classrooms, or the lack of motivation of some of the teachers, and so on. The aspect of the curriculum, however, can and should be improved. To teach mathematics successfully, the teacher must be aware of the difficulties arising from the conflict between healthy intuition and the logical structure of mathematical discussion. Special teaching methods should be devised to impart the technical ability to analyze the logical, nonintuitive aspects of mathematics, and the students should not be expected at the same time to develop an intuitive ability to use such material. Together with achieving broad recognition of the many facets of mathematics and its role in human culture, it will be possible at last to shake off mathematics' ill-deserved reputation as the hardest subject in school, and it certainly does not need to be the least interesting.

* * *

AFTERWORD

The reader will certainly have noticed the oft-repeated claim in the book that evolution did not prepare us for entirely error-free rigorous analysis and discussion. I would ask readers to treat any errors that they will almost certainly find in the book in that spirit.

SOURCES

The author is wholly and solely responsible for the facts and opinions in the book. Nonetheless, the book could not have been written without reference to numerous sources, including books mentioned in the following bibliography. In the bibliography, interested readers will find popular essays, as well as books that will expand on some of the subjects and on the personalities mentioned herein.

Aczel, Amir D. *The Mystery of the Aleph: Mathematics, the Kabbalah, and the Search for Infinity*. New York: Washington Square Press, 2000.

———. *Descartes's Secret Notebook: A True Tale of Mathematics, Mysticism, and the Quest to Understand the Universe*. New York: Broadway Books, 2005.

Adams, William J. *The Life and Times of the Central Limit Theorem*. 2nd ed. Providence, RI: American Mathematical Society, 2009.

Bertsch McGrayne, Sharon. *The Theory That Would Not Die: How Bayes' Rule Cracked the Enigma Code, Hunted down Russian Submarines, and Emerged Triumphant from Two Centuries of Controversy*. New Haven, CT: Yale University Press, 2011.

Bochner, Salomon. *The Role of Mathematics in the Rise of Science*. Princeton, NJ: Princeton University Press, 1966.

Blackmore, Susan. *The Meme Machine*. Oxford: Oxford University Press, 1999.

Boyer, Carl B. *The History of the Calculus and Its Conceptual Development*. New York: Dover Publications, 1959.

Cohen, Bernard I. *The Birth of a New Physics*. Revised and updated. New York: W. W. Norton, 1985.

Coyne, Jerry A. *Why Evolution Is True*. New York: Viking Penguin Group, 2009.

Crease, Robert P. *The Great Equations: Breakthroughs in Science from Pythagoras to Heisenberg*. New York: W. W. Norton, 2009.

Davis, Philip J., and Reuben Hersh. *The Mathematical Experience*. Boston: Houghton Mifflin, Birkhäuser, 1981.

Dehaene, Stanislas. *The Number Sense: How the Mind Creates Mathematics*. Oxford: Oxford University Press, 1997.

Devlin, Keith. *The Math Gene: How Mathematical Thinking Evolved and Why Numbers Are Like Gossip*. New York: Basic Books, Perseus Books Group, 2000.

———. *The Unfinished Game: Pascal, Fermat, and the Seventeenth-Century Letter That Made the World Modern*. New York: Basic Books, Perseus Books Group, 2008.

Dixit, Avinash K., and Barry J. Nalebuff. *The Art of Strategy: A Game Theorist's Guide to Success in Business and Life*. New York: W. W. Norton, 2008.

Drake, Stillman. *Galileo: A Very Short Introduction*. Oxford: Oxford University Press, 1980.

du Sautoy, Marcus. *The Music of the Primes: Searching to Solve the Greatest Mystery in Mathematics*. New York: HarperCollins, 2003.

———. *Symmetry: A Journey into the Patterns of Nature*. New York: HarperCollins, 2008.

Ekeland, Ivar. *Mathematics and the Unexpected*. Chicago: University of Chicago Press, 1988.

Eves, Howard. *An Introduction to the History of Mathematics*. 3rd ed. New York: Holt, Rinehart and Winston, 1969.

Forbes, Nancy, and Basil Mahon. *Faraday, Maxwell, and the Electromagnetic Field: How Two Men Revolutionized Physics*. Amherst, NY: Prometheus Books, 2014.

Gessen, Masha. *Perfect Rigor: A Genius and the Mathematical Breakthrough of the Century*. Boston: Houghton Mifflin Harcourt, 2009.

Gigerenzer, Gerd. *Reckoning with Risk: Learning to Live with Uncertainty*. London: Penguin Books, 2002.

Hacking, Ian. *The Emergence of Probability*. 2nd ed. Cambridge: Cambridge University Press, 2006.

Harel, David. *Computers Ltd.: What They Really Can't Do*. Oxford: Oxford University Press, 2000.

Harel, David, with Yishai Feldman. *Algorithmics: The Spirit of Computing*. 3rd ed. Harlow, UK: Addison-Wesley, Pearson Education, 2004.

Hoffman, Paul. *The Man Who Loved Only Numbers: The Story of Paul Erdős and the Search for Mathematical Truth*. New York: Hyperion, 1998.

Huntly, H. E. *The Divine Proportion: A Study in Mathematical Beauty*. New York: Dover Publications, 1970.

Isaacson, Walter. *Einstein: His Life and Universe*. New York: Simon and Schuster, 2007.

Israel, Georgio, and Ana Millán Gasca. *The World as a Mathematical Game: John von Neumann and Twentieth Century Science*. Boston: Birkhäuser, 2009.

Kahneman, Daniel. *Thinking, Fast and Slow*. New York: Farrar, Straus and Giroux, 2011.

Kline, Morris. *Mathematical Thought from Ancient to Modern Times*. Oxford: Oxford University Press, 1972.

———. *Why Johnny Can't Add: The Failure of the New Math*. New York: St. Martin's Press, 1973.

———. *Mathematics: The Loss of Certainty*. Oxford: Oxford University Press, 1980.

———. *Mathematics and the Search for Knowledge*. Oxford: Oxford University Press, 1985.

Koestler, Arthur. *The Watershed: A Biography of Johannes Kepler*. London: Heinemann Educational, 1961.

Lanczos, Cornelius. *The Einstein Decade (1905–1915)*. New York: Academic Press, 1974.

Liberman, Varda, and Amos Tversky. *Statistical Reasoning and Intuitive Judgment* (in Hebrew). Tel Aviv: Open University of Israel, 1996.

Livio, Mario. *The Golden Ratio: The Story Phi, the World's Most Astonishing Number*. New York: Broadway Books, 2002.

Magee, Bryan. *The Great Philosophers: An Introduction to Western Philosophy*. Oxford: Oxford University Press, 1987.

Mahon, Basil. *The Man Who Changed Everything: The Life of James Clerk Maxwell*. Chichester, UK: John Wiley and Sons, 2003.

Mangel, Marc, and Colin W. Clark. *Dynamic Modeling in Behavioral Ecology*. Princeton, NJ: Princeton University Press, 1988.

Monk, Ray. *Russell*. London: Phoenix, 1987.

Nagel, Ernest, and James R. Newman. *Gödel's Proof*. New York: New York University Press, 1960.

Ne'eman, Yuval, and Yoram Kirsh. *The Particle Hunters: The Search after the Fundamental Constituents of Matter* (in Hebrew). Tel Aviv: Massada, 1983.

Netz, Reviel, and William Noel. *The Archimedes Codex: Revealing the Secrets of the World's Greatest Palimpsest*. London: Phoenix, 2007.

Rudman, Peter S. *How Mathematics Happened: The First 50,000 Years*. Amherst, NY: Prometheus Books, 2007.

Ruelle, David. *The Mathematician's Brain*. Princeton, NJ: Princeton University Press, 2007.

Saari, Donald G. *Decisions and Elections: Expecting the Unexpected*. Cambridge: Cambridge University Press, 2001.

Singh, Simon. *Fermat's Last Theorem: The Story of a Riddle That Confounded the World's Greatest Minds for 358 Years*. London: Fourth Estate, 1998.

———. *The Code Book: The Science of Secrecy from Ancient Egypt to Quantum Cryptography*. New York: Anchor Books, Random House, 2000.

Swetz, Frank J., and T. I. Kao. *Was Pythagoras Chinese?* University Park: Pennsylvania State University Press, 1977.

van der Warden, B. L. *Science Awakening*. Translated into English by Arnold Dresden. Groningen: Noordhoff, 1954.

Wilson, Robin. *Four Colors Suffice: How the Map Problem Was Solved*. Princeton, NJ: Princeton University Press, 2002.

Yavetz, Ido. *Wandering Stars and Ethereal Spheres: Landmarks in the History of Greek Astronomy* (in Hebrew). Or Yehuda, Israel: Kinneret, Zmora-Bitan, Dvir, 2010.

INDEX OF NAMES

INDEX OF SUBJECTS